Nanomagnetism: Fundamentals and Applications

FRONTIERS OF NANOSCIENCE

Series Editor: Richard E. Palmer
The Nanoscale Physics Research Laboratory,
The School of Physics and Astronomy,
The University of Birmingham, UK

Vol. 1 **Nanostructured Materials** edited by Gerhard Wilde

Vol. 2 **Atomic and Molecular Manipulation** edited by Andrew J. Mayne and Gérald Dujardin

Vol. 3 **Metal Nanoparticles and Nanoalloys** edited by Roy L. Johnston and J.P. Wilcoxon

Vol. 4 **Nanobiotechnology** edited by Jesus M. de la Fuente and V. Grazu

Vol. 5 **Nanomedicine** edited by Huw Summers

Vol. 6 **Nanomagnetism: Fundamentals and Applications** edited by Chris Binns

Nanomagnetism: Fundamentals and Applications

Edited by

Chris Binns
Department of Physics and Astronomy,
University of Leicester,
Leicester, United Kingdom

ELSEVIER

Amsterdam • Boston • Heidelberg • London • New York • Oxford
Paris • San Diego • San Francisco • Singapore • Sydney • Tokyo

Elsevier
The Boulevard, Langford Lane, Kidlington, Oxford, OX5 1GB, UK
Radarweg 29, PO Box 211, 1000 AE Amsterdam, The Netherlands

British Library Cataloguing in Publication Data
A catalogue record for this book is available from the British Library

Library of Congress Cataloging-in-Publication Data
A catalog record for this book is available from the Library of Congress

ISSN: 1876-2778
ISBN: 978-0-08-098353-0

For information on all Elsevier publications
visit our web site at store.elsevier.com

Contents

Contributors

Numbers in Parentheses indicate the pages on which the author's contributions begin.

Chris Binns (1, 85, 217), Department of Physics and Astronomy, University of Leicester, Leicester, United Kingdom

Jonathan Bramble (85), Department of Chemistry, University of Sheffield, Sheffield, United Kingdom

J. Dorantes-Dávila (33), Instituto de Física, Universidad Autónoma de San Luis Potosí, San Luis Potosí, Mexico

Andrey A. Fraerman (189), Department of Magnetic Nanostructures, Institute for Physics of Microstructures, Russian Academy of Science, and Lobachevsky State University of Nizhny Novgorod, Gagarin Avenue, Nizhny Novgorod, Russia

Joaquín C. García-Martínez (85), Pharmaceutical and Organic Chemistry. Facultad de Farmacia, Edificio Polivalente, Universidad de Castilla-La Mancha, Albacete, Spain

Petra E. Jönsson (129), Department of Physics and Astronomy, Uppsala University, Box 516, Uppsala, Sweden

Sara Laureti (129), Institute of Structure of Matter, National Research Council (CNR) Via Salaria, Monterotondo Scalo, Rome, Italy

Victor L. Mironov (189), Department of Magnetic Nanostructures, Institute for Physics of Microstructures, Russian Academy of Science, and Lobachevsky State University of Nizhny Novgorod, Gagarin Avenue, Nizhny Novgorod, Russia

G.M. Pastor (33), Institut für Theoretische Physik, Universität Kassel, Kassel, Germany

Davide Peddis (129), Institute of Structure of Matter, National Research Council (CNR) Via Salaria, Monterotondo Scalo, Rome, Italy

Andrew Pratt (259), Department of Physics, University of York, York, United Kingdom

Andrea Rawlings (85), Department of Chemistry, University of Sheffield, Sheffield, United Kingdom

Sarah S. Staniland (85), Department of Chemistry, University of Sheffield, Sheffield, United Kingdom

Juan Tolosa (85), Pharmaceutical and Organic Chemistry. Facultad de Farmacia, Edificio Polivalente, Universidad de Castilla-La Mancha, Albacete, Spain

Gaspare Varvaro (129), Institute of Structure of Matter, National Research Council (CNR) Via Salaria, Monterotondo Scalo, Rome, Italy

Orla Wilson (85), Department of Materials Science and Engineering, Johns Hopkins University, Baltimore, Maryland, USA

Preface

Magnetism at the nanoscale is a burgeoning area that has attracted intense activity in both fundamental and applied research. The nanoscale is interesting fundamentally because it defines the size scale when the basic properties of matter start to diverge from those of the bulk and also become size dependent, this latter feature being alien in the macroscopic world. Arguably, the first true novel nanoscale behavior is the formation of single-domain magnetic particles below a critical size. Modern nanoparticle synthesis techniques also enable the study of how magnetism develops in matter as it is built atom by atom from the monomer.

In addition, the special properties of magnetic nanoparticles have made them central to a diverse range of technologies that span magnetic recording, cancer treatment and diagnosis, high-performance magnetic materials for a new generation of generators and motors, and environmental applications.

This volume begins in Chapter 1 with a tutorial description of nanomagnetism useful for nonexperts that will enable them to engage with the rest of the book. Chapter 2 deals with spin and orbital magnetism in nanoparticles, how they can be calculated and how they vary with size for different elements. Chapter 3 describes synthesis methods for magnetic nanoparticles, with a special focus on new and emerging techniques that can produce a new generation of nanoparticles required in technology. Chapter 4 presents the properties of nanostructured materials built by bottom-up synthesis of nanoparticles and how their properties can be controlled. Chapter 5 looks at magnetic states in patterned ferromagnetic nanostructures produced by top-down synthesis methods. Chapter 6 describes medical applications of magnetic nanoparticles with a focus on cancer treatment and diagnosis. Finally, Chapter 7 explores environmental applications of magnetic nanoparticles such as water purification and soil remediation.

The volume has a wide scope and will provide the reader with an overview of the current state of the art in nanomagnetism in both fundamental research and technological applications.

Chris Binns

Colour versions of the figures in this book can be found on the companion website: http://booksite.elsevier.com/9780080983530/.

Chapter 1

Tutorial Section on Nanomagnetism

Chris Binns

Department of Physics and Astronomy, University of Leicester, Leicester, United Kingdom

1 WHY IS THE NANOMETER SCALE SPECIAL IN MAGNETISM?

Given that distance scales in Physics range from 10^{-35} m (the Planck length) to 10^{26} m (the observable Universe), it is interesting to reflect on why the nanoscale (10^{-9} m) is so important in materials. There are two main reasons. One is that for pieces of matter smaller than ~100 nm their fundamental properties are different to those of the bulk material. The other is that those fundamental properties become dependent on the size of the piece. This is quite alien to our macroscopic view of the world in which we take it for granted that when we cut a piece of material in half its fundamental properties remain unchanged. At the nanometer scale, a number of factors come into play to modify this behaviour. The proportion of atoms that are on the surface atomic layer, whose electronic states are modified relative to interior layers, becomes significant. For example in a 10-nm particle, 10% of the atoms are on the surface layer and this proportion increases to 50% in a 2-nm particle. Also in a metallic particle, the energy separation of conduction electron quantum states becomes significant relative to other energy parameters such as thermal and Zeeman energies. These and other effects mean that as the size of a piece of matter is reduced to ~100 nm and below, its electronic, magnetic, optical and chemical properties all start to evolve with size. This evolution is smooth near the upper boundary of the nanoscale region but at small sizes significant changes in behaviour can occur with the removal or addition of a single atom.

The detailed quantum mechanical theory of orbital and spin magnetic moments in nanoparticles is presented in Chapter 2, and here, the focus is on understanding the fundamental changes in magnetism at the nanoscale compared with bulk materials. To begin this tutorial, let us examine how intrinsic magnetic properties change as we reduce the size of a piece of material from the bulk through to a single atom. Interestingly, it was this thought process that led Leucippus and his student Demokritos to originally propose the concept of the atom around 400–450 BC. They argued that if matter

Frontiers of Nanoscience, Vol. 6. http://dx.doi.org/10.1016/B978-0-08-098353-0.00001-4

was truly a continuum, then it could be cut into ever-smaller pieces *ad infinitum* so that in principle it would be possible to cut it into pieces of nothing that could then not be reassembled. To avoid this paradox, they hypothesised that there must be a smallest uncuttable piece—the *atomon*. With modern techniques, it is possible to synthesise and study materials at any size scale down to the atom, and if we carry out the Demokritos thought experiment, it turns out that the properties of the material start to change long before we reach the size of an atom. With respect to magnetic properties, arguably the first true size affect in materials is when the dimension of a magnet becomes too small to support the formation of domains. So let us begin by examining why magnetic domains form in bulk materials.

2 FORMATION OF DOMAINS IN MAGNETIC MATERIALS

Magnetism fundamentally arises from the exchange interaction between atoms, whose origin is the requirement that electron wavefunctions are anti-symmetric and can be understood qualitatively as follows. Consider a two-electron system with space and spin co-ordinates, $\mathbf{r_1}$, σ_1 and $\mathbf{r_2}$, σ_2 with a wavefunction $\psi(\mathbf{r_1},\sigma_1,\mathbf{r_2},\sigma_2)$. We know that this must be anti-symmetric with respect to the co-ordinates of the two electrons so if we exchange the co-ordinates ψ changes sign, that is, $\psi(\mathbf{r_1},\sigma_1,\mathbf{r_2},\sigma_2) = -\psi(\mathbf{r_2},\sigma_2,\mathbf{r_1},\sigma_1)$. It follows that the wavefunction is zero if the co-ordinates of both electrons are identical (this is the Pauli principle) so there is no probability of finding two electrons with the same spin at the same point in space. The same is not true for electrons with opposite spins however. The fundamental nature of the wavefunction thus tends to keep electrons with the same spin apart, which lowers their Coulomb energy as illustrated in Figure 1.1. The system energy is therefore lower if the electrons have parallel spins, and the difference in energy between the parallel and anti-parallel alignments is the exchange energy. Note that it is this energy that is the source of Hunds first rule, which states that intra-atomic electrons try to maintain parallel alignment of their spins.

The exchange is effectively a correction to the Coulomb energy required by the anti-symmetric nature of the wavefunction and produces a difference in energy between the parallel and anti-parallel alignment of neighbouring atomic spin moments. For two neighbouring atoms with spins $\mathbf{S_1}$ and $\mathbf{S_2}$, the interaction can be represented by the energy term

$$E_{ex} = -2J\mathbf{S}_1 \cdot \mathbf{S}_2, \tag{1.1}$$

where J is the exchange constant. If we move beyond the simplistic argument presented above, it turns out that J can be positive (parallel alignment favoured—ferromagnetism) or negative (anti-parallel alignment favoured—anti-ferromagnetism). The exchange interaction is orders of magnitude stronger than the magnetic dipolar interactions and is the source of ferromagnetism.

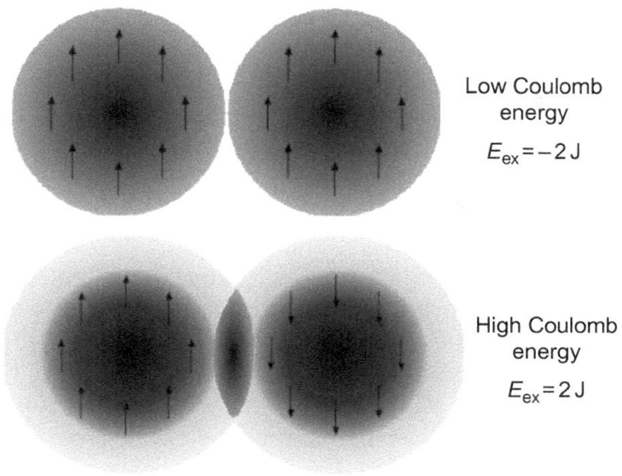

FIGURE 1.1 Schematic representation of the exchange interaction. The requirement that the electron wavefunction is anti-symmetric tends to separate electrons with the same spin, thus lowering their Coulomb energy.

The exchange energy of an entire crystal is represented by the *Heisenberg* Hamiltonian

$$H=-\sum_i\sum_{j\neq i}J_{ij}\mathbf{S_i}\cdot\mathbf{S_j}, \tag{1.2}$$

where $2J_{ij}\mathbf{S_i}\mathbf{S_j}$ is the exchange energy of atoms i and j and the factor 2 has been omitted to avoid double counting. Note that $\hbar\mathbf{S_i}$ is the total angular momentum of atom i but throughout the discussion below and in most books it is referred to as a spin (hence the symbol). In transition metals, because of orbital momentum quenching (see below) the total angular momentum is almost entirely due to spin.

The exchange energy per atom is several orders of magnitude stronger than the direct magnetic dipolar interaction between atoms, as can be shown by an elementary calculation. If we put magnetic moments of strength 1 μ_B a distance of 3 Å apart, the magnitude of the field, B, at one moment due to its neighbour is

$$B=\frac{\mu_0\mu_B}{4\pi r^3}, \tag{1.3}$$

so the interaction energy, ΔE, is

$$\Delta E=\mu_B B=\frac{\mu_0\mu_B^2}{4\pi r^3}, \tag{1.4}$$

which is approximately 3×10^{-25} J or an energy k_BT for $T\sim0.03$ K. So any magnetic order would be destroyed at temperatures well below 1 K, yet Fe, for example, remains magnetic up to over 1000 K. In addition, the classical dipolar interaction would tend to produce an *antiferromagnetic* interaction, that is, the moments would be aligned anti-parallel. The dipolar interaction does however have an important role to play in the formation of domains and also becomes significant in assemblies of magnetic nanoparticles, which have large magnetic moments.

In response to the exchange energy, a magnetic material should be fully magnetised, but it is an elementary observation that unless some special effort is made to stop it a magnetic material demagnetises so that it generates no external field. This is in order to minimise the magnetic self-energy of the atomic dipoles in the material. If they obey the exchange force and line up, then the magnetisation generates a field within the material in which the dipoles are aligned such that they have maximum magnetostatic energy. Reversing is of no use because the generated field reverses and again they are aligned in the least favourable direction to minimise energy. The magnetostatic self-energy can be written

$$\varepsilon_M = \frac{1}{2}\int_{\text{solid}} M \cdot H \, d\tau. \tag{1.5}$$

This represents the interaction energy of each dipole with the field H generated by all the other dipoles integrated throughout the material. The factor 1/2 avoids double counting.

The exchange interaction and the magnetostatic self-energy are competing, and as demonstrated earlier, the energy of the direct magnetic interaction of atomic dipoles is insignificant compared with the exchange energy per atom. Bearing in mind however that the dipolar interaction is long range while the exchange interaction only operates between atomic neighbours, there is a compromise that will minimise the energy relative to the totally magnetised state. If the material organises its magnetisation into domains of opposite magnetisation, ε_M is reduced relative to the state of uniform magnetisation. For example, Figure 1.2A shows a bar that is uniformly magnetised and its magnetostatic energy is reduced if it is subdivided into domains of opposite magnetisation (Figure 1.2B and C). The magnetisation in the domains will be along the easy axis, taken to be along the bar in Figure 1.2. A material with cubic anisotropy can also form closure domains at the ends at right-angles to the main domains to reduce the external field to zero. Figure 1.3 shows an image of the domain pattern at zero field in a single-crystal Fe whisker with a thickness of 50 μm.

3 DOMAIN WALLS

Minimisation of the magnetostatic energy demands that the material forms as many domains as possible but every new domain introduces a boundary of oppositely aligned atomic spins thereby increasing the exchange energy

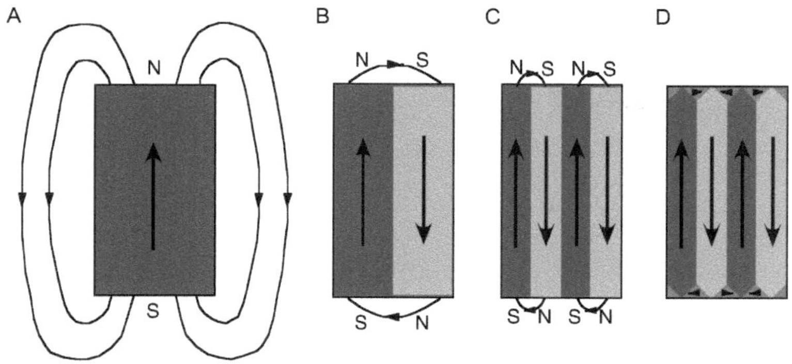

FIGURE 1.2 Formation of domains in a bar of magnetic material. (A–C) The magnetostatic energy is reduced by the formation of oppositely magnetised domains. (D) In a material with cubic anisotropy, closure domains will also form to reduce the external field to zero.

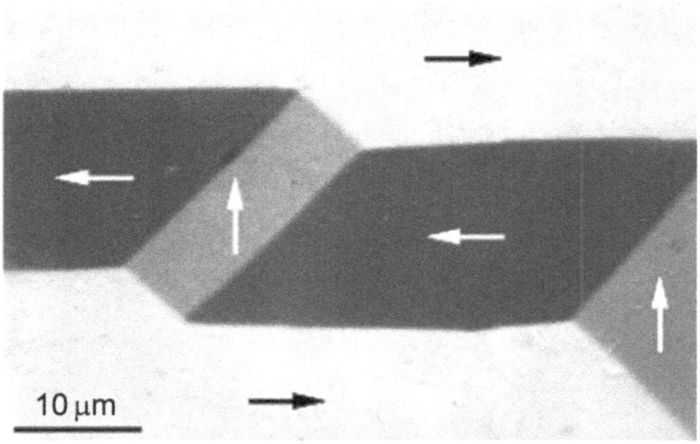

FIGURE 1.3 Domain structure in an FeSi(100) single crystal imaged by Kerr microscopy. *Image obtained from Prof. Dr. Rudolf Schäfer, Leibnitz institute for Solid State and Materials Research, Dresden and reproduced with permission from Springer.*

within the material. The energy balance works because only atoms at the boundary interact via the powerful exchange force whereas a much larger number of atoms benefits from the reduction in the dilute magnetostatic energy. Increasing the number of domains brings diminishing returns and eventually it becomes energetically unfavourable to form an extra domain because of the exchange energy of the extra boundary.

In reality, the energy balance still would not work if the boundaries occurred abruptly going from one atomic plane to the next. The formation of a domain structure relies on the fact that the exchange energy of the boundary can be lowered by spreading the reversal over many spins as the following analysis shows.

Consider n atoms in contact at a domain boundary. The exchange energy *difference* between perfect alignment and spins canted at a small angle ϕ is (from Equation 1.1)

$$\delta E_{\mathrm{ex}} = 2JS^2 n(1-\cos\phi) = 4JS^2 n \sin^2\left(\frac{\phi}{2}\right) \tag{1.6}$$

So, for small ϕ

$$\delta E_{\mathrm{ex}} = JS^2 n\phi^2 \tag{1.7}$$

If the domain wall between oppositely magnetised domains spreads the spin reversal over N atomic planes (Figure 1.4B), then the angle, ϕ, between adjacent planes is π/N and the total difference, that is the energy of the domain wall, is

$$\delta E_{\mathrm{ex}} = JS^2 nN\left(\frac{\pi}{\mathrm{N}}\right)^2 = \frac{JS^2 n\pi^2}{N}. \tag{1.8}$$

The energy of an abrupt boundary (Figure 1.4A) is $2JS^2n$ so spreading the boundary over N planes has reduced its energy by a factor $\pi^2/2N$.

This result predicts that the wall will be arbitrarily wide—eventually getting back to a uniform magnetisation, so Equation (1.8) apparently predicts that domain walls do not exist. As with the formation of domains, there is a law of diminishing returns and when N becomes large there is little to gain by making it yet larger so that eventually much weaker forces than exchange can limit the wall thickness. In this case, it is the magnetocrystalline anisotropy. Within the domains, the magnetisation will point along the local easy axis but this cannot be the case within the wall and a thick wall will force a lot of spins to point along a hard direction. The wall thickness, t, is given by

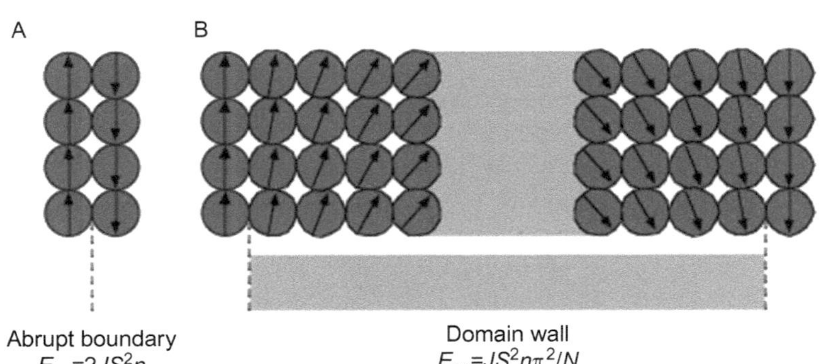

FIGURE 1.4 (A) Boundary with abrupt reversal of atomic spins. (B) Reversal spread over N atomic planes.

TABLE 1.1 Role of Magnetic Energy Terms

Energy Term	Order of Magnitude Energy Per Atom (eV)	Role in Bulk Magnetic Configuration
Exchange energy	1	Responsible for ferromagnetism
Magnetocrystalline anisotropy energy	10^{-3}	Limits thickness of domain wall
Magnetic dipolar energy	10^{-6}	Responsible for formation of domains

$$t = Na = \pi a \left(\frac{W}{3K}\right)^{1/2}, \tag{1.9}$$

where K is the anisotropy energy density and W is the exchange energy/unit volume. For example, in a simple cubic crystal

$$W = \frac{6JS^2}{a^3}. \tag{1.10}$$

Typically, $W \sim 10^9 \text{J m}^{-3}$ while $K \sim 10^5 \text{J m}^{-3}$ so from Equation (1.10) $t \sim 100$ nm.

Thus in the bulk, the three energy terms, the exchange energy, the anisotropy energy and the magnetostatic energy, which have very different energies all play a role in the final magnetic configuration of the material. The role and the order of magnitude energy per atom for each of the terms is listed in Table 1.1

4 SINGLE-DOMAIN PARTICLES

The average size of the magnetic domains is a function of the three parameters listed in Table 1.1 and as the volume of a piece of magnetic material is reduced the number of domains decreases. It is clear that when the volume drops below a certain critical value, it becomes energetically unfavourable to include a single-domain wall and the uniformly magnetised state illustrated in Figure 1.5 becomes the lowest energy configuration. Thus, a piece of magnetic material below the critical size stays permanently magnetised at close to its saturation magnetisation. It may not have the full saturation magnetisation in remanence due to canting of spins at the particle surface.

A rough estimate of the critical diameter for the formation of a single domain can be made by assuming it is approximately the thickness of a domain wall, that is about 100 nm but more rigorous estimates can be derived.[1,2] Consider a particle with a uniaxial anisotropy whose anisotropy energy is given by

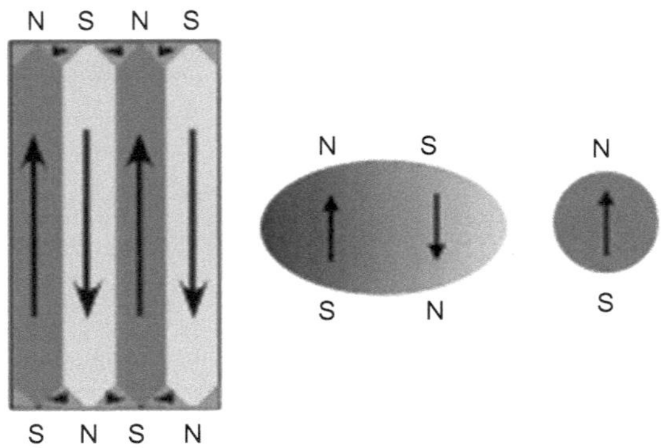

FIGURE 1.5 The number of domains decreases with sample size and below a critical value, the energy balance favours just a single domain and a piece of magnetic material stays permanently and fully magnetised.

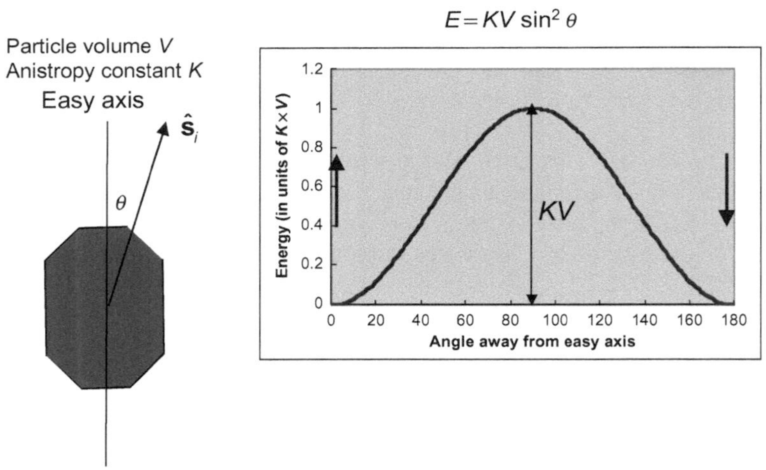

FIGURE 1.6 Dependence of anisotropy energy on magnetisation direction in a nanoparticle with a uniaxial anisotropy.

$$E = KV \sin^2 \theta, \tag{1.11}$$

where K is the anisotropy energy density (in J m^{-3}), V is the particle volume and θ is the angle between the magnetisation vector and the easy axis. The system therefore has two minimal energy states separated by an energy barrier of height KV between them as illustrated in Figure 1.6. In this case, it can be shown that for a spherical particle with a large anisotropy, which satisfies the condition

$$K > \frac{\mu_0 M_S^2}{6}, \tag{1.12}$$

where M_S is the saturation magnetisation (in $A\,m^{-1}$), that the critical diameter, d_C, for single-domain behaviour is given approximately by[2]

$$d_C \approx \frac{18\sqrt{A_{ex}K}}{\mu_0 M_S^2}. \tag{1.13}$$

The term A_{ex} in equation (1.13) is the exchange stiffness, defined by $zJ_{ex}S^2/a$, where z is the number of nearest neighbours, J_{ex} is the exchange integral and a is the atomic spacing. Note that J_{ex} is not the same as the exchange constant, J, discussed above though it also has units of energy and for the interaction between orthogonal orbitals, $J = J_{ex}$.

In most materials, $A_{ex} \sim 10^{-11}\,J\,m^{-1}$ and so, for example, for Co particles with $K \sim 7 \times 10^5\,J\,m^{-3}$ (see below) and $M_S = 1.3 \times 10^6\,A\,m^{-1}$, $d_C \sim 22\,nm$. Equation (1.13) shows that d_C increases with the anisotropy since from Equation (1.9), the domain wall thickness is reduced so that the energy of a domain wall (from Equation 1.8) is larger and it becomes energetically unfavourable to form one at a larger particle size. The opposite is true for the saturation magnetisation.

The important point as far as nanomagnetism is concerned is that as the particle size is reduced to below d_C the behaviour changes radically and the particle magnetisation switches from approximately zero to approximately the full saturated value. Within the Demokritos thought experiment, this could signal the appearance of the *atomon* though in fact it is just the transition into nanoscale behaviour and the properties continue to change as the size is reduced further as discussed below.

Evolution has made good use of this particular nanoscale magnetic behaviour within magnetotactic bacteria, which form internal chains of magnetite (Fe_3O_4) nanoparticles with a diameter of around 50 nm as illustrated in Figure 1.7.[3] According to Equation (1.13), the critical diameter for single-domain behaviour in magnetite ($M_S \sim 2.8 \times 10^5\,A\,m^{-1}$, $K \sim 1.1 \times 10^4\,J\,m^{-3}$) is above 60 nm so the nanoparticles within the bacteria are single domains. Thus, there is a guarantee that they are permanently magnetised and a chain of them as shown in Figure 1.6 will be a permanently magnetised needle. The bacterium uses the structure for simple navigation as it can follow the Earth's magnetic field angle of dip to the bottom of its liquid environment where the food is. The system would not work if the needle was a single piece of mineral as the domain structure would demagnetise it. Only by forming the nanoparticle structure is permanent magnetism in the needle assured.

5 THE BLOCKING TEMPERATURE

As pointed out in the previous section for a particle with a uniaxial anisotropy (which is mostly the case for nanoparticles), the energy barrier separating the

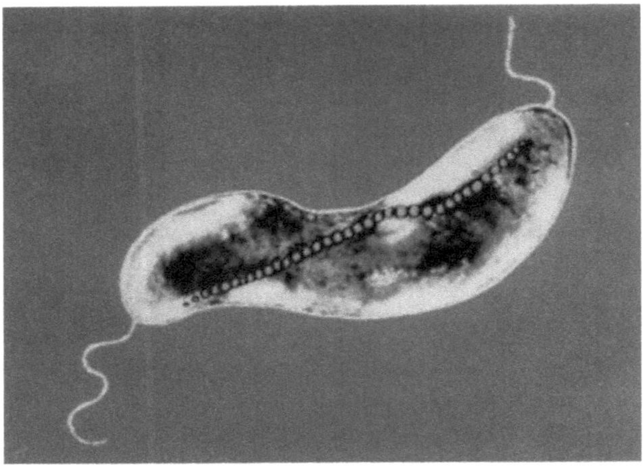

FIGURE 1.7 Electron microscope image of a magnetotactic bacterium. It forms nanoparticles of magnetite with a diameter of ~50 nm in a chain (visible in the interior). According to Equation (1.13) d_C for magnetite is ~60 nm so the particles are all single domains thus guaranteeing that the structure is permanently magnetised. *Reproduced with permission from Ref. 3.*

different magnetisation directions is KV. As the volume decreases at a specific temperature, there comes another critical diameter (well below d_C) at which the magnetisation of the nanoparticle is unstable against thermal fluctuations, that is, $KV \sim k_B T$ and the time averaged magnetisation goes to zero. Thus for a given particle size, there is a temperature that marks the transition from a permanent static moment to one that is fluctuating in a nanoparticle.

The blocking temperature can be estimated by writing the lifetime of the magnetisation vector along a particular direction, τ, in the form of an Arrhenius law

$$\tau = \tau_0 \exp\left(\frac{KV}{k_B T}\right), \tag{1.14}$$

where τ_0 is a natural lifetime or the value at the high temperature limit. This has been measured to be 10 ns in 2-nm diameter Fe nanoparticles embedded in Ag matrices[4] and 1 ns in 35-nm diameter × 10-nm thickness CoPt discs.[5] It is also possible to derive an expression for τ_0, which according to the model by Brown[6] is given by

$$\tau_0 = \frac{1 + (\eta \gamma_0 M_S)^2 KV}{\eta \gamma_0^2} \frac{\sqrt{\pi}}{4K} \left(\frac{k_B T}{KV}\right)^{1/2}, \tag{1.15}$$

where γ_0 is the gyromagnetic ratio and η is a dissipation constant. Equation (1.15) indicates that there is a weak dependence of τ_0 on temperature but as shown below, both τ and τ_0 have a minor influence on the blocking temperature so it is reasonable to treat τ_0 as a constant.

In any given experiment, the moments will appear to be blocked when the lifetime τ is longer than the measurement time. If we pick an arbitrarily long measurement time, for example, 1000 s, then the blocking temperature, T_B, can be estimated from

$$\log_e 1000 = \log_e\left(10^{-9}\right) + \left(\frac{KV}{k_B T_B}\right), \tag{1.16}$$

that is,

$$T_B = \left(\frac{KV}{k_B log_e\left(10^{12}\right)}\right). \tag{1.17}$$

For example, as shown in Section 8, non-interacting dilute assemblies of Fe nanoparticles with a diameter of 2 nm in Ag matrices have $K = 2.6 \times 10^5$ J m^{-3} so from Equation (1.17) their blocking temperature is $T_B = 2.1$ K. The specific time chosen for the lifetime in estimating the blocking temperature is not critical. For example if we specify, instead of 1000 s, one year (3.2×10^7 s), $T_B = 1.5$ K, or the lifetime of the universe (10^{18} s), $T_B = 1$K. So between 1 and 2 K, the 2-nm particles change from essentially, permanent frozen moments to fluctuating moments. The blocking temperature shows a similar insensitivity to τ_0.

It is sometimes useful to determine the particle diameter at which the magnetic moment becomes blocked at a given temperature. From Equation (1.14), this is given, for a spherical particle, as

$$d_{bl} = \left(\frac{6k_B T \ln(\tau f_0)}{\pi K}\right)^{1/2}, \tag{1.18}$$

where $f_0 = 1/\tau_0$. Equation (1.18) is used in Chapter 6 when describing magnetic nanoparticle hyperthermia.

6 MAGNETISATION DYNAMICS IN NANOPARTICLES ABOVE THE BLOCKING TEMPERATURE—SUPERPARAMAGNETISM

Let us first consider the magnetisation in an assembly of single-domain nanoparticles at a temperature above T_B (Equation 1.17) so that the magnetic moments fluctuate freely in zero applied field. In a magnetic nanoparticle, the atomic magnetic moments are locked together by the exchange interaction to form a single 'giant' moment. The total angular momentum quantum number, J, for the cluster is very large compared with an atom and so the quantised μ_z states for the whole nanoparticle form a quasi-continuum with tiny increments between the allowed pointing directions. The nanoparticle can thus be considered as a classical magnetic particle whose magnetic vector can point freely as illustrated in Figure 1.8.

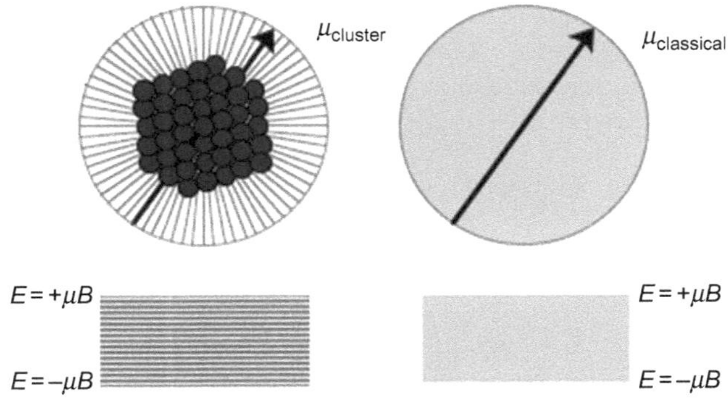

FIGURE 1.8 Atomic moments of the atoms in a nanoparticle are locked together by the exchange interaction to form a single 'giant' cluster moment. The energy states of the different μ_z values (different pointing directions of the moment) form a quasi-continuum so the moment can be treated as a classical vector.

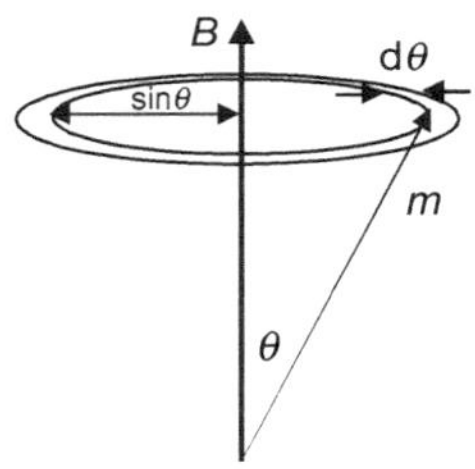

FIGURE 1.9 Suitable shell element to integrate over spherical shell defined by $\theta=0–\pi$.

In an applied field, **B**, the energy of the nanoparticle magnetic moment, $\boldsymbol{\mu}$, is

$$E=-\boldsymbol{\mu}\cdot\mathbf{B}=-\mu B\cos\theta, \tag{1.19}$$

which is minimised when the moment and the field are aligned but perfect alignment is prevented by thermal excitations away from the minimum energy direction. This is a standard problem in classical statistics, which predicts that the probability of the moment pointing along a direction, θ, is proportional to

$$\exp\left(\frac{-E}{k_{\mathrm{B}}T}\right)=\exp\left(\frac{\mu B\cos\theta}{k_{\mathrm{B}}T}\right)=e^{m\cos\theta}, \tag{1.20}$$

where $m=\mu B/k_{\mathrm{B}}T$. The angle, θ, can vary between 0 and π and for an ensemble of particles, the average angle made with the field, $\langle\cos\theta\rangle$, is obtained by integrating the factor $\langle\cos\theta\rangle=\cos\theta e^{m\cos\theta}/e^{m\cos\theta}$ over a spherical shell defined by $\theta=0-\pi$. A suitable shell element is shown in Figure 1.9.

The average angle the moment makes with the field is thus

$$= \frac{\int_0^{\pi} \cos\theta e^{m\cos\theta} \sin\theta d\theta}{\int_0^{\pi} e^{m\cos\theta} \sin\theta d\theta}. \tag{1.21}$$

using $\sin\theta d\theta = d(\cos\theta)$ and writing $x = \cos\theta$, this can be reduced to

$$\langle x \rangle = \frac{\int_0^1 x e^{mx} dx}{\int_0^1 e^{mx} dx} \tag{1.22}$$

and integrating by parts gives

$$\langle x \rangle = \frac{e^m + e^{-m}}{e^m - e^{-m}} - \frac{1}{m} \frac{e^m - e^{-m}}{e^m - e^{-m}} \tag{1.23}$$

that is

$$\langle \cos\theta \rangle = \coth m - \frac{1}{m} \equiv L(m). \tag{1.24}$$

So, the average magnetic moment per particle along the field direction is

$$\overline{\mu} = \mu \langle \cos\theta \rangle = \mu L(m). \tag{1.25}$$

$L(m)$ is the *Langevin Function* and is the equivalent in the nanoparticle system, treated classically, to the Brillouin function $B_J(m)$ in the quantised atomic system. In fact taking the limit $S \rightarrow \infty$ in the Brillouin function produces the Langevin function. It has similar limits, that is $L(m) = 0$ at $m = 0$ and $L(m) \rightarrow 1$ as $m \rightarrow \infty$ (saturation). Figure 1.10 shows the Langevin function plotted for Fe nanoparticles containing 1000 atoms, in

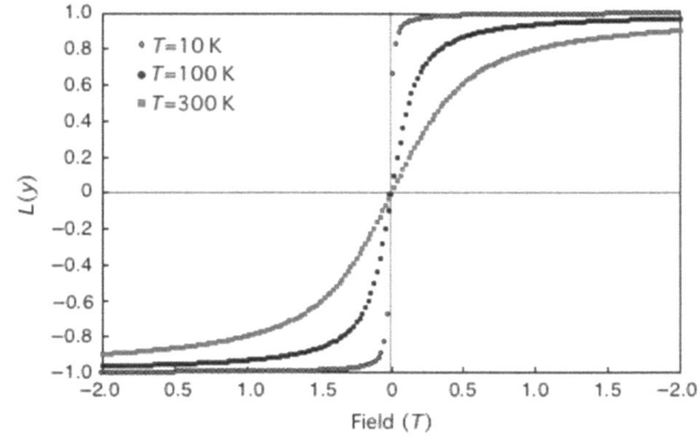

FIGURE 1.10 Langevin functions at $T = 10$, 100 and 300 K for 1000 atom Fe nanoparticles ($\mu = 2.04 \times 10^{-20}$ A m^2), $m = 1478.5 \times B/T$.

which $\mu = 2.04 \times 10^{-20}$A m^2 so, for example at $T = 10$ K, $B = 1$ T, $m = 148$. It is clear that the nanoparticles are much easier to magnetically saturate than atoms because of their large moments. For example even at 300 K, a field of 2 T produced 90% of saturation. Since the paramagnetic behaviour is derived from giant or 'super' moments, it is described as *superparamagnetism*.

The low m limit of the Langevin function is

$$L(m) \approx \frac{m}{3} = \frac{\mu B}{3kT} \tag{1.26}$$

giving

$$\chi = \frac{N\overline{\mu}}{H} = \frac{\mu_0 N \mu^2}{3kT} \tag{1.27}$$

so at sufficiently low fields (much lower than in the case of atoms) the superparamagnetic system also obeys the Curie law, $\chi = C/T$.

7 OBSERVATION OF SUPERPARAMAGNETIC BEHAVIOUR

Superparamagnetism in nanoparticle assemblies has been observed in a variety of systems but some of the clearest data have been obtained from pure Fe and Co nanoparticles embedded in non-magnetic matrices using the ultra-high vacuum (UHV) gas-phase method described in Chapter 3, section 3.5. With this synthesis technique, there is independent control over the nanoparticle size and volume fraction so it is possible to prepare dilute assemblies of nanoparticles of controlled size in which there is no insignificant interaction between them so that single-particle behaviour can be observed. It is then straightforward to increase the volume fraction so the modification of the behaviour due to dipolar–dipolar and at high volume fractions, exchange interactions can be studied. The effect of inter-particle interactions on magnetic behaviour is described in Section 10.

Figure 1.11A shows magnetisation data (dots) measured from 2-nm diameter Fe nanoparticles in Ag matrices with a volume fraction of 1% as a function of temperature in the range of 50–300 K. The blocking temperature of these nanoparticles estimated in Section 5 was around 2 K so in the temperature range used the assembly should show superparamagnetic behaviour. The lines drawn through the data in Figure 1.11A, are fits using the Langevin function (Equation 1.24) and the agreement is excellent in every case. Since the argument in the Langevin function is $m = \mu B/k_B T$, plotting the data against B/T should result in all the magnetisation curves measured at different temperatures lying on top of each other and this is demonstrated in Figure 1.11B. An *in situ* STM image of the Fe nanoparticles produced under the same source conditions and deposited on Si(111) in UHV[7,8] is shown in Figure 1.11C. With low-noise magnetisation data such as that obtained in this experiment, it is feasible to fit more than one Langevin function to each curve

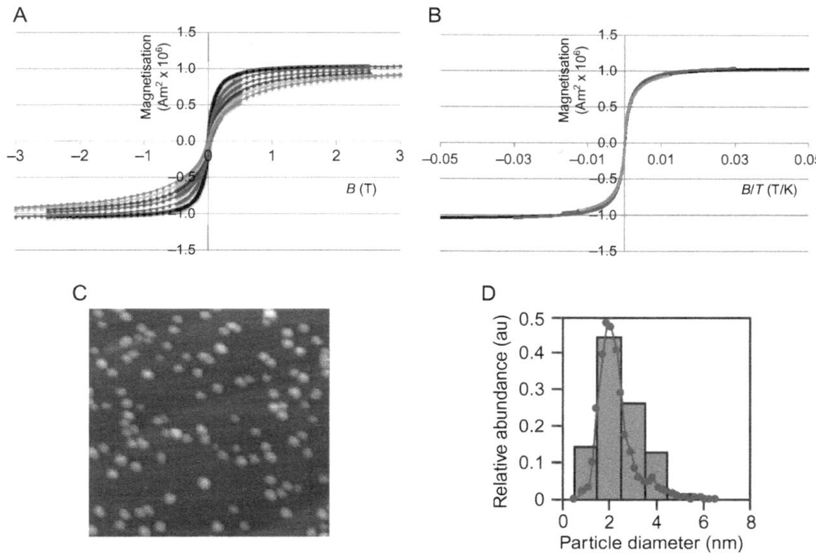

FIGURE 1.11 (A) Magnetisation cruves measured (dots) from 2-nm diameter Fe nanoparticle assemblies in Ag matrices with a volume fraction of 1% as a function of temperature (in the range 50 K to 300 K). The lines are fits using the Langevin function (Equation 1.24) at different temperatures. (B) The same data as in (A) plotted versus B/T to demonstrate the coincidence of all curves as required for a superparamagnetic system. (C) STM image of Fe nanoparticles produced under the same conditions deposited on Si(111). (D) Size distribution obtained from STM images compared to a size histogram produced by fitting multiple Langevin functions to the magnetisation data. *Panels (C) and (D) are reproduced with permission from Ref. 7.*

and treat the amplitude of each as a fitting parameter to obtain the size distribution as well as the mean size. The histogram obtained using this procedure is plotted in Figure 1.11D and superimposed on the size distribution obtained from the STM image (dots/line). It is evident that with well-characterised samples the magnetic behaviour of a nanoparticle assembly above the blocking temperature is exactly as predicted in Section 5.

8 MAGNETISATION IN ASSEMBLIES OF BLOCKED NANOPARTICLES

As shown in Section 6, superparamagnetism is a result of the competition between the magnetic energy trying to align the particle magnetic moments with the applied field and thermal fluctuations that tend to demagnetize them. The magnetisation behaviour of an assembly of nanoparticles below their blocking temperature is determined by the competition between the particle moment aligning with the direction of the applied field or the direction of the local anisotropy axis. To predict the behaviour, we can generalise the uniaxial anisotropy energy expression (1.11) to that in a particle whose anisotropy axis is at an angle θ relative to the magnetisation direction as shown in Figure 1.12. If ϕ is the angle that the particle magnetisation vector makes relative to the direction in which the field is applied, then in zero field, the energy above the ground state is

$$E_\phi = KV\sin^2(\theta - \phi) \tag{1.28}$$

and when the field is applied

$$E_\phi = KV\sin^2(\theta - \phi) - \mu B\cos\phi. \tag{1.29}$$

So the particle moment will rotate to minimise E_ϕ and this will be at some angle between the directions of the field and the anisotropy axis. For small fields and/or large anisotropies, the moment will be close to the anisotropy

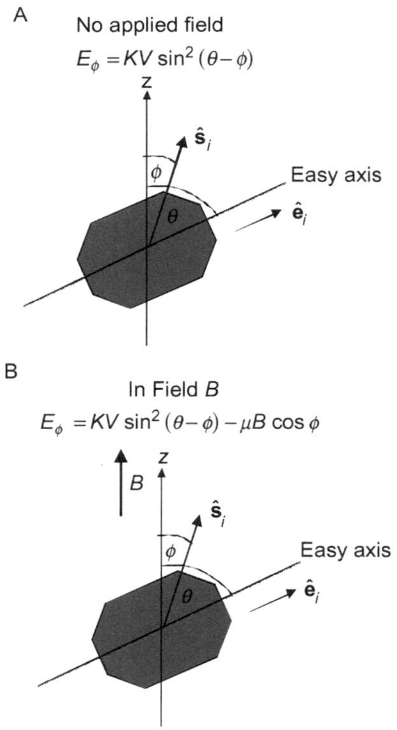

FIGURE 1.12 (A) Dependence of uniaxial anisotropy energy on magnetisation direction in zero field in a nanoparticle whose axis is at an angle to the direction in which the field will be applied. (B) After the field is applied.

axis, but as the field is increased, the magnetic moment will align more closely with it.

To calculate the magnetisation of an ensemble in which the particles (i.e. their anisotropy axes) are randomly oriented, for each field and particle orientation θ, the pointing direction (ϕ) is calculated by minimising the energy in Equation (1.29). This determines the measured magnetic moment along the field direction, $\mu_z = \mu \cos \phi$, where μ is the magnitude of the particle magnetic moment. The values of μ_z are then averaged over all possible particle orientations θ to get the magnetisation of the ensemble.

The magnetisation curves measured at 2 K and the fits using the above model are shown in Figure 1.13 for 2-nm diameter Fe and Co nanoparticles in Ag matrices at 1% and 2% volume fraction, respectively. At this temperature, the samples are expected to be below the blocking temperature for the relevant measurement time ($\sim$100 s) and it is observed that both sets of data display hysteresis. In these films, the particles are randomly oriented so when the samples are first cooled, their magnetisation vectors point randomly in three dimensions and the sample is unmagnetised (Figure 1.13A). Then if the sample is saturated and the field is removed, all the magnetic moments will relax back to the direction along the anisotropy axes that is closest to saturation and will thus be randomly oriented over a hemisphere (insets in Figure 1.13C and E). The remanence in the sample is thus expected to be 50% of saturation. In both samples, the remanence is observed to be less than 50%, which is attributed to the smaller nanoparticles in the ensemble not being blocked at 2 K.[9]

For fields between zero and saturation, the model described above can be used to fit the data and the only unknown in Equation (1.29) is K so this can be treated as an adjustable fitting parameter. The fits are shown in Figure 1.13C and E for Fe and Co nanoparticles, respectively, and are observed to closely model the data. This is clearly a good way of extracting the anisotropy constant in nanoparticle ensembles, and the quality of the fit indicates that the initial assumption that the nanoparticle anisotropy is uniaxial is correct. For both Fe and Co, the value of K is significantly larger than the bulk magnetocrystalline anisotropy.

The vast majority of synthesised nanoparticles display uniaxial anisotropy even if they are composed of materials like Fe that have cubic magnetocrystalline anisotropy in the bulk. An anisotropy symmetry that is higher than uniaxial would only occur in clusters containing magic numbers of atoms as according to Xie and Blackman[10] adding a few atoms to one of the facets is sufficient to induce uniaxial anisotropy along an axis. Only a small proportion of nanoparticles in an assembly contain magic numbers, so the vast majority of the clusters are expected to show uniaxial anisotropy. Even for the nanoparticles that contain magic numbers, the process of depositing and embedding the clusters is likely to induce stresses that lower the symmetry of the anisotropy.

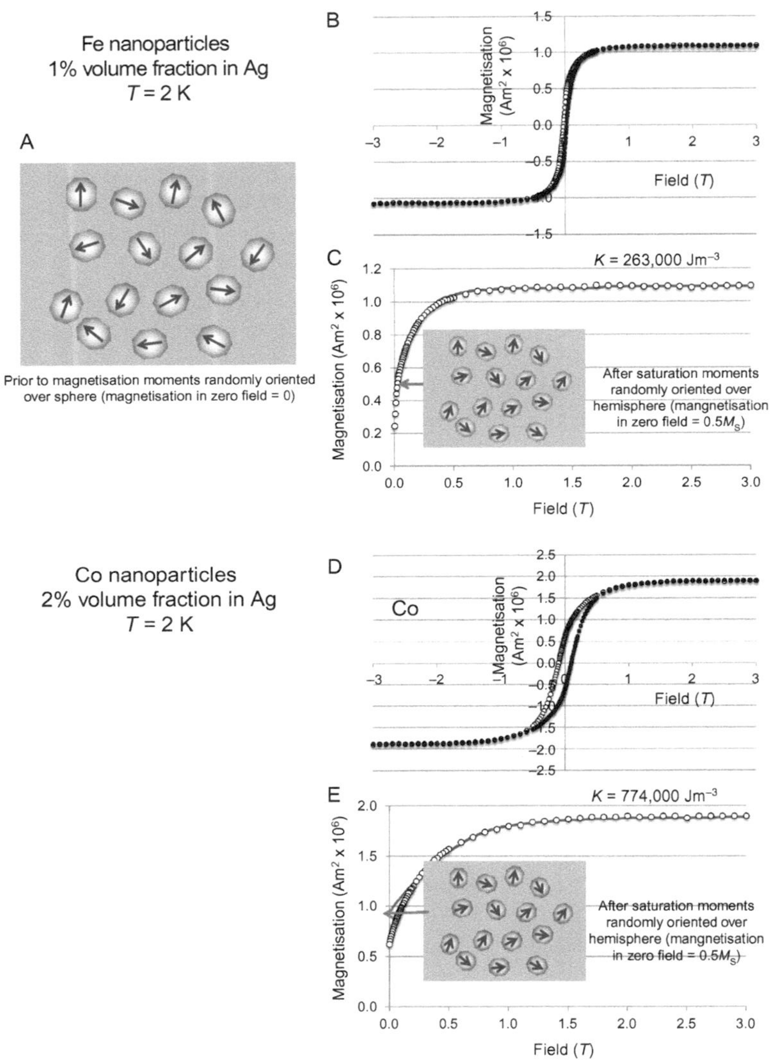

FIGURE 1.13 Magnetic behaviour of 2-nm diameter Fe and Co nanoparticles with a volume fraction of 1% and 2%, respectively, in Ag matrices at 2 K. (A) After deposition, the nanoparticle anisotropy axes are randomly oriented over a sphere so prior to applying a field the sample magnetisation is zero. (B) Magnetisation of Fe nanoparticles for the field swept down (empty circles) and the field swept up (filled circles). (C) Fit to the data (field swept down) using the model described in the text with the K value required to optimise the fit. The model remanence is 50% as required but the sample resonance is less. (D) and (E) as (B) and (C) but for Co nanoparticles.

9 ENHANCED MAGNETIC MOMENTS IN NANOPARTICLES

Chapter 2 deals with the rigorous theory of orbital and spin moments in nanoparticles but here is presented a simplistic and general overview of how magnetic moments are expected to be modified relative to bulk values. The discussion will be restricted to transition metal nanoparticles, which are the most relevant to applications and for which the vast majority of studies have been done. Rare-earth nanoparticles show a rich variety of behaviour as a function of particle size including spin canting and ferrimagnetic structures and need to be considered on an element-by-element and size-by-size basis. Readers are referred to existing reviews.[11]

9.1 Orbital Magnetic Moments

It is well known that the orbital moment of a free transition metal atom is quenched when it is within a solid or any matrix with which it interacts. This is shown most starkly by the effective magneton number, p, defined by

$$p \equiv g\sqrt{J(J+1)} \tag{1.30}$$

If we compare the value of p obtained by Hund's rules with the measured values for transition metal ions in paramagnetic salts shown in Table 1.2, there appears to be no correlation as demonstrated more clearly by the data plotted in Figure 1.14. On the other hand, there is quite good agreement between the data and the p value obtained by arbitrarily setting $L=0$ after applying Hund's rules to get S, L and J. Thus, it appears that when the transition metal ions interact with the ligands in the salts the orbital moment is quenched.

The same phenomenon does not occur in rare-earth ions since the 4f orbitals responsible for magnetism in the rare earths are shielded from their environment by the outermost 6s and 6p electrons of the atom. The magnetic 3d electrons in transition metals interact strongly with their environment and it is this interaction that quenches the orbital moment. In general terms, the reason is that the interaction lifts degeneracies of the d orbitals and it is well known that a non-degenerate quantum state must be represented by a real wavefunction, say ψ_{real}. On the other hand, the operator for the z-component of the orbital angular momentum is

$$\langle L_z \rangle = -i\hbar\frac{\partial}{\partial\varphi}, \tag{1.31}$$

that is, purely imaginary. Thus, the expectation value $\langle \psi_{\text{real}}|L_z|\psi_{\text{real}}\rangle$ vanishes. In the bulk magnetic transition metals, the orbital moment has a small residual value due to spin–orbit coupling.

A nanoparticle has a high proportion of atoms at the surface atomic layer (e.g. ~50% for a nanoparticle with a diameter of 2 nm) and the interaction of

TABLE 1.2 Measured and Derived Magneton Numbers for the Transition Metal Series

Ion	Configuration	State	p (Hund's Rules) $= g\sqrt{J(J+1)}$	p (Measured)	$p\ (L=0) = 2\sqrt{S(S+1)}$
V^{4+}	$3d^1$	$^2D_{3/2}$	1.55	1.8	1.73
V^{3+}	$3d^2$	$^3F_2$	1.63	2.8	2.83
V^{2+}	$3d^3$	$^4F_{3/2}$	0.77	3.8	3.87
Cr^{3+}	$3d^3$	$^4F_{3/2}$	0.77	3.7	3.87
Mn^{4+}	$3d^3$	$^4F_{3/2}$	0.77	4.0	3.87
Cr^{2+}	$3d^4$	$^5D_0$	0	4.8	4.90
Mn^{3+}	$3d^4$	$^5D_0$	0	5.0	4.90
Mn^{2+}	$3d^5$	$^6S_{5/2}$	5.92	5.9	5.92
Fe^{3+}	$3d^5$	$^6S_{5/2}$	5.92	5.9	5.92
Fe^{2+}	$3d^6$	$^5D_4$	6.70	5.4	4.90
Co^{2+}	$3d^7$	$^4F_{9/2}$	6.54	4.8	3.87
Ni^{2+}	$3d^8$	$^3F_4$	5.59	3.2	2.83
Cu^{2+}	$3d^9$	$^2D_{3/2}$	3.55	1.9	1.73

There is no agreement except for the d^5 configuration. The final column shows the value of p that would be obtained if we arbitrarily set $L=0$. These values are close to the measured ones.

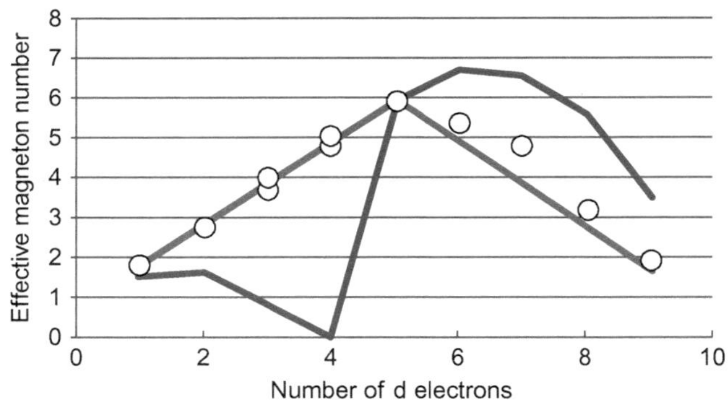

FIGURE 1.14 Effective magneton number, p (Equation 1.30), obtained using Hunds rules (thin line) compared to the measured values from various transition metal ions as a function of the number of d electrons (circles). There appears to be no correlation but the data appear to follow closely the thick line, which is the effective magneton number obtained by arbitrarily setting $L=0$.

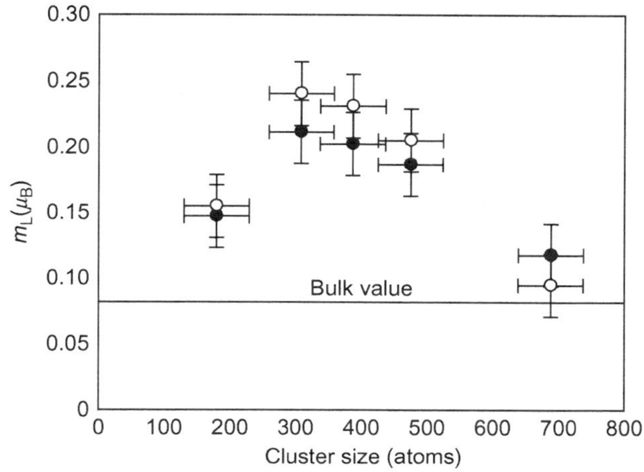

FIGURE 1.15 Orbital magnetic moment per atom in Fe nanoparticles deposited on graphite in UHV as a function of the number of atoms in the particles. The bulk value is indicated by the grey line. The data were taken at X-ray incidence angles of 0° (filled circles) and 55° (open circles) and shows little anisotropy of the moment.

these with their neighbours has a lower symmetry than in the case of an atom in the bulk. Thus, the lifting of degeneracy and the quenching of the orbital moment is reduced, giving the nanoparticles a higher orbital moment per atom than the bulk material. This is borne out by the detailed calculations presented in Chapter 2 and has been confirmed by measurements. The orbital magnetic moment of free Co clusters containing 8–22 atoms was measured by X-ray Magnetic Circular dichroism (XMCD) to be four to six times larger than the bulk value.[12] Figure 1.15 shows the orbital moment measured by XMCD in Fe nanoparticles deposited onto graphite surfaces in UHV as a function of nanoparticle size.[13] The largest particles have an orbital moment close to the bulk value indicated by the grey line but the moment increases with decreasing size reaching a peak about three times the bulk value. The data were taken at two different incidence angles of the X-ray beam and show no significant anisotropy of the orbital moment. The orbital moment is observed to decrease for nanoparticles containing less than 300 atoms, but it is believed that this is part of an oscillatory behaviour of the orbital moment versus size. It is observed to increase again for very small cluster sizes (<100 atoms)[14] and also for larger cluster sizes of 6 nm (>9000 atoms)[15] before settling back to the bulk value.

9.2 Spin Magnetic Moments

As with the orbital moment, the behaviour of the spin moment at the nanoscale can be understood in general terms by considering the behaviour of a nanoparticle to be intermediate between the atom and the bulk. In the atom, the spin moment is obtained by Hund's rules and applying these to Fe, Co and Ni atoms gives the orbital spin and total magnetic moments shown in Table 1.3. By comparison, the measured magnetic moments per atom in the metal are shown in the final column and indicate a large reduction compared with the free atom. This is partly explained by orbital moment quenching discussed in the previous section, but in addition, the spin moment is reduced to just over half the atomic value in all three metals.

The reason for the reduction is that the Hund's rule ground state is broadened by interactions with atomic neighbours in the bulk and overlaps with similarly broadened excited states with anti-parallel spin moments. In a metal, it is more appropriate to consider the electrons filling overlapping bands with majority and minority spin moments as illustrated in Figure 1.16. Some of the majority spin band states lie above the Fermi level and are not filled, while some of the minority spin band states lie below the Fermi level and are filled cancelling that proportion of majority spins. Thus, the spin moment is reduced below that of the atom and is a non-integer multiple of Bohr magnetons per atom. The high proportion of surface atoms in a nanoparticle reduces the average atomic co-ordination and thus the 3d bandwidth (in transition metals). If the bandwidth is reduced and the valence band exchange splitting remains the same, the imbalance between the majority and minority spin bands increases, giving a higher spin moment per atom. In the atomic limit, the bands become sharp energy levels producing a complete spin polarisation and the Hund's rule value for the spin moment.

Figure 1.17 shows the spin moment measured by XMCD in Fe nanoparticles deposited on graphite[13] as a function of the number of atoms in the particles compared with the bulk value indicated by the grey line. In general, the pure spin moment cannot be determined directly from the XMCD measurements as the sum rules yield the value of $m_S + 7m_T$ where m_T is the dipole term, which is related to the asymmetry of the spin distribution around the atom.[16] In bulk systems, the value of m_T is very small compared with the spin moment and is often ignored during data analysis. This assumption cannot be made in the case of low-dimensional systems and nanoparticles however it has been shown that, within certain assumptions, the dipole term goes to zero for data taken at an X-ray incidence angle of 55° (the 'magic angle').[16] Figure 1.17 shows the spin moments obtained from data at X-ray incidence angles of 0° and 55° revealing the magnitude of the dipole contribution. In principle, the pure spin moment (and the value to be compared with measurements by other methods) is obtained at the magic angle and this is evidently enhanced relative to the bulk value for cluster sizes below 700 atoms.

The discussion above pertains to the transition metals. In rare-earth metals, there is little interaction between the 4f levels responsible for magnetism producing no overlap between the majority and minority spin states; thus, the full Hund's rule spin moment is preserved in the bulk. For the same reason, there

TABLE 1.3 Spin and Orbital Magnetic Moments of Fe, Co and Ni Atoms Obtained Using Hunds Rules

Element	No. of d Electrons	m_l −2	−1	0	1	2	|S|	|L|	|J|	Orbital Moment (μ_B)	Spin Moment (μ_B)	Total Moment in Atom (μ_B)	Total Moment/Atom in Metal (μ_B)
Fe	6	↓↑	↓	↓	↓	↓	2	2	4	2	4	6	2.22
Co	7	↓↑	↓↑	↓	↓	↓	3/2	3	9/2	3	3	6	1.7
Ni	8	↓↑	↓↑	↓↑	↓	↓	1	3	4	3	2	5	0.6

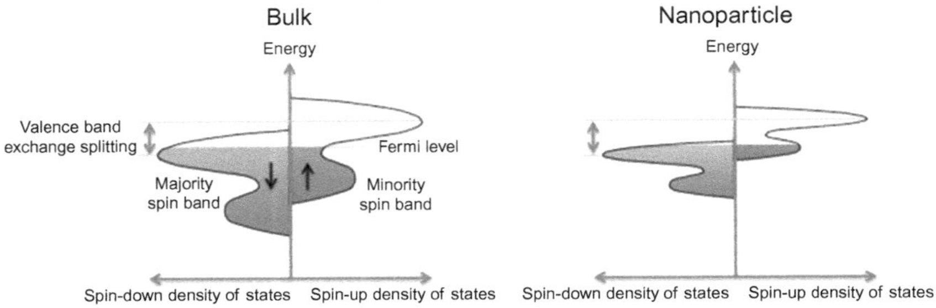

FIGURE. 1.16 Increased spin polarisation in a nanoparticle due to the narrowing of the 3d bands.

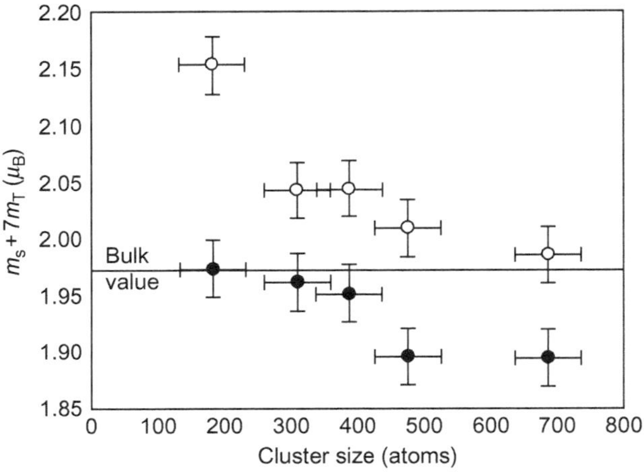

FIGURE 1.17 Spin magnetic moment per atom in Fe nanoparticles deposited on graphite in UHV as a function of the number of atoms in the particles. The bulk value is indicated by the grey line. The data were taken at X-ray incidence angles of 0° (filled circles) and 55° (open circles). The data at 55° (the magic angle) show the pure spin moment with no dipole contribution (see text).

is no orbital moment quenching so the total magnetic moment per atom in the metal is similar to that of the free atom. The situation in nanoparticles is complicated however since the modified interactions at the surface can produce canted spins and antiferromagnetism, depending on the cluster material, size and temperature. The magnetic moment in rare-earth nanoparticles is often smaller than the bulk value and must be obtained on a temperature-by-temperature and size-by-size basis.[11]

10 THE EFFECT OF NANOPARTICLE INTERACTIONS ON MAGNETIC BEHAVIOUR

The discussion so far has focused on the native properties of individual nanoparticles or non-interacting assemblies. Thus, for example, the magnetisation loops presented in Figures 1.11 and 1.13 have the same shape as would be obtained from one particle but with the signal multiplied by the number of particles (leaving aside the effect of the size distribution). In this section, we will consider how the behaviour is modified when the particles have a sufficiently high density in the assembly to interact.

10.1 Volume Fractions Below the Percolation Threshold

The cluster deposition technique described in Chapter 3, section 3.5 is best suited to producing samples to study interaction effects as the deposited particle size can be kept constant while the density is adjusted by altering the deposition rates of the nanoparticles and matrix material. Thus, the volume fraction and nanoparticle size can be controlled independently. The evolution of the magnetisation curves at 300 K for Fe nanoparticles with the size distribution shown in Figure 1.11 in Ag matrices as a function of the Fe volume fraction is shown in Figure 1.18 for nanoparticle densities below the percolation threshold. Included is a sample at the dilute limit (Fe_1Ag_{99}), which shows superparamagnetic behaviour (see Figure 1.11), and the set of Langevin functions weighted with the particle size distribution shown in Figure 1.11 has been fitted to this data (green line).

The most noticeable change as the nanoparticle volume fraction increases is the rise in the low-field susceptibility. This would occur if the assembly contained larger particles and indeed it is possible to get visually good fits to the data using Langevin functions with larger particle diameters in the

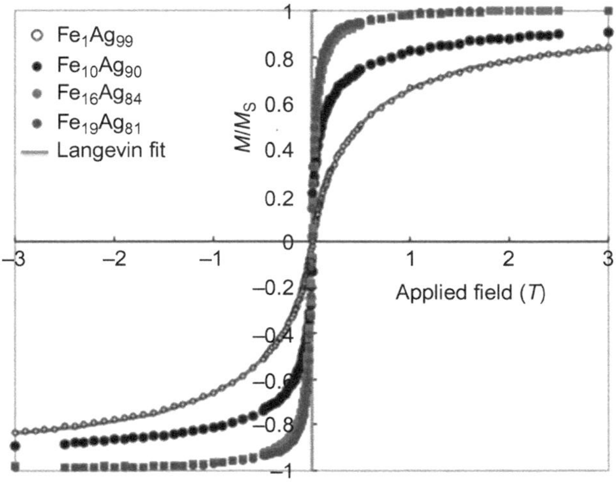

FIGURE 1.18 Magnetisation curves for Fe nanoparticles with the size distribution shown in Figure 1.11 in Ag as a function of the Fe volume fraction. The line through the open circles is a fit using a set of Langevin functions weighted by the size distribution of the deposited particles. *Reproduced with permission from Ref. 7.*

argument. It is erroneous however to treat these films as superparamagnetic assemblies of larger particles than the deposited ones because the data do not have other characteristics required to demonstrate superparamagnetism. For example, plotting the data against B/T does not cause the curves taken at different temperatures to line up with each other as demonstrated for the dilute sample in Figure 1.11B. It is also found that the particle size required in the Langevin function to get a good fit to the measured data depends on the temperature at which the data were taken. As pointed out in Ref. 17, both these features violate the criteria for superparamagnetism. The reason is that, while at the higher densities there is aggregation so that the magnetic inclusions are larger than the deposited particles, these aggregates interact with each other. A detailed Monte Carlo simulation of the $Fe_{10}Ag_{90}$ film[7] showed that it consists of aggregates of particles, whose magnetic moments are strongly coupled within the aggregate by exchange forces, which interact with each other via dipolar forces.

A simple model for such a system has been presented by Allia *et al.*[17], which enables one to determine the approximate size of the aggregates at a given volume fraction and parameterise the strength of the dipolar coupling. The approach is to include within the Langevin function an interaction parameter, T^*, with the units of temperature similarly to the Curie–Weiss theory of ferromagnetism. That is, the magnetisation of an assembly of N interacting aggregates in a field B at temperature T is given by

$$M = N\mu L\left(\frac{\mu B}{k(T+T^*)}\right) \tag{1.32}$$

where μ is the total magnetic moment of an aggregate. It was shown that if one fits simple Langevin functions (i.e. including no interaction) to the magnetisation data using a size parameter d_a in the arguments, this is related to the true diameter, d, by the expression

$$d_a = \left(\frac{1}{1+\frac{T^*}{T}}\right)^{1/3} d \tag{1.33}$$

Thus by plotting d_a versus temperature and fitting the curve predicted by Equation (1.33) using d and T^* as parameters, it is possible to obtain the true aggregate size (d) and the strength of the dipolar interactions.

The result of applying this procedure to the magnetisation data of the Fe_1Ag_{99} and $Fe_{10}Ag_{90}$ films at different temperatures is shown in Figure 1.19. In the case of the dilute Fe_1Ag_{99} film, which has already been shown to be superparamganetic, the procedure gives a true diameter that is the same as that of the deposited nanoparticles and an interaction parameter of $T^* = 0$ as expected. In the case of the $Fe_{10}Ag_{90}$ film, the true aggregated diameter is 5.7 nm (so around 6–7 of the deposited nanoparticles) and the dipolar interactions are characterised by an equivalent temperature of 94 K. The true

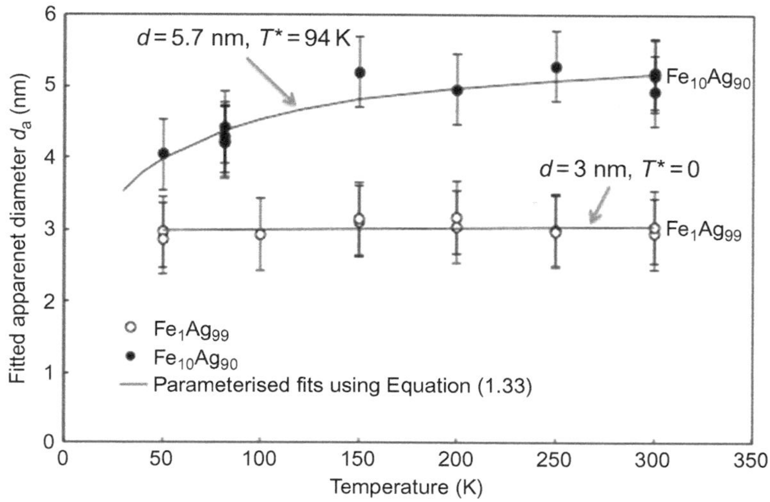

FIGURE 1.19 Fitted particle (or aggregate) diameter as a function of temperature for Fe_1Ag_{99} films (open circles) and $Fe_{10}Ag_{90}$ films (filled circles). *Reproduced with permission from Ref. 7.*

aggregate diameter and interaction parameter both increase with the volume fraction but the basic assumption of the model, that is aggregates interacting via dipolar forces, breaks down as the volume fraction approaches the percolation threshold (~25% for a three-dimensional assembly).

10.2 Volume Fractions Above the Percolation Threshold

In a three-dimensional assembly of random sites such as in samples of deposited nanoparticles within a matrix, the percolation threshold is in the range of 25–30%[18] and for volume fractions above this there is metal-to-metal contact between particles in chains extending across the sample. Thus, the entire nanoparticle assembly becomes exchange coupled. In order to understand the behaviour of this type of interacting nanoparticle system, it is easiest to start at the volume fraction extreme of 100%, that is, pure nanoparticle films with no matrix.

For this type of film, a suitable model developed by Chudnovsky and co-workers[19–23] is based on a random anisotropy formalism in which the anisotropy changes on a length scale much smaller than the domain width (see Section 2). In their approach, the magnetic ground state in a granular film is determined by the relative strength of a random anisotropy field

$$H_r = \frac{2K_r}{M_s} \tag{1.34}$$

and an exchange field

$$H_{\mathrm{ex}} = \frac{2A}{M_{\mathrm{S}} R_{\mathrm{a}}^2} \tag{1.35}$$

Here, K_{r} is the (randomly oriented) anisotropy of the grains, M_{S} is their saturation magnetisation, A is the exchange constant for the interaction between the grains (see Equation 1.13) and R_{a} is the size of the region over which the local anisotropy axis is correlated, that is the characteristic grain size. The ratio of the fields is given by the dimensionless parameter

$$\lambda_{\mathrm{r}} = \frac{H_{\mathrm{r}}}{H_{\mathrm{ex}}} \tag{1.36}$$

The RA model predicts that the approach to saturation of the magnetisation of the films in an applied field H is given by

$$M = M_{\mathrm{s}} \left(1 - \frac{1}{30} \frac{\lambda_{\mathrm{r}}^2}{\sqrt{h_{\mathrm{ex}}}} \int_0^{\infty} \mathrm{d}x C(x) x^2 \exp\left[x\sqrt{h_{\mathrm{ex}}} \right] \right), \tag{1.37}$$

where $h_{\mathrm{ex}} = H/H_{\mathrm{ex}}$, and $C(x)$ is the correlation function for the anisotropy axes with x in units of R_{a}. In a cluster-assembled film with monodisperse clusters, $C(x)$ can be taken to be a simple step function with a cut-off at $x = 1$. A key feature of the RA model is that it describes how pure cluster films can be magnetically soft when composed of clusters with a high anisotropy constant by an averaging of the anisotropy over all directions.

The model was originally applied to amorphous films in which a local, randomly oriented, anisotropy is due to local atomic disorder. It is thus perfectly adapted to describing the magnetisation in films of deposited nanoparticles in which the distance R_{a} over which an anisotropy axis is correlated is well defined (i.e. the particle diameter). For $\lambda_{\mathrm{r}} > 1$, the magnetic correlation length at zero field is R_{a}, and the magnetic vector in each particle points along the local intra-particle anisotropy axis. With increasing inter-particle exchange (or decreasing intra-particle anisotropy), the configuration becomes a correlated spin glass in which the magnetisation vector in neighbouring particles is nearly aligned but the random deviation of the moments from perfect alignment produces a smooth rotation of the magnetisation throughout the system with a magnetic correlation length that is a factor $1/\lambda_{\mathrm{r}}^2$, larger than the particle diameter. Since the film is composed of single-domain particles with super-moments, it is more appropriate to call this state a correlated super-spin glass (CSSG). Thus at $\lambda_{\mathrm{r}} = 1$, the ground state changes from a simple spin glass where the magnetisation in each nanoparticle points in a random direction to a CSSG and the difference between the two configurations is illustrated in Figure 1.20. The RA model has been used successfully to model the magnetisation data in several magnetometry studies of deposited-nanoparticle films.[7,9,24–28]

A simpler intuitive argument for the loss of long-range order emerges from a consideration of the average displacement from perfect alignment

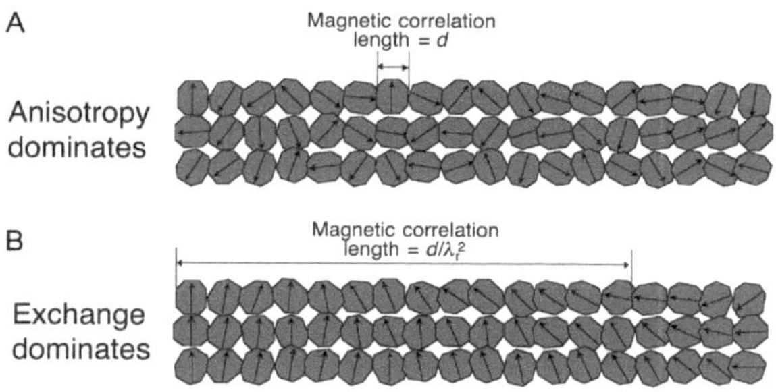

FIGURE 1.20 A stack of randomly oriented nanoparticles with the slight elongation of each one representing the anisotropy axis. In (A), $\lambda_r \geq 1$ and the magnetisation vector points along the local anisotropy axis so the magnetic correlation length is a single-particle diameter (simple spin glass). In (B), $\lambda_r < 1$ and the magnetic vectors are nearly aligned. The random perturbation from perfect alignment results in a finite magnetic correlation length that is a factor $1/\lambda_r^2$ larger than a single particle (correlated super-spin glass). *Reproduced with permission from Ref. 9.*

between two neighbouring nanoparticle moments. If the magnetisation vector of a particle is taken as a reference axis relative to which the anisotropy axis of a neighbouring particle makes an angle ϕ, then if the angle between the magnetisation vectors of the two particles is θ the value of θ will be given by the minimum of the energy term

$$H_{ex}\langle\cos\theta\rangle + H_r\langle\sin^2(\phi-\theta)\rangle \qquad (1.38)$$

Along a line of N particles, the value of N required for the average displacements to add up to 90° (when correlation with the original spin is lost) is proportional to $(H_{ex}/H_r)^2$, that is, proportional to $1/\lambda_r^2$, as predicted by the RA model.

The conclusion of this simple argument is that in the absence of an applied field, a film of deposited nanoparticles should exhibit no domain structure. The uniformly rotating magnetisation of the CSSG state produces no external field and thus removes the competition between the magnetostatic term and the exchange interaction, which in a normal film produces domains (see Section 2). Magnetisation patterns above a deposited nanoparticle film have been obtained using magnetic force microscopy (MFM) and show a randomised patchwork of fields at the nanoscale as expected for a CSSG.[28,29] In all cases studied so far however the films were composed of core–shell Co/CoO particles and in addition the stray field from an MFM tip can be more than 1000Oe,[30] depending on the imaging conditions and the tip coating. This may be strong enough to alter the magnetisation pattern of the ground state. An alternative method to image domain structure is X-ray photoelectron microscopy (XPEEM) using circularly polarised X-rays around the Fe L edge in which circular dichroism is used to provide contrast between different magnetisation directions.[31] Generally, XPEEM does not have as good a spatial resolution at MFM but it provides images in a field-free environment,

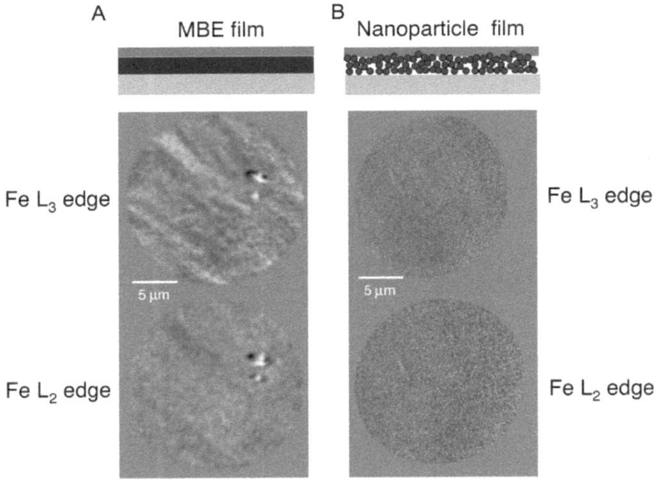

FIGURE 1.21 (A) XPEEM images at the Fe L_3 and L_2 edges of the domain structure in a 200-Å thick MBE-grown film on Si(111). (B) Similar XPEEM images from the cluster-assembled film showing no sign of a domain structure. *Reproduced with permission from Ref. 31.*

and the lack of a domain structure in pure nanoparticle films is clearly demonstrated in Figure 1.21. This compares images taken from a conventional Fe thin film (Figure 1.21A) and a film of deposited Fe nanoparticles of the same thickness (Figure 1.21B).[31] While the conventional film shows a clear domain pattern (with opposite contrast at the L3 and L2 edges as expected), the images from the nanostructured film are a uniform grey indicating that any magnetic correlation is less than the resolution of the probe (~50 nm). The lack of a domain structure in nanoparticle films makes them magnetically soft and also it has been shown that their magnetisation switches faster than in conventional films.[32]

The discussion has focused on pure nanoparticle films with no matrix but the essential physics remains the same in any exchange coupled array above the percolation including the case when there is a matrix material filling the gaps. For example, in the case of cluster deposited $Fe_{70}Ag_{30}$ films, the approach to saturation was well described by the random anisotropy model but with an exchange field strength (Equation 1.35) lower than that for a pure nanoparticle film.[7]

REFERENCES

1. Kittel C. Theory of the structure of ferromagnetic domains in films and small particles. *Phys Rev* 1946;**70**:965–71.
2. O'Handley RC. *Modern magnetic materials: principles and applications.* New York, NY: Wiley; 2000.

3. Williams RJP. Iron and the origin of life. *Nature* 1990;**343**:213–4.
4. Jackson TJ, Binns C, Forgan EM, Morenzoni E, Niedermayer Ch, Glückler H, et al. Superparamagnetic relaxation in Fe nanoclusters measured by low energy muon spin rotation. *J Phys Condens Matter* 2000;**12**:1399–411.
5. Mironov VL, Gribkov BA, Vdovichev SN, Gusev SA, Fraerman AA, Ermolaeva OL, et al. Magnetic force microscope tip-induced remagnetization of CoPt nanodisks with perpendicular anisotropy. *J Appl Phys* 2009;**106**:053911.
6. Brown Jr WF. Thermal fluctuations of a single domain particle. *Phys Rev* 1963;**130**:1677–86.
7. Binns C, Maher MJ, Pankhurst QA, Kechrakos D, Trohidou KN. Magnetic behavior of nanostructured films assembled from preformed Fe clusters embedded in Ag. *Phys Rev B* 2002;**66**:184413.
8. Upward MD, Cotier BN, Moriarty P, Beton PH, Baker SH, Binns C, et al. Deposition of Fe clusters on Si surfaces. *J Vac Sci Technol B* 2000;**18**:2646–9.
9. Binns C, Maher MJ. Magnetic behaviour of thin films produced by depositing pre-formed Fe and Co nanoclusters. *New J Phys* 2002;**4**:85.1–85.15.
10. Xie Y, Blackman J. Magnetic anisotropy of nanoscale cobalt particles. *J Phys Condens Matter* 2004;**16**:3163–72.
11. Bucher JP. Magnetism of free and supported metal clusters. In: Khanna SN, Castleman AW, editors. *Quantum phenomena in clusters and nanostructures*. Berlin: Springer-Verlag; 2003 [chapter 5].
12. Peredkov Neeb SM, Eberhardt W, Meyer J, Tombers M, Kampschulte H, Niedner-Schatteburg G. Spin and orbital magnetic moments of free nanoparticle. *Phys Rev Lett* 2011;**107**:233401.
13. Baker SH, Binns C, Edmonds KW, Maher MJ, Thornton SC, Louch S, et al. Enhancements in magnetic moments of exposed and co-coated Fe nanoclusters as a function of cluster size. *J Magn Magn Mater* 2002;**247**:19–25.
14. Lau JT, Föhlisch A, Nietubyc R, Reif M, Wurth W. Size-dependent magnetism of deposited small iron clusters studied by X-ray magnetic circular dichroism. *Phys Rev Lett* 2002;**89**:057201.
15. Kleibert A, Meiwes-Broer K-H, Bansmann J. Size-dependent magnetic spin and orbital moments of Fe nanoparticles deposited onto Co/W(110). *Phys Rev B* 2009;**79**:125423.
16. Binns C, Blackman JA. Magnetism in supported and embedded clusters. In: Blackman JA, editor. *Metallic nanopaarticles*. Hungary: Elsevier; 2008.
17. Allia P, Coisson M, Tiberto P, Vinai F, Knobel M, Novak MA, et al. Granular Cu-Co alloys as interacting superparamagnets. *Phys Rev B* 2001;**64**:144420.
18. Wang J, Zhou Z, Zhang W, Garoni TM, Deng Y. Bond and site percolation in three dimensions. *Phys Rev E* 2013;**87**:052107.
19. Chudnovsky EM. A theory of two-dimensional amorphous ferromagnet. *J Magn Magn Mater* 1983;**40**:21–6.
20. Chudnovsky EM, Saslow WM, Serota R. Ordering in ferromagnets with random anisotropy. *Phys Rev B* 1986;**33**:251–6.
21. Saslow WM. Ferromagnets with weak random anisotropy. *Phys Rev B* 1987;**35**:3454–60.
22. Chudnovsky EM. Magnetic properties of amorphous ferromagnets. *J Appl Phys* 1988;**64**:5770.
23. Chudnovsky EM. Random anisotropy in amorphous alloys. In: Fernandez-Baca JA, Ching W-Y, editors. *The magnetism of amorphous metals and alloys. Springer series in solid-state sciences*, vol. 80. Singapore: World Scientific; 1995.
24. Loffler JF, Meier JP, Doudin B, Ansermet J-P, Wagner W. Random and exchange anisotropy in consolidated nanostructured Fe and Ni: role of grain size and trace oxides on the magnetic properties. *Phys Rev B* 1998;**57**:2915–24.

25. Loffler JF, Braun H-B, Wagner W. Magnetic correlations in nanostructured ferromagnets. *Phys Rev Lett* 2000;**85**:1990–3.
26. Thomas L, Tuaillon J, Perez JP, Dupuis V, Perez A, Barbara B. Approach to saturation in nanocrystallized films of iron and nickel. *J Magn Magn Mater* 1995;**140**:437–8.
27. Perez JP, Dupuis V, Tuaillon J, Perez A, Paillard V, Melinon P, et al. Magnetic properties of nanostructured iron films obtained by low energy neutral cluster beam deposition. *J Magn Magn Mater* 1995;**145**:74–80.
28. Koch SA, Palasantzas G, Vystavel T, De Hosson JTM, Binns S, Louch S. Magnetic and structural properties of Co nanocluster thin films. *Phys Rev B* 2005;**71**:085410.
29. Koch SA, Velde RHT, Palasantzas G, De Hosson JTH. Magnetic force microscopy on cobalt nanocluster films. *Appl Surf Sci* 2004;**226**:185.
30. Kleiber M, Kümmerlen F, Löhndorf M, Wadas A, Weiss D, Wiesendanger R. Magnetization switching of submicrometer Co dots induced by a magnetic force microscope tip. *Phys Rev B* 1998;**58**:5563–7.
31. Binns C, Howes PB, Baker SH, Marchetto H, Potenza A, Steadman P, et al. Loss of long-range magnetic order in a nanoparticle assembly due to random anisotropy. *J Phys Condens Matter* 2008;**20**:055213.
32. Binns C, Sirotti F, Gruegel H, Prieto P, Baker SH, Thornton SC. Static and dynamic magnetic behaviour of iron nanoclusters on magnetic substrates. *J Phys Condens Matter* 2003;**15**:4287–300.

Chapter 2

Spin and Orbital Magnetism in Free Nanoparticles: Size, Composition, and Temperature Effects

J. Dorantes-Dávila* and G.M. Pastor†
*Instituto de Física, Universidad Autónoma de San Luis Potosí, San Luis Potosí, Mexico
†Institut für Theoretische Physik, Universität Kassel, Kassel, Germany

1 INTRODUCTION

Magnetism is probably the most well known, everyday tangible manifestation of the quantum nature of matter. In transition metals (TMs) and their alloys, magnetism originates in the valence electrons belonging to the open atomic d shell, which are fairly localized and thus tend to develop a non-vanishing spin polarization. The formation of local magnetic moments corresponds to an uneven occupation of electronic states having different spin projections. It is the result of a subtle interplay between the electron–electron Coulomb interactions, which favour spin polarizations, and the kinetic and potential energies, which favour electronic delocalization throughout the lattice and band broadening. The diversity of magnetic behaviours that result from these competing effects can hardly be overstated. Depending on the strength of the intra-atomic Coulomb interactions, the *d* band width, and band filling, one may find TM compounds showing, for example, strong ferromagnetic (FM) order with robust saturated spin moments, weak ferromagnetism with non-saturated moments, paramagnetic (PM) behaviour, antiferromagnetic (AF) arrangements, or even more complex ferrimagnetic and non-collinear orders. Clusters, nanoparticles (NPs), and the nanostructures made out of them display a far richer behaviour as a function of size, structure, and composition.

The magnetic properties of 3d TM NPs have been investigated with a variety of experimental techniques, which originate from different related disciplines such as molecular, surface, and solid-state physics.[1–15] Stern–Gerlach (SG) deflection measurements on size-selected beams have provided information on the average magnetization per atom $\overline{\mu}_N$ of isolated Fe_N, Co_N, and Ni_N

Frontiers of Nanoscience, Vol. 6. http://dx.doi.org/10.1016/B978-0-08-098353-0.00002-6

clusters, including their dependences as a function of temperature T.[13–16] These measurements show that Fe, Co, and Ni clusters are FM with low-temperature values of $\overline{\mu}_N$ that are significantly larger than the ground-state magnetization of the corresponding solids. In this way, some of the early ground-state theoretical predictions could be confirmed.[17,18] In addition, a very interesting temperature dependence of the SG deflection profiles has been reported from which information on the temperature dependence of $\overline{\mu}_N$ could be inferred.[9–14] Putting aside some important differences in the shape of the magnetization curves and in the characteristic temperature scales (cluster Curie-like temperatures),[19] the SG results demonstrate that the FM order in 3d TM clusters is stable at finite temperatures. More recent studies based on an electronic spin-fluctuation theory have confirmed the observed strong stability of the FM order within small clusters.[20] Nevertheless, a detailed understanding the temperature-dependent SG deflection profiles remains a challenge.

Spectroscopic experiments such as X-ray magnetic circular dichroism (XMCD)[21,22] provide a more detailed information on the magnetic properties of clusters. While XMCD requires the clusters to be embedded in a matrix or deposited on a surface, it has the central advantage of allowing the separate determination of both spin and orbital contributions μ_S and μ_L to the cluster magnetic moments $\overline{\mu}_N$.[23] Moreover, since the XMCD originates in a 2p inner-shell absorption, it is element specific. This is crucial for investigations of heterogeneous systems, such as magnetic interfaces and nanoalloys, in which the magnetic behaviour of different components can be very different. Furthermore, the direct experimental access to the average orbital moments gives a clear insight on the role of spin–orbit interactions in small NPs and on the origin of magnetic anisotropy. Therefore, its description sets a new challenge to the theory description of the transition-metal magnetism at the nanoscale.

The XMCD results for μ_S of Fe, Co, and Ni clusters confirmed the enhancement of spin magnetic moments and the enhancement of $\overline{\mu}_N$ found in previous SG measurements and electronic theories.[9–14,17,18] In addition, remarkably large values of μ_L have been revealed.[3,5,24] These experimental studies motivated theoretical investigations of orbital magnetism in TM clusters, which disclosed the dependence of μ_L as a function of size, geometry, and composition. In this way, the crossover from the atomic Hund's rule to the bulk-like quenching of L has been uncovered.[25] The ensemble of these studies demonstrated the importance of orbital contributions to the total magnetic moments of TM clusters, particularly in order to achieve an accurate comparison between theory and experiment.

Another property deserving considerable attention is the magnetic anisotropy energy (MAE), which can be defined as the energy involved in rotating the magnetization from a low-energy direction, or easy axis, to a high-energy direction, or hard axis. The magnetic anisotropy is one of the main characteristics of a magnetic material. It determines the low-temperature orientation of the magnetization with respect to the structure of the system and the stability of the magnetization direction in single-domain particles. These are properties of crucial importance in technological applications (e.g. magnetic recording or memory devices) where the magnetization must be pinned to a given direction in space. The MAE also determines the blocking temperature above which superparamagnetic behaviour sets in. It controls the nature of the magnetization reversal process and the energy dissipation involved. Consequently, numerous studies have been devoted to the magneto-anisotropic properties of clusters, nanostructures, and low-dimensional systems.[26–28] These investigations are also relevant in the context of SG cluster-beam experiments,[1,2,13,15] since the magnetic anisotropy is responsible for the coupling between the magnetization and the cluster rotations and vibrations. The underlying spin–orbit interactions provide a natural mechanism for the spin relaxation observed in isolated clusters in an external magnetic field.[2]

A number of independent experimental studies, including VSM and SQUID magnetometry, Mössbauer spectroscopy, and magneto-optical Kerr effect, have shown that the MAE of TM clusters is larger than in the corresponding periodic crystals.[26] These trends have been confirmed by theoretical investigations, which quantified the effects of the spin–orbit coupling on the spin-polarized charge distribution and on the electronic spectrum for different orientations of the magnetization.[27] A complicated, non-perturbative behaviour as a function of cluster size, structure, bond length, and *d*-band filling has been revealed. In addition, depending on the cluster symmetry, one observes strong multiaxial behaviours, in which in-plane anisotropies are comparable or even larger than the usually considered uniaxial terms.

Remarkable effects are expected in TM-alloy clusters as a function of composition and chemical order. In order to tailor the materials' properties, one often mixes 3d TMs with 4d and 5d elements which, being heavier, are subject to stronger spin–orbit (SO) interactions. In this context, it is important to recall that large NPs and three-dimensional solids of 4d and 5d elements are non-magnetic. Only for very small sizes, the 4d and 5d clusters can develop a spontaneous low-temperature magnetization, due to the reduction of local coordination and the associated *d*-band narrowing.[29–33] This remarkable finite-size effect has been first observed experimentally in Rh clusters by Cox *et al.*, who determined the average magnetic moments per atom $\overline{\mu}_N = 0.15-0.80\mu_B$ of Rh_N having $N \leq 30$–50 atoms by means of SG deflection.[33] These measurements, together with theoretical results,[29–32] confirm that the spin polarizability of clusters of 4d and 5d elements can be very important. Taking into account the contrasting behaviours of 3d, 4d, and 5d clusters, one expects that 3d–4d and 3d–5d nanoalloys should show very interesting structural, electronic, and magnetic properties.

The remainder of the chapter is organized as follows. In Section **2**, the main theoretical methods used in the study of the magnetic properties of small clusters and NPs are outlined. First, we focus on ground-state approaches, such as density-functional theory (DFT) from the first principles perspective, and self-consistent tight-binding (SCTB) theory from the point of view of many-body electronic models. In this section, we also present a finite-temperature theory of cluster magnetism, which allows us to investigate the effects of spin fluctuations from an electronic perspective. Sections **3–8** are devoted to applications of the theory to representative specific problems. In Section **3**, the size- and environment-dependent ground-state properties of small single-component clusters are discussed. This section should allow the non-specialist to familiarize himself with the dominant spin contributions to cluster magnetism. In Section **4**, the discussion is extended to binary-alloy clusters. This incorporates a number of new features such as composition and chemical environment dependences, as well as proximity and charge-transfer effects on spin magnetism. In Section **5**, we address the fundamental crossover from atomic Hund's rule to bulk-like quenching of orbital magnetism as a function of cluster size. Ni clusters are considered in some detail. Trends for other TMs are also discussed. Section **5**, as well as Section **6** on the magnetic anisotropy, sets the spin–orbit interactions at the centre of our attention. The microscopic origin of the magneto-anisotropic behaviour is analysed, in particular by correlating the MAE to the anisotropy of the orbital moments. Section **7** investigates the possibilities of tailoring the magnetic anisotropy of 3d–4d nanoalloys by varying cluster size and composition. The last specific presented in Section 8 concerns the magnetic properties of 3d TM clusters at finite temperatures. First, the stability of FM order is explored by computing the low-temperature spin-fluctuation energies. A discussion of the short-range magnetic order above the cluster Curie temperature then follows. Finally, the temperature dependence of the average cluster magnetization per atom and of the correlations between the local magnetic moments is analysed. The chapter is closed with a brief outlook on some important research directions of current interest.

2 THEORETICAL BACKGROUND

The purpose of this section is to present the theoretical methods used for describing the magnetic behaviour of itinerant electrons in nanostructures. Three main complementary approaches are introduced. Two of them concern ground-state properties, while the other applies to finite temperatures. In the first place, the basic concepts of DFT are recalled, giving emphasis to the aspects which are specific to magnetism. Second, we outline a self-consistent tight-binding approach, which is particularly performant in studies of orbital magnetism and magnetic anisotropy of large NPs. Finally, we present a functional-integral spin-fluctuation theory of cluster magnetism at the finite temperatures.

2.1 Density-Functional Theory

The first-principles calculations to be reviewed in this chapter have been performed in the framework of Hohenberg–Kohn–Sham's (KS) DFT.[34,35] The foundation of this approach relies on replacing the wave function $|\Psi\rangle$ by the electronic density $n(\vec{r}) = \left\langle \Psi | \Sigma_\alpha \hat{\Psi}_\alpha(\vec{r})^\dagger \hat{\Psi}_\alpha(\vec{r}) | \Psi \right\rangle$ as the basic variable of the quantum many-body problem. As usual, the operator $\hat{\Psi}_\alpha(\vec{r})^\dagger \left[\hat{\Psi}_\alpha(\vec{r})\right]$ creates (annihilates) an electron with spin projection $\alpha = \pm 1/2$ at the position $\vec{r}$. Hohenberg and Kohn demonstrated that the ground-state energy E^0 and density $n^0(\vec{r})$ of an N_e-electron system, under the action of an arbitrary external potential $v(\vec{r})$, can be obtained by minimizing the energy functional

$$E\left[n(\vec{r})\right] = F\left[n(\vec{r})\right] + \int n(\vec{r}) v(\vec{r}) \mathrm{d}r^3 \tag{2.1}$$

with respect to the electronic density $n(\vec{r})$.[34] The functional $F\left[n(\vec{r})\right]$ is universal in the sense that it is independent of $v(\vec{r})$. It represents the minimum value of the sum of the kinetic and Coulomb energies among all N_e-electron states $|\Psi\rangle$ yielding the density $n(\vec{r})$. The latter must satisfy of course the condition $N_e = \int n(\vec{r}) \mathrm{d}r^3$. For spontaneously spin-polarized systems, or in the presence of an external magnetic field $\vec{B}(\vec{r})$, the theory is suitably extended[36] by introducing the spin-density matrix $n_{\alpha\beta}(\vec{r}) = \left\langle \hat{\Psi}_\alpha(\vec{r})^\dagger \hat{\Psi}_\beta(\vec{r}) \right\rangle$ or, equivalently, the charge density

$$n(\vec{r}) = \mathrm{Tr}\left[n_{\alpha\beta}(\vec{r})\right] = \sum_\alpha n_{\alpha\alpha}(\vec{r}) \tag{2.2}$$

and the magnetization density

$$\vec{m}(\vec{r}) = \sum_{\alpha,\beta=+,-} n_{\alpha\beta}(\vec{r}) \vec{\sigma}_{\alpha\beta}. \tag{2.3}$$

Here, $\vec{\sigma} = (\sigma_x, \sigma_y, \sigma_z)$ stands for the vector of Pauli matrices, and α and β refer to the spin quantum numbers. In the following, we focus for simplicity on collinear magnetic states, for which the external magnetic field and magnetization have only one non-vanishing component: $\vec{B}(\vec{r}) = \left[0, 0, B(\vec{r})\right]$ and $\vec{m}(\vec{r}) = \left[0, 0, m(\vec{r})\right]$. This is equivalent to requiring $n_{\alpha\beta}(\vec{r}) = \delta_{\alpha\beta} n_\alpha(\vec{r})$ and implies $n(\vec{r}) = n_+(\vec{r}) + n_-(\vec{r})$ and $m(\vec{r}) = n_+(\vec{r}) - n_-(\vec{r})$.

The actual minimization of $E\left[n(\vec{r})\right]$ is performed within the KS scheme. To this aim, the spin-resolved density is expressed in terms of N_α normalized single-particle spin orbitals $\varphi_{i\alpha}(\vec{r})$ as

$$n_\alpha\left(\vec{r}\right) = \sum_{i=1}^{N\alpha} |\varphi_{i\alpha}\left(\vec{r}\right)|^2, \tag{2.4}$$

where N_α is the number of electrons having spin α. Thus, $N_e = N_+ + N_-$ gives the total number of electrons in the cluster and $S_z = (N_+ - N_-)/2$ the z component of the total spin. The applicability of this approach presupposes that the actual ground-state spin-density $n_\alpha^0\left(\vec{r}\right)$ of the system under study is non-interacting v-representable, that is, that it coincides with the ground-state density of N_e non-interacting electrons, having a total spin projection S_z, in the presence of some spin-polarized single-particle potential $v_\alpha^S\left(\vec{r}\right)$. Minimization of $E\left[n\left(\vec{r}\right)\right]$ with respect to $\varphi_{i\alpha}\left(\vec{r}\right)$ under the normalization constraint yields the KS equations[35]

$$\left[\frac{-\hbar^2}{2m}\nabla^2 + v'\left(\vec{r}\right) + \int \frac{e^2}{\left|\vec{r} - \vec{r'}\right|} n\left(\vec{r'}\right) \mathrm{d}r'^3 + v_\alpha^{\mathrm{XC}}\left(\vec{r}\right)\right] \varphi_{i\alpha}\left(\vec{r}\right) = \varepsilon_{i\alpha}\varphi_{i\alpha}\left(\vec{r}\right), \tag{2.5}$$

where $v_\alpha^{XC}\left(\vec{r}\right) = \delta E_{\mathrm{XC}}/\delta n_\alpha\left(\vec{r}\right)$ is known as the exchange and correlation (XC) potential. It is given by the functional derivative of the XC energy

$$\begin{aligned} E_{\mathrm{XC}}\left[n\left(\vec{r}\right)\right] = F\left[n\left(\vec{r}\right)\right] - \frac{1}{2}\int\int \frac{e^2}{\left|\vec{r} - \vec{r'}\right|} n\left(\vec{r}\right) n\left(\vec{r'}\right) \mathrm{d}r^3 \mathrm{d}r'^3 \\ + \frac{\hbar^2}{2m}\sum_\alpha \sum_{i=1}^{N\alpha} \int \varphi_{i\alpha}\left(\vec{r}\right)^* \nabla^2 \varphi_{i\alpha}\left(\vec{r}\right) \mathrm{d}r^3 \end{aligned} \tag{2.6}$$

with respect to the spin-resolved density $n_\alpha\left(\vec{r}\right)$. By definition, the functional $E_{\mathrm{XC}}\left[n\left(\vec{r'}\right)\right]$ contains the exchange and correlation contributions to the Coulomb energy, as well as the correlation contributions to the kinetic energy. The KS equations are usually solved numerically by iteration or minimization methods.

In order to explore as thoroughly as possible the magnetic solutions of the KS equations, it is useful to consider a large number of different spin configurations as starting points for the numerical iterations. A systematic search of the collinear solutions can be performed by varying the total spin moment S_z, in the full range relevant for the given d-shell filling n_d and number of atoms N_a. This approach known as fixed-spin moment method was first introduced in the context of condensed matter.[37] It is important to remark that the cluster geometry should be optimized for all S_z and for all starting magnetic configurations by following the forces obtained from the self-consistent solution of the KS equations. This is necessary in order to determine the most stable structure, total spin moment S_z, and internal magnetic order, since the rearrangements of geometric, electronic, and magnetic degrees of freedom are interrelated. They should be treated self-consistently on the same footing.

Once the optimization with respect to structural and magnetic degrees of freedom is achieved, one derives the binding energy per atom E_B by referring the total energy E to the corresponding energy of the isolated atoms. For each stationary point of the total energy surface (i.e. for each relaxed structure having vanishing $\|\vec{\nabla} E\|$), one determines the vibrational frequencies from the diagonalization of the dynamical matrix. In this way, saddle points can be ruled out, to which the local optimization procedure sometimes happens to converge. Only configurations corresponding to true minima are discussed in the following. Finally, a number of electronic and magnetic properties—for example, the magnetic energy $\Delta E_m = E(S_z = 0) - E(S_z)$, the local magnetic moments μ_i integrated within the Wigner–Seitz (WS) or Bader atomic cells of atom i,[38] and the spin-polarized density of electronic states $\rho_\alpha(\varepsilon)$—are derived from the self-consistent spin-polarized density and KS spectrum.

2.2 Self-Consistent Tight-Binding Theory

The magnetic properties of TM compounds are known to be dominated by the valence s-, p-, and d-electron states. It is therefore physically reasonable to focus one's attention on these orbitals. Instead of predicting the cluster properties from first principles, one is mainly interested in understanding the trends in the properties of nanostructures as a function of size, structure, and composition. From this perspective, the SCTB theory is a reliable approach, which combines remarkable flexibility and high numerical accuracy at an affordable computational cost. In contrast to *ab initio* methods, SCTB theory is based on many-body models whose parameters characterize the electronic hybridizations, Coulomb interactions, and spin–orbit interactions for the different elements. Rather than being a limitation, the model simplifications and the possibility of varying the interaction parameters yield truly new insights on the microscopic origin of the observed size-, structure-and composition-dependent magnetic phenomena. Moreover, SCTB allows us to perform a transparent analysis of the magnetic behaviour from a local perspective. Therefore, it is well suited for investigations of nanostructures, where the interplay of different local and chemical environment is crucial. The practical possibilities offered by this method are exemplified in the following sections.

We consider an spd-band model Hamiltonian

$$\hat{H} = \hat{H}_0 + \hat{H}_1 + \hat{H}_{SO}, \tag{2.7}$$

which is given by sum of three main contributions.[25,27,39,40] The first term

$$\hat{H}_0 = \sum_{lm,km'} t_{lk}^{mm'} \hat{c}_{lm\sigma}^{\dagger} \hat{c}_{km'\sigma} \tag{2.8}$$

takes into account interatomic hybridizations between the orbitals m and m' of atoms l and k. In the usual notation, $\hat{c}_{lm\sigma}^{\dagger}, \hat{c}_{lm\sigma}$ and $\hat{n}_{lm\sigma} = \hat{c}_{lm\sigma}^{\dagger} \hat{c}_{lm\sigma}$ refer to the creation, annihilation, and number operator for a spin-σ electron at the orbital m of atom l. The second term

$$\hat{H}_I = \frac{1}{2} \sum_{\substack{l,m,m' \\ \sigma,\sigma'}} U_{mm'}^{\sigma\sigma'}{}' \hat{n}_{lm\sigma} \hat{n}_{lm\sigma'} \tag{2.9}$$

approximates the interactions among the electrons by an intra-atomic multi-band Hubbard-like model, where the orbital-dependent $U_{mm'}^{\sigma\sigma'}$ refers to the Coulomb repulsion between electrons with spin σ and σ'. The prime in the summation indicates that the terms with $m = m'$ and $\sigma = \sigma'$ are excluded. Thus, we have direct Coulomb integrals $U_{mm'}^{\uparrow\downarrow} = U_{mm'}^{\downarrow\uparrow} = U_{mm'}$ and $U_{mm'}^{\uparrow\uparrow} = U_{mm'}^{\downarrow\downarrow} = U_{mm'}^{\uparrow\downarrow} - J_{mm'}$, where $J_{mm'}$ is the exchange integral. In most applications, the exchange interactions are taken into account only for the more localized d orbitals.

The many-body Coulomb interaction term H_{I} is often simplified by applying the mean-field unrestricted Hartree–Fock approximation. One therefore replaces Equation (2.9) by

$$\hat{H}_I = \sum_{lm\sigma} \varepsilon_{lm\sigma} \hat{n}_{lm\sigma}, \tag{2.10}$$

where

$$\varepsilon_{lm\sigma} = \varepsilon_{lm}^{0} + \sum_{m'} \left[\left(U_{mm'} - \frac{J_{mm'}}{2} \right) \nu_{lm'} - \frac{\sigma}{2} \sum_{m'} J_{mm'} \mu_{lm'} \right] + z_l \Omega_m \tag{2.11}$$

are the orbital- and spin-dependent energy levels, which depend on the local occupations $\nu_{lm} = \langle \hat{n}_{lm\uparrow} \rangle + \langle \hat{n}_{lm\downarrow} \rangle$ and spin polarizations $\mu_{lm} = \langle \hat{n}_{lm\uparrow} \rangle + \langle \hat{n}_{lm\downarrow} \rangle$. In Equation (2.11), the spin index σ takes the value 1 for spin up and -1 for spin down, while ε_{lm}^{0} stands for the energy level of the orbital m of atom l in the PM solution of the corresponding periodic solid. The two-centre direct and exchange Coulomb integrals $U_{mm'}$ and $J_{mm'}$ for d electrons are expressed in terms of the three independent radial integrals $F^{(0)}$, $F^{(2)}$, and $F^{(4)}$ given by atomic symmetry.[41] In this way, the first and second Hund's rules are naturally fulfilled. The last term in Equation (2.11) takes into account the environment-dependent energy-level shifts due to non-orthogonality effects[39,42–44] and due to the crystal-field potential of the neighbouring atoms,[45] which are approximately proportional to the local coordination number z_l.

Finally, the third term

$$H_{SO} = -\xi \sum_{l,m\sigma,m'\sigma'} \left(\vec{L} \cdot \vec{S} \right)_{m\sigma,m'\sigma'} \hat{c}_{lm\sigma}^{\dagger} \hat{c}_{lm'\sigma'} \tag{2.12}$$

takes into account SO interactions, where $\left(\vec{L}\cdot\vec{S}\right)_{m\sigma,m'\sigma'}$ refers to the intra-atomic matrix elements of $\vec{L}\cdot\vec{S}$. H_{SO} couples the up and down spin-manifolds and introduces the dependence of the magnetic properties on the relative orientation between the magnetization direction and the structure of the cluster. Its strength is characterized by spin–orbit coupling (SOC) parameter ξ, which can be derived from atomic calculations.[46]

The mean-field approximation of Coulomb interactions (Equations 2.10 and 2.11) requires a self-consistent iterative solution of the relation

$$\langle \hat{n}_{lm\sigma}\rangle = \int_{-\infty}^{\varepsilon F} \rho_{lm\sigma}(\varepsilon)\mathrm{d}\varepsilon \tag{2.13}$$

giving the average occupation numbers of the orbital $lm\sigma$ in terms of the local density of states $\rho_{lm\sigma}(\varepsilon) = -\frac{1}{\pi}\mathrm{Im}\{G_{lm\sigma}(\varepsilon)\}$. The latter can be obtained most efficiently from the local Green's functions $G_{lm\sigma}(\varepsilon) = \left\langle lm\sigma\middle|\left(\varepsilon-\hat{H}\right)^{-1}\middle|lm\sigma\right\rangle$ by using Haydock–Heine–Kelly's recursion method.[47] Once self-consistency is achieved, the most relevant electronic and magnetic properties can be derived from $\rho_{lm\sigma}(\varepsilon)$. The effects of non-collinear magnetic moments can be investigated by performing the mean-field approximation in a rotationally invariant form.[48,49]

The orbital moment operator is given by

$$\hat{\vec{L}} = \sum_{\sigma}\int \mathrm{d}\vec{r}\,\hat{\Psi}_{\sigma}^{\dagger}\left(\vec{r}\right)\vec{r}\times\vec{p}\,\hat{\Psi}_{\sigma}\left(\vec{r}\right), \tag{2.14}$$

where the field operator $\hat{\Psi}_{\sigma}\left(\vec{r}\right) = \sum_{lm}\psi_{lm}\left(\vec{r}\right)\hat{c}_{lm\sigma}$ is expanded in a d-orbital basis $\psi_{lm}\left(\vec{r}\right) = u_l^{\mathrm{d}}(r)Y_{2m}(\theta,\phi)$ centred at each atom l. Neglecting the overlaps between the ψ_{lm} on different atoms, one obtains

$$\hat{L}_{\delta} = \sum_{lm\sigma} m\hat{c}_{lm\sigma}^{\dagger}\hat{c}_{lm\sigma}, \tag{2.15}$$

where δ refers to the orbital quantization direction (e.g. $\delta = x, y$, and z). The average local orbital moment $\langle L_{l\delta}\rangle = \sum_{m\sigma} m\langle \hat{n}_{lm\sigma}\rangle$ at atom l is calculated from

$$\langle L_{l\delta}\rangle = \sum_{\sigma}\sum_{m=-2}^{2}\int_{-\infty}^{\varepsilon F} m\rho_{lm\sigma}^{\delta}(\varepsilon)\mathrm{d}\varepsilon, \tag{2.16}$$

where m indicates the magnetic quantum number. The quantization axis of the orbital momentum is thereby taken to be the same as the spin quantization axis.

The electronic energy per atom

$$E_{\delta} = \frac{1}{N}\sum_{l} E_{\delta}(l) \tag{2.17}$$

can be written as the sum of local contributions

$$E_\delta(l) = \sum_{m\sigma} \left[\int_{-\infty}^{\varepsilon F} \varepsilon \rho_{lm\sigma}^{\delta}(\varepsilon) \mathrm{d}\varepsilon - E_{lm\sigma}^{\mathrm{dc}} \right] \tag{2.18}$$

corresponding to each atom l of the cluster. Here $E_{lm\sigma}^{\mathrm{dc}} = (1/2)\Delta\varepsilon_{lm\sigma}\langle \hat{n}_{lm\sigma}\rangle$ stands for the double-counting correction.

The magnetic anisotropy energy (MAE) is defined as the change $\Delta E_{\delta\gamma} = E_\delta - E_\gamma$ in the electronic energy associated to a change in the orientation of the magnetization from the direction γ to the direction δ. For example, if $\gamma = z$ refers to the surface normal and $\delta = x$ to an in-plane direction, one has that positive (negative) values of $\Delta E_{xz} = E_x - E_z$ correspond to a perpendicular (in-plane) easy axis. The calculation of the MAE as the difference between electronic energies is a non-perturbative approach, which includes the effects of spin-density redistributions resulting from SO interactions for different magnetization directions. Consequently, a very precise solution of the self-consistent equations is required, since ΔE_{xz} is usually a small quantity of the order of a few meVs per atom. The MAE thus provides a most delicate test on the effects of SO interactions and orbital polarizations. One should finally note that the mean-field equations (2.11) and (2.13) do not satisfy rotational invariance, except in the case of constant $U_{mm'} = U$ and $J_{mm'} = J$. This implies that the self-consistent solutions may in fact depend on the choice of the quantization axis of the orbital momentum, even in the absence of SO interactions. Any artefacts that could result from such a symmetry breaking can be avoided by averaging the calculated electronic energies over the different orbital quantization directions δ and γ involved in the MAE.

2.3 Spin-Fluctuation Theory

Previous studies of the ground-state properties of TM clusters have shown that the magnetic behaviour is very sensitive to the size and structure of the system. This is a consequence of the itinerant character of the dominant d-electron states, which depend on the local environment of the atoms. Taking this into account is also important in order to derive reliable conclusions at finite temperatures. Consequently, a sound electronic theory at $T > 0$ should incorporate both the fluctuations of the magnetic moments and the itinerant character of the d-electron states on the same footing. Simple spin models, for example, based on the Heisenberg or Ising model, are not expected to be very predictive, at least until they integrate in some effective way the electronic effects responsible for the size dependence of the local magnetic moments and their effective interactions.

The finite-temperature magnetic properties of clusters are derived from the canonical partition function $Q = \exp\{-\beta \hat{H}\}$, where $\hat{H}$ stands for the many-body Hamiltonian given by Equations (2.7)–(2.9) and $\beta = 1/k_B T$. In the case of isolated clusters, the temperature T corresponds to the source where the clusters are formed. This defines the macroscopic thermal bath with which the small clusters are in equilibrium before expansion.[50] Thermal average refers then to the ensemble of clusters in the beam. For deposited clusters the temperature is defined by the support. Keeping the number of atoms N and the number of electrons n_d fixed (canonical ensemble) matches the experimental situation found in charge- and size-selected beams or at inert supports.

The partition function is solved by extending the functional-integral formalism developed by Hubbard and Hasegawa for periodic solids[51–53] to the case of finite systems with arbitrary symmetry.[54,55] To this aim we rewrite the many-body interaction $\hat{H}_\mathrm{I}$ given by Equation (2.9) as

$$\hat{H}_I = \sum_l \left(\frac{U}{2} \hat{N}_l^2 - J \hat{S}_{lz}^2 \right), \tag{2.19}$$

where $\hat{N}_l = \sum_{m\sigma} \hat{n}_{lm\sigma}$ is the electron number operator at atom l, and $\hat{S}_{lz} = (1/2)\sum_m (\hat{n}_{lm\uparrow} - \hat{n}_{lm\downarrow})$ is the z component of the spin operator. $U = (U_{\uparrow\downarrow} + U_{\uparrow\uparrow})/2$ represents an average direct Coulomb repulsion. Notice that Equation (2.19) includes the self-interaction terms $U_{\uparrow\uparrow}\hat{n}_{lm\sigma}^2/2 = U_{\uparrow\uparrow}\hat{n}_{lm\sigma}/2$, which are cancelled out by redefining the d-energy levels as $\varepsilon_l^0 - U_{\uparrow\uparrow}/2$. For the calculation of the canonical partition function Q, the quadratic terms in Equation (2.19) are linearized by means of a two-field Hubbard–Stratonovich transformation within the static approximation.[56] Thus, a charge field η_l and an exchange field ξ_l are introduced at each cluster site l, which describe the finite-temperature fluctuations of the d-electron energy levels and local exchange splittings, respectively. Using the notation $\vec{\xi} = (\xi_1, \ldots, \xi_N)$ and $\vec{\eta} = (\eta_1, \ldots, \eta_N)$, we have

$$Q \propto \int \mathrm{d}\vec{\eta}\, \mathrm{d}\vec{\xi}\, \exp\left\{ -\beta F'\left(\vec{\xi}, \vec{\eta}\right) \right\}, \tag{2.20}$$

where the free energy F' associated to the exchange fields $\vec{\xi}$ and $\vec{\eta}$ is given by

$$F'\left(\vec{\xi}, \vec{\eta}\right) = \frac{1}{2} \sum_l \left(U\eta_l^2 + \frac{J}{2}\xi_l^2 \right) \} - \frac{1}{\beta} \ln\left\{ \mathrm{Tr}\left[\exp\left\{ -\beta \hat{H}' \right\} \right] \right\}. \tag{2.21}$$

The effective Hamiltonian

$$\hat{H}' = \sum_{l,m,\sigma} \varepsilon'_{l\sigma} \hat{n}_{lm\sigma} + \sum_{\substack{l \neq k \\ m,m',\sigma}} t_{lk}^{mm'} \hat{c}_{lm\sigma}^\dagger \hat{c}_{km'\sigma} \tag{2.22}$$

describes the dynamics of the d electrons as if they were independent particles moving in a random alloy with energy levels

$$\varepsilon'_{l\sigma} = \varepsilon_l^0 + Ui\eta_l - \sigma \frac{J}{2} \xi_l. \tag{2.23}$$

The thermodynamic properties of the system are obtained as a statistical average over all possible distributions of the energy levels $\varepsilon'_{l\sigma}$ throughout the cluster. This is known as the static approximation, which is exact in the atomic limit ($t_{lm}^{mm'}=0,\ \forall l\neq m$), where no fluctuations are present, and in the non-interacting limit ($U_{\sigma\sigma'}=0$).

For $T\to 0$, the dominating field configuration $\left(\vec{\xi}^{\,0},\vec{\eta}^{\,0}\right)$ corresponds to the saddle point in the free energy $F'\left(\vec{\xi},\vec{\eta}\right)$. This is determined from the self-consistent equations

$$\left.\frac{\partial F'}{\partial \xi_l}\right|_0=\frac{J}{2}\left(\xi_l^0-2\langle\hat{S}_{lz}\rangle'\right)=0 \tag{2.24}$$

and

$$\left.\frac{\partial F'}{\partial \eta_l}\right|_0=U\left(\eta_l^0+i\langle\hat{N}_l\rangle'\right)=0, \tag{2.25}$$

where $\langle\ldots\rangle'$ indicates average with respect to the single-particle Hamiltonian $\hat{H}'$. Replacing Equations (2.24) and (2.25) in Equation (2.23) yields the known mean-field approximation to the energy levels $\varepsilon'_{l\sigma}$.[18] The present approach provides therefore a proper finite-temperature extension of the SCTB theory developed in the previous section.[18]

In this chapter, we are interested in the temperature dependence of the magnetic properties, which are dominated by the low-lying spin fluctuations. Moreover, we know that $J\ll U$, which implies that the energy involved in local charge fluctuations is much larger than the spin-fluctuation energies. It is, therefore, reasonable to neglect the thermal fluctuations of the charge fields η_l. For each exchange-field configuration $\vec{\xi}$, we set η_l equal to the saddle point of $F'\left(\vec{\xi},\vec{\eta}\right)$ which is given by $i\overline{\eta}_l=\nu_l=\langle\hat{N}_l\rangle'$. Physically, this means that the charge distribution ν_l is calculated self-consistently for each exchange-field configuration $\vec{\xi}$. Since the ν_l are implicit functions of $\vec{\xi}$, one may write

$$Q\propto\int d\vec{\xi}\,\exp\left\{-\beta F'\left(\vec{\xi}\right)\right\}, \tag{2.26}$$

where the free energy

$$F'\left(\vec{\xi}\right)=-\frac{1}{2}\sum_l\left(U\nu_l^2-\frac{J}{2}\xi_l^2\right)-\frac{1}{\beta}\ln\left\{\mathrm{Tr}\left[\exp\left\{-\beta\hat{H}'\right\}\right]\right\} \tag{2.27}$$

depends only on the exchange variables ξ_l, which describe the relevant fluctuations of the spin degrees of freedom. Notice that $F'\left(\vec{\xi}\right)$ in Equations (2.26) and (2.27) is actually a shorthand for $F'\left(\vec{\xi},\overline{\vec{\eta}}\left(\vec{\xi}\right)\right)$, where $\overline{\vec{\eta}}\left(\vec{\xi}\right)$ refers to the saddle-point value of $\vec{\eta}$ for the exchange configuration $\vec{\xi}$. The integrand in Equation (2.26) is proportional to the probability $P(\vec{\xi}) = \left[\exp\left\{-\beta F'(\vec{\xi})\right\}\right]/Q$ of the exchange-field configuration $\vec{\xi}$.

The thermodynamic properties are obtained by averaging over all possible $\vec{\xi}$ with $\exp\left\{-\beta F'(\vec{\xi})\right\}$ as a weighting factor. For example, the local spin magnetization $m_l = 2\langle \hat{S}_{lz}\rangle$ at atom l is given by

$$m_l(T) = \frac{2}{Q}\int \mathrm{d}\vec{\xi}\, \exp\left\{\frac{\beta}{2}\sum_l \left(Uv_l^2 - \frac{J}{2}\xi_l^2\right)\right\} \mathrm{Tr}\left[\hat{S}_{lz}\exp\{-\beta\hat{H}'\}\right] \tag{2.28}$$

$$= \frac{2}{Q}\int \mathrm{d}\vec{\xi}\, \langle \hat{S}_{lz}\rangle' \mathrm{e}^{-\beta F'(\vec{\xi})}, \tag{2.29}$$

where $\langle \hat{S}_{lz}\rangle'$ is the average spin moment at atom l, according to the effective single-particle Hamiltonian $\hat{H}'$, which depends on the fluctuating $\vec{\xi}$. Taking into account that

$$\frac{\partial F}{\partial \xi_l} = \frac{J}{2}\left(\xi_l - 2\langle \hat{S}_{lz}\rangle'\right) \tag{2.30}$$

one may rewrite Equation (2.29) as

$$m_l(T) = \frac{1}{Q}\int \mathrm{d}\vec{\xi}\, \xi_l \mathrm{e}^{-\beta F'(\vec{\xi})}. \tag{2.31}$$

Thus, the local magnetization at atom l is equal to the average of the exchange field at the same atom. Equation (2.31) justifies the intuitive though not quite rigorous association between the fluctuations of the local moment $2\langle \hat{S}_{lz}\rangle$ at atom l and those of the exchange field ξ_l. Notice that in Equations (2.29) and (2.31), the restriction $\sum_l \xi_l \geq 0$, or positive total cluster moment $\sum_l \langle \hat{S}_{lz}\rangle \geq 0$, must be enforced in order to avoid trivially vanishing results for the average magnetization due to time-inversion symmetry. This applies to any finite-system calculation and corresponds to the experimental situation, where the cluster moment is aligned along an external magnetic field. We therefore compute the local magnetizations from

$$m_l(T) = \frac{1}{Q}\int \mathrm{d}\vec{\xi}\, \mathrm{sgn}\left(\sum_l \xi_l\right) \xi_l \mathrm{e}^{-\beta F'(\vec{\xi})}. \tag{2.32}$$

The cluster magnetization per atom $\overline{m}_N$ is determined by averaging the z component of the total spin operator $\hat{S}_z = \sum_l \hat{S}_{lz}$ under the constraint $\langle \hat{S}_z\rangle' \geq 0$. This is given by

$$\overline{m}_N(T) = \frac{2}{N}\langle |\hat{S}_z|\rangle = \frac{1}{N}\sum_l m_l(T), \tag{2.33}$$

which coincides with the cluster average of the local magnetizations m_l. The previous definitions of local and average magnetizations are equivalent to the constraint of positive magnetization used in Monte Carlo (MC) simulations of phenomenological spin models like the Ising model.[57] In this way, the local magnetizations $m_l(T)$ can be determined for different local environments, in analogy to the layer by layer magnetizations in thin films.

An alternative definition of the average magnetic moment per atom is provided by

$$\overline{\mu}_N(T) = \frac{2}{N}\sqrt{\left\langle \hat{S}_z^2 \right\rangle}, \tag{2.34}$$

where

$$\left\langle \hat{S}_z^2 \right\rangle = \frac{1}{Q}\int d\vec{\xi} \left\langle \hat{S}_z^2 \right\rangle' e^{-\beta F'(\vec{\xi})} \tag{2.35}$$

denotes the average of the square total spin. Using Equation (2.30) one can express $\left\langle \hat{S}_z^2 \right\rangle$ in terms of the exchange-field averages as[51]

$$4\left\langle \hat{S}_z^2 \right\rangle = -\frac{2N}{\beta J} + \frac{1}{Q}\int d\vec{\xi}\, \xi^2 e^{-\beta F'(\vec{\xi})}, \tag{2.36}$$

where $\xi^2 = (\sum_l \xi_l)^2$. The first term in Equation (2.36) cancels the trivial contribution to $\langle \xi^2 \rangle$, which is present even if $\hat{H}'$ is independent of $\vec{\xi}$. The magnetic order within the cluster and its stability at finite T are characterized by the correlation functions

$$\gamma_{lk} = 4\left\langle \hat{S}_{lz}\hat{S}_{kz} \right\rangle = -\frac{2}{\beta J}\delta_{lk} + \frac{1}{Q}\int d\vec{\xi}\, \xi_l \xi_k e^{-\beta F'(\vec{\xi})} \tag{2.37}$$

between the magnetic moments at atoms l and k. Notice that $4\left\langle \hat{S}_z^2 \right\rangle = \sum_{lk}\gamma_{lk}$. Positive (negative) values of γ_{lk} for $l \neq k$ indicate FM (AF) correlations, which tend to enhance (reduce) the total magnetization per atom $\overline{\mu}_N(T)$. The diagonal elements γ_{ll} give the local magnetic moments

$$\mu_l = 2\sqrt{\left\langle \hat{S}_{lz}^2 \right\rangle} = \sqrt{\gamma_{ll}} \tag{2.38}$$

at atom l. In this context, it is important to note that temperature fluctuations in itinerant-electron magnets not only affect the spin–spin-correlation functions, for example, by destroying the ground-state FM order in Fe_N but also modify the size of the local spin polarizations μ_l. This contrasts with the case of localized magnetism, where spin and charge degree of freedoms are well separated. In the applications, it is interesting to investigate the different temperature scales yielding changes of the various γ_{lk}, in order to infer which local moments fluctuate more or less easily and which are the moments that trigger the breakdown of the cluster magnetic order.

3 SIZE AND ENVIRONMENT DEPENDENCE ON SPIN MAGNETISM IN TRANSITION-METAL CLUSTERS

One of the fundamental goals in the theory of transition-metal cluster magnetism is to understand how the physical properties change as the electrons of a single atom become part of a group of several atoms and delocalize, and how bulk-like behaviour is achieved. The first theoretical studies in this context were performed in the framework of DFT by considering high-symmetry structures and small sizes[58]. In agreement with experiment,[1] one obtains[58] for FeN and CrN ($N=9$ and 15) larger magnetic moments than for the bulk. Later on, a systematic study[18] of 3d transition-metal clusters was performed by using the theory outlined in Section 2.2. The main results of this work can be summarized as follows: (i) Fe_N, Cr_N, and Ni_N clusters show an important enhancement of the spin magnetic moments as compared to the bulk material. (ii) The local magnetic moments μ (l) at atom l become more important as we go from Ni to Cr, that is, as we approach half-band filling. In fact, Cr clusters show even larger magnetic moments than Fe clusters, in contrast to the bulk. (iii) An interesting dependence of the magnetic order within the cluster is obtained as a function of structure and band filling: bcc-Fe_N is ferromagnetic, whereas bcc-Cr_N is AF, and fcc-Ni_N is FM, whereas fcc-Fe_N is AF. This behaviour can be related to the well-known increasing stability of the antiferromagnetism as we approach half-band filling. (iv) Several interesting local environment effects are observed. For example, the variations of the interatomic distance d_N versus N cause important changes in the size dependence of the average magnetic moment $\overline{\mu}_N$, which qualitatively improve the agreement between theory and experiment for Fe_n.[1]

An important step towards the understanding the cluster magnetic properties has been made by studying the onset of magnetism as a function of the exchange integral J for different sizes, including the comparison with the bulk.[30] The local magnetic moments μ (l) and average magnetic moment $\overline{\mu}_9$ of Fe_9 having a bcc-like structure are given in Figure 2.1 as a function of J/W, where W is the bulk band width. One observes that magnetism sets in at a critical value of the exchange integral J_c, which is smaller than Stoner's bulk $J_c=1/\rho$ (E_F). This implies that clusters can be magnetic even though the corresponding solid is non-magnetic, a phenomenon experimentally observed in the case of Rh.[33] Moreover, the average magnetic moment $\overline{\mu}_9(J)$ is always larger than the corresponding bulk value $\overline{\mu}_b(J)$.[18] The larger stability of magnetism in small 3d TM clusters reflects the increasing importance of Coulomb interactions, relative to kinetic energy, as the local coordination number decreases. Furthermore, one observes that for sufficiently large J, the highest occupied minority-spin states become unoccupied (lies above E_F). The resulting spin-down to spin-up charge transfer causes an increase of $\overline{\mu}_N$ given by $\Delta\overline{\mu}_N=2n_e/N$, where n_e is the number of electrons, whose spin is flipped. Such redistributions of spin density are usually accompanied by

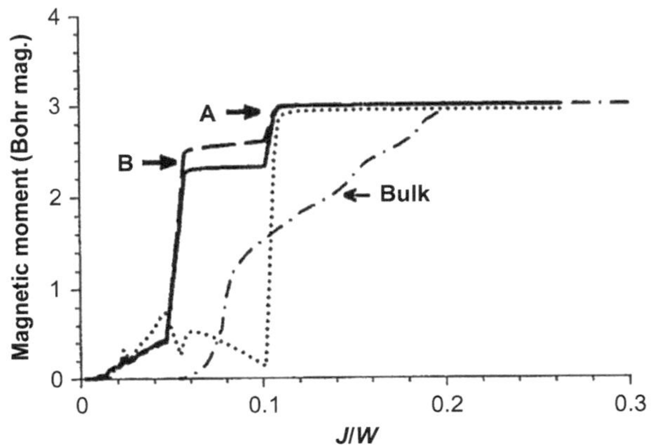

FIGURE 2.1 Local magnetic moments $\mu(l)$ of bcc Fe_9 as a function of the interaction parameter J/W. The dotted curve refers to the central atom and the dashed curve to the surface shell of its first neighbours. Results for the average magnetic moment $\overline{\mu}_N = (1/N)\sum_l \mu(l)$ (solid curve) and of bulk Fe (dashed–dotted curve) are also given. A (B) indicates the valued of J/W for which the electronic densities of states are shown in Figure 2.2 (3). *After Ref. 30.*

strong changes in the electronic structure. This is demonstrated in Figures 2.2 and 2.3, where the local density of states for representative values of J close to the spin transitions is shown.

The above discussion motivates exploring the onset of magnetism in clusters of elements which are non-magnetic in the bulk.[29,31,32] It is well known that atoms having an open shell, and TM atoms in particular, are all magnetic. This is a consequence of the Coulomb repulsion among the electrons which tends to maximize the total spin S, as stated by the first Hund's rule.[59] In the corresponding solids, however, the kinetic energy associated to d-electron delocalization and band formation tends to favour a non-magnetic ground state. In the end, the elements being magnetic in the macroscopic limit are rather few. In fact, none of the 4d or 5d TM solids are magnetic under equilibrium conditions.

Rhodium was the first element for which magnetism is observed in clusters of a non-magnetic solid.[33] The average magnetic moment per atom $\overline{\mu}_N$ of Rh_N derived from SG experiments shows a remarkable size dependence (see Figure 2.4). Within a reduced size range ($N \leq 60$–90) $\overline{\mu}_N$ decreases from rather large values for the smallest clusters (e.g. $\overline{\mu}_9 = (0.8 \pm 0.2)\mu_B$) to very small or vanishing values (e.g. $\overline{\mu}_{30} = (0.13 \pm 0.14)\mu_B$ and $\overline{\mu}_{60} = (0.05 \pm 0.1)\mu_B$). Moreover, $\overline{\mu}_N$ oscillates as a function of N showing maxima and minima at certain cluster sizes. For example, $\overline{\mu}_N$ is particularly large for $N = 15-16$ and 19 and drops for $N = 13-14$, $17-18$, and 20.[33] This behaviour is different from what is observed in ferromagnetic TMs such as Fe_N, Co_N, and Ni_N, where the variations of $\overline{\mu}_N$ are relatively less important and extend over a much wider range of cluster sizes.

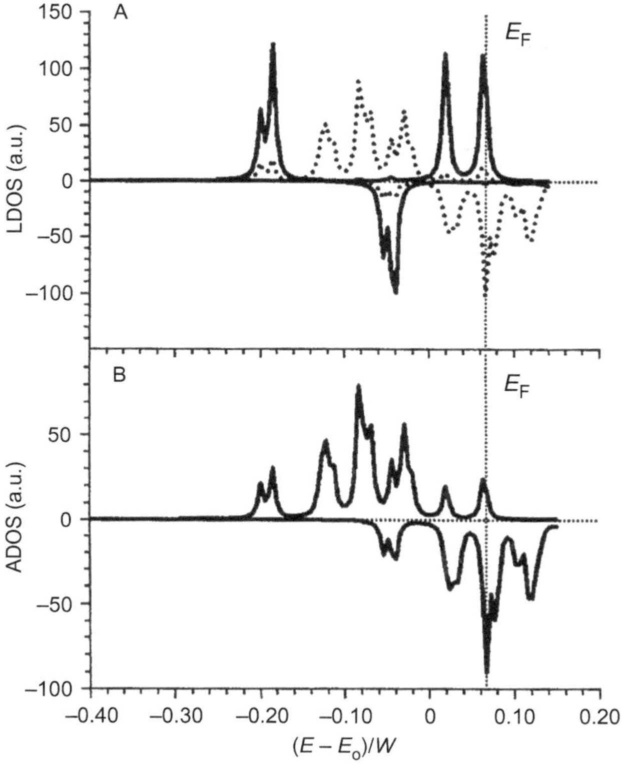

FIGURE 2.2 (A) Local densities of states $\rho_{l\sigma}$ (E) and (B) average density of states $\rho_\sigma(E) = 1/N\sum_l \rho_{l\sigma}(E)$ of bcc Fe_9 for J/W corresponding to the label "A" in Figure 2.1. Positive (negative) values stand for majority (minority) spin. In (A), the solid curve refers to the central atom, while the dotted curve refers to the surface shell. A Lorentzian was used to broaden the cluster energy levels ($\gamma = 0.1$ eV). *After Ref. 30.*

A first attempt to shed light on the behaviour of Rh_N clusters has been to consider a few representative types of cluster geometries (e.g. fcc, bcc, and icosahedral structures) and to compare their relative stabilities and corresponding magnetic behaviours.[32] The size and structural dependence of the magnetic properties of Rh clusters is remarkably rich and quite complex. Depending on the interatomic bond length and on the cluster structure, one finds important variations of spin density within the cluster, leading to both FM and AF orders. As shown in Figure 2.4, this results in oscillations of the average magnetic moment per atom $\overline{\mu}_N$ as a function of N. Deriving simple trends, for example, relating the local magnetic moments to the local environment of the atoms, seems far from obvious.

Concerning the relative structural stabilities, one observes that the geometries yielding the largest cohesive energy change as a function of N. These

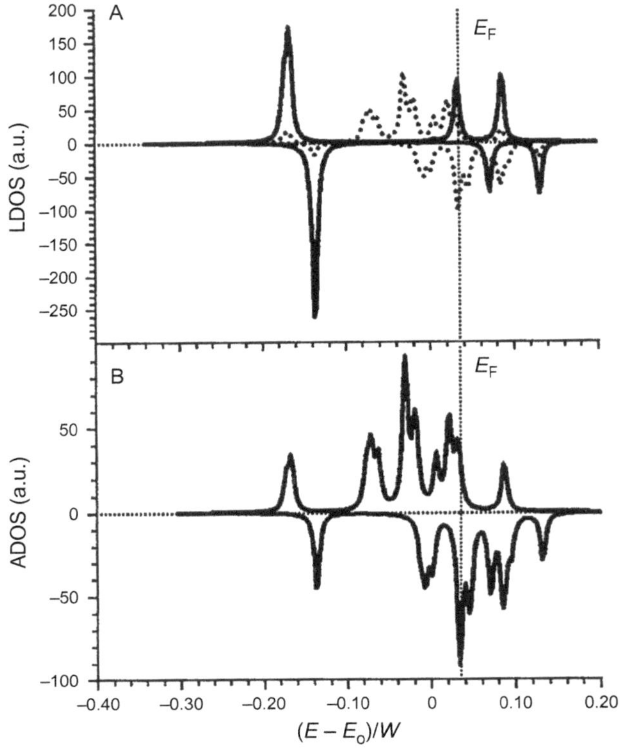

FIGURE 2.3 (A) Local densities of states $\rho_{l\sigma}(E)$ and (B) average density of states $\rho_\sigma(E) = 1/N \sum_l \rho_{l\sigma}(E)$ of bcc Fe_9 for J/W corresponding to the label B in Figure 2.1. See the caption of Figure 2.2. *After Ref. 30.*

structural rearrangements play a crucial role in the determination of the average magnetic moment $\overline{\mu}_N$ of Rh_N. The calculated size dependence of $\overline{\mu}_N$ corresponding to the most stable geometries presents oscillations, which are in good qualitative agreement with experiment. Such strong structural dependences, as found in Rh_N, are characteristic of weak unsaturated itinerant ferromagnetism.

4 FeRh, CoRh, AND CoPt ALLOY CLUSTERS

In past years, the focus of attention of the research activities on cluster magnetism has been progressively shifting from mono-element clusters to nanoscale alloys.[50,60–68,7,69–72] This is largely motivated by the prospects of tuning the magnetic properties by varying their composition, concentration, and chemical order. The study of binary-element clusters is particularly challenging, since it offers numerous ways of exploring competing FM, PM, and

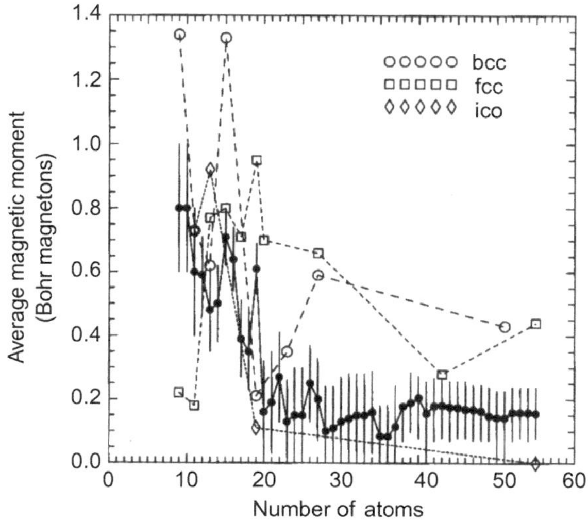

FIGURE 2.4 Average magnetic moment $\overline{\mu}_N$ of Rh_N clusters as a function of N. The experimental results from Ref. 33 are indicated by dots and error bars. The open symbols are obtained from SCTB calculations: circles correspond to bcc-like clusters, squares to fcc-like clusters, and diamonds to icosahedral-like clusters. The lines connecting the symbols are simply a guide to the eye. *After Ref. 32.*

antiferromagnetic (AF) behaviours. In this context, alloying a strongly FM 3d element (e.g. Fe, Co, or Ni) with highly spin-polarizable 4d or 5d element (e.g. Rh, Pd, Pt) deserves special attention. Indeed, besides the finite-size and surface effects discussed in the previous section, one expects to observe very interesting proximity and interface effects, which should depend critically on the actual distribution of the different chemical species within the NP. Alloying also opens the possibility of inducing significant magnetic moments on the otherwise PM 4d or 5d elements, by introducing ferromagnetic 3d atoms in their immediate environment.[61,73] In this way, the 4d and 5d atoms would become magnetic components on their own and therefore condition the properties of the clusters as a whole.

The theoretical study of binary-metal clusters is hindered by the diversity of geometrical conformations, ordered and disorder arrangements, as well as segregation tendencies. This poses a serious challenge to both first-principles and model approaches. Exploring the ground-state energy landscape is a formidable task, since one needs to consider a large, most possibly complete and unbiased set of initial structures. Such a thorough geometry optimization must include not only the representative cluster geometries or topologies but also all relevant chemical orders. This requires taking into account all distributions of both elements for any given size and composition. In the following, we discuss the interplay between local environment, chemical order, and magnetism in FeRh, CoRh, and CoPt clusters in the framework of DFT.

The DFT calculations discussed in this section have been performed for the most part by using the Vienna *ab initio* simulation package (VASP).[74] The exchange and correlation energy is described by using both the spin-polarized local density approximation (LDA) and the generalized gradient approximation (GGA).[75] VASP solves the spin-polarized KS equations in an augmented plane-wave basis set, taking into account the core electrons within the projector augmented wave method.[76] This is an efficient frozen-core approach which allows to incorporate the proper nodes of the KS orbitals in the core region, and the resulting effects on the electronic structure, total energy, and interatomic forces. The wave functions are expanded in a plane-wave basis set with the kinetic energy cut-off $E_{\rm max}$. In order to improve the convergence of the solution of the self-consistent KS equations, the discrete energy levels are broadened by using a Gaussian smearing λ. For calculations on finite systems, one usually considers a simple cubic supercell with periodic boundary conditions. The size of this cell is taken enough so that any pair of the cluster images is well separated. In this way, the interaction between them is negligible. As long as one is interested in finite systems, the reciprocal-space summation is restricted to the Γ point. The actual choice of computational parameters $E_{\rm max}$, λ, and supercell size depends somewhat on the specific application. It is a compromise between numerical accuracy and computational cost, the validity of which can be explicitly verified by comparing the calculated properties.

The interplay of magnetic and chemical order in small Fe_mRh_n clusters ($N = n + m$) has been investigated by performing fully unconstrained structural relaxations in the entire range of compositions and spin polarizations. A very rich variety of structural and magnetic behaviour has been revealed.[77,78] The optimized geometries are in general compact with a strong tendency to intermixing in order to maximize the number of FeRh nearest-neighbour (NN) bonds. For very small sizes, the low-lying isomers have usually a different topology than the optimal structure. However, in larger clusters, the lowest-energy isomerizations involve mainly changes in the chemical order, keeping the underlying topology essentially unchanged. The calculations often predict very small energy differences among the low-lying isomers, suggesting that different chemical orders might coexist in a cluster beam. This could lead to an enhanced dispersion of the average magnetic moments per atom, since different isomers many times show somewhat different magnetic properties.

The average magnetic moments $\overline{\mu}_N$ of Fe_mRh_n increase monotonously with the number of Fe atoms.[77] This is a consequence of the larger Fe local moments and the underlying FM-like order. The change in $\overline{\mu}_N$ per additional Fe atom tends to increase with decreasing N, since the resulting change in Fe concentration is more important the smaller the size is. The typical increase in $\overline{\mu}_N$ obtained per Fe substitution is about $(1/N)\mu_{\rm B}$.

The local magnetic moments $\mu_{\rm Fe}$ and $\mu_{\rm Rh}$ in the WS spheres of the Fe and Rh atoms provide further insight on the interplay between 3d and 4d magnetism in Fe_mRh_n. The Fe spin moments are essentially given by the saturated

3d contributions. For pure Fe clusters, the actual values of μ_{Fe} are somewhat lower than $3\mu_B$ due to a partial spill-off of the spin-polarized density. Remarkably, the Fe moments increase as we replace Fe by Rh atoms, showing some weak oscillations as a function of Rh content m. This effect can be traced back to a d-electron charge transfer from Fe to Rh which, together with the extremely low coordination number, yields a full polarization of the larger number of Fe 3d holes. On the other side, the Rh moments are not saturated. Therefore, they are more sensitive to size, structure, and composition. The values of μ_{Rh} are in the range of 1.0–$1.5\mu_B$ showing some oscillations as a function of concentration. They are less affected by charge transfer, since no systematic enhancement of μ_{Rh} is observed with increasing Fe content.

Another appealing example of magnetic nanoalloy is CoRh. Experiments on CoRh NPs indicate not only the presence of important Co moments but also the development of significant spin polarizations at the Rh atoms.[61] Small clusters are also very interesting in this context, particularly as we approach the size range where Rh_N becomes magnetic. As discussed in the previous section, several calculations have shown that Rh_N with $N \simeq 40$ are PM or weakly ferromagnetic. They have nearly vanishing magnetic moments but a high magnetic susceptibility. Therefore, CoRh clusters in this size range should show a remarkable dependence of the magnetic properties as a function of concentration and chemical order. In the following, we focus on compact fcc-like Co_nRh_m clusters having $n+m=43$ atoms.[79] This chemical order is assumed to be core-shell-like, which corresponds to the situation found in experiment when the dissociation times of the Co and Rh organometallic precursors are very different.[61] In Figure 2.5, the local magnetic moments of different CoRh clusters with $N=43$ atoms are shown as a function of the shell number l, where $l=1$ corresponds to the central atom and $l=4$ to the outermost shell. The results are obtained by integrating the spin density $m\left(\vec{r}\right)$ within the WS atomic spheres.

In order to discuss the properties of the alloys, it is most interesting to start from the highly spin-polarizable pure Rh clusters and analyse the effects of replacing progressively Rh by Co. We consider first the case where Rh is in the core. As a result of the substitution of the outer shell by Co atoms, one obtains the cluster $Rh_{19}Co_{24}$, which shows an important enhancement of μ_l at all Rh atoms (see Figure 2.5). The Co moments (outermost shell) are also slightly larger than in pure Co_{43}. This is probably related to a slight increase in the number of 3d holes at the Co atoms in contact with Rh. The enhancement of the average magnetic moment per atom—from $\overline{\mu}_N = 0.54\mu_B$ in Rh_{43} to $\overline{\mu}_N = 1.61\mu_B$ in $Rh_{19}Co_{24}$—is not only due to the larger Co contributions ($\overline{\mu}_{Co}^{WS} = 2.02\mu_B$). The enhancement of the Rh local moments is equally important. In fact, $\overline{\mu}_{Rh}^{WS} = 0.48\mu_B$ in Rh_{43}, while $\overline{\mu}_{Rh}^{WS} = 0.91\mu_B$ in $Rh_{19}Co_{24}$. This demonstrates the high spin polarizability of Rh clusters in this size range.

A further increase of Co content in $Rh_{13}Co_{30}$ does not yield a very significant change in the local Rh moments calculated in the GGA. Actually, a small

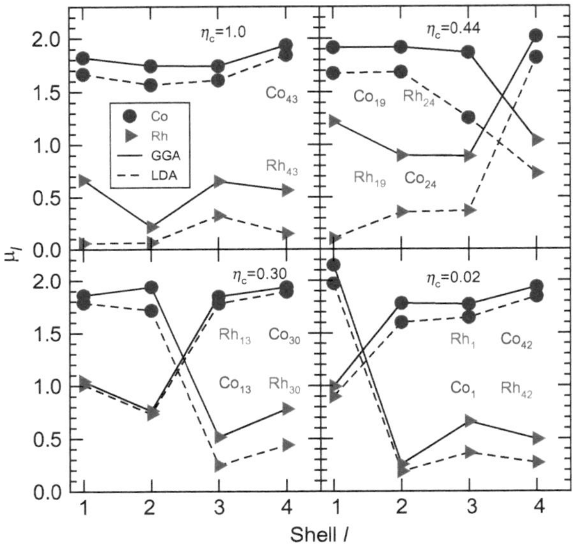

FIGURE 2.5 Local spin magnetic moments μ_l within the Wigner–Seitz spheres of the Co atoms (circles) and Rh atoms (triangles) in core-shell CoRh clusters having $N=43$ atoms. Results are given for the average μ_l at each NN shell l surrounding the central atom $l=1$. Co_nRh_m (Rh_nCo_m) designates a cluster having n Co atoms (Rh atoms) in the core and m Rh atoms (Co atoms) in the outer shell. The relative core sizes η_c are indicated in the insets. The lines connecting the points are a guide to the eye, where full (dashed) lines refer to the generalized gradient (local density) approximation. *After Ref. 79.*

reduction of μ_l^{Rh} is observed. However, the LDA results for μ_l^{Rh} are significantly enhanced, now becoming very similar to the GGA ones (see Figure 2.5 for $\eta_c=0.30$). The average magnetic moments per atom in $Rh_{13}Co_{30}$ is nearly the same as in $Rh_{19}Co_{24}$. This is the result of a compensation between a larger contribution of Co atoms and a small reduction of the Rh local moments. Finally, for very high Co content as in Rh Co_{42} ($\eta_c=0.02$), we find a magnetization profile which is very similar to pure Co_{43}, as far as the Co shells are concerned. The local moment $\mu_1^{Rh}=0.99\mu_B$ at the central Rh atom in $RhCo_{42}$ and in $Rh_{13}Co_{30}$ is very similar.

Co substitution at the cluster core, which we denote by Co_nRh_m, yields a qualitatively different behaviour. In this case, one observes $\overline{\mu}_{Rh}^{WS}=0.48\mu_B$, $0.45\mu_B$, $0.73\mu_B$, and $1.04\mu_B$ in Co_nRh_{43-n} for $n=0$, 1, 13, and 19, respectively. Even in the very dilute limit (e.g. $CoRh_{42}$), the Co spin moments are strong ($\overline{\mu}_{Co}^{WS}\simeq 2\mu_B$). Interestingly, $\overline{\mu}_{Co}^{WS}$ tends to decrease with increasing Co concentration. This suggest that the enhancement of $\overline{\mu}_{Co}^{WS}$ is not due to a reduction of the local coordination numbers—the Co atoms are in the core—but rather to a small charge transfer from Co to Rh, which causes an increase in the number of polarizable 3d holes at the Co atoms. Similar charge-transfer effects have been found in FeRh clusters.[77]

Concerning the spatial distribution of the spin polarization, we observe that more than 95% of the total magnetization originates in the WS spheres of the atoms, when the Co content is important. In the Rh-rich limit (e.g. in Rh_{43} and Co_1Rh_{42}), the local contributions drop to about 90%. This indicates a more important spill-off of the spin-polarized density, although the local atomic contributions remain largely dominant. Comparing d and sp moments within the WS spheres, we observe, as expected, that the latter are quite small. In most cases, the sp moments are parallel to the d moments. However, when Rh is at the core (e.g. in $Rh_{19}Co_{24}$ and $Rh_{13}Co_{30}$), we find an appreciable antiparallel sp polarization $\mu_{sp}^{Rh} \simeq -(0.02-0.05)\mu_B$ at the Rh atoms. Antiparallel sp moments are also found at the Co atoms of some clusters having a Co core (e.g. in $Co_{13}Rh_{24}$). This effect could be investigated experimentally by performing Knight-shift measurements.

Experiments and simulations[80] have motivated calculations on larger clusters having fcc-like structures with segregated core–shell Co and Rh components. As an example, we consider fcc clusters with $N=273$ atoms, which contain a central atom and the successive shells of its closest 15 NNs. In these nanoalloys, the calculated average magnetic moments per Co atom $\overline{\mu}_{Co}^{WS} = \left(\sum_{i=1}^{m+n} \mu_i\right)/n$ are in most cases similar, regardless of the type of core–shell arrangement. This holds as long as the concentrations are comparable and the Rh content is not very large. However, the magnetization profiles, that is, the distributions of the spin-polarized density μ_l(Rh) as a function of l, are most sensitive to the chemical order.[79] In the case of Rh-core NPs, the Co moments are largest at the cluster surface ($9 \leq l \leq 15$) and decrease slightly at the CoRh interface. The induced Rh moments are quite significant, in particular, at the interface ($\mu_l(\text{Rh}) \simeq 0.4$–$0.5\mu_B$) where they couple parallel to the Co moments. In addition, important oscillations and changes of sign of μ_l(Rh) are observed as one moves away from the interface towards the cluster centre. This AF-like order, combined with the increase of the Co moments, yields average magnetic moments per atom which are similar to those of the Co-core case. Concerning the role of electronic correlations, it is worth noting that in most cases, LDA and GGA give very similar qualitative behaviours for almost all considered alloy clusters. This is remarkable taking into account the observed wide variety of magnetization profiles that is observed. The main difference is a systematic tendency of the LDA to underestimate the absolute value of μ_l. Nevertheless, the predicted magnetic order coincides in general with the GGA.

To conclude this section, we would like to comment on recent experimental and theoretical investigations on local atomic order in CoPt NPs, which revealed atypical lattice-parameter ratios c/a in chemically ordered structures.[81] Density functional calculations indicate that CoPt clusters obtained by size-selected cluster-beam deposition with diameters around 3 nm transform upon annealing into a chemically ordered $L1_0$-like phase. The observed anomalous c/a ratios are an intrinsic property of CoPt NPs. This finite-size effect is expected to be very important for the relative stability of the different structures. Quite generally, and in contrast to the bulk structure, the Co and Pt layers appear to relax independently from each other, despite the strong CoPt hybridization. In fact, these different relaxations of Co and Pt atoms are responsible for the different *c/a* ratios observed in experiment. For example, significant in-plane tetramerizations are observed in the outer Co layers, which do not match the underlying Pt layer. These unusual rearrangements derived from the theory have been used to explain the experimentally observed strong dispersion in the Co–Co interatomic distances ($2.33 \leq d_{\mathrm{CoCo}} \leq 3.23$ Å).

5 FROM HUND'S RULES TO BULK QUENCHING OF ORBITAL MAGNETISM

In atoms, Hund's rules predict maximum orbital angular moment L compatible with maximum spin multiplicity, while in TM solids, electron delocalization and band formation result in an almost complete quenching of $\langle L \rangle$. Such intrinsic differences between atomic and bulk behaviours are characteristic of systems developing itinerant-electron magnetism. Consequently, investigations of orbital magnetism in TM clusters—in the way from the atom to the solid—should reveal novel size-dependent phenomena that are important both from a fundamental standpoint and in view of applications of cluster-based magnetic nanostructures. These general considerations already indicate that $\langle L \rangle$ should be very sensitive to the local environment of the atoms. Indeed, recent experiments on chains[6] and clusters at surfaces[3,5,24] show that $\langle L \rangle$ is typically a factor 2–5 larger than in the corresponding solids. Electronic calculations on small supported clusters confirm these conclusions.[82] In the case of free clusters, however, very little is known about this problem, particularly from the point of view of theory which has been so far concerned only with the spin contributions.[83] This is quite remarkable since a size-dependent enhancement of $\langle L \rangle$ would have direct consequences on the average magnetic moments per atom and for the comparison with experiment.[9,11,14] In this section, we present and discuss results on orbital magnetism in free TM clusters, including the crossover from the atom to the bulk. Emphasis is given to the dependence on size, structure, local atomic environment, and *d*-band filling.

In order to examine the crossover from atomic L to bulk-like quenching of $\langle L \rangle$ with increasing cluster size, we focus on Ni_N clusters. Besides its experimental relevance,[9,11,14] Ni is very interesting, since the difference between the atomic orbital moment L_{a} and bulk moment L_{b} is dramatic ($L_{\mathrm{a}} = 2\mu_{\mathrm{B}}$ for $\mathrm{s}^1\mathrm{d}^9$ and $L_{\mathrm{b}} = 0.05\mu_{\mathrm{B}}$). Moreover, the spin moments in Ni_N are the smallest among the ferromagnetic TMs, and therefore the contribution of $\langle L \rangle$ to the magnetic moment $\langle \vec{M} \rangle = 2\langle \vec{S} \rangle + \langle \vec{L} \rangle$ should be particularly significant. Figure 2.6 shows the calculated average orbital moment per atom $\langle L_\delta \rangle = [\sum_{i=1}^{N} L_\delta(i)]/N$

of Ni_N with $N \leq 165$, having face-centred-cubic (fcc) or icosahedral-like geometries.[25] Results for low-symmetry structures and two-dimensional (2D) islands are also given. These are representative of different local symmetries and shapes and allow to quantify the role of cluster structure on orbital magnetism. First of all, one observes that the reduction of system size causes a remarkable enhancement of $\langle L_\delta \rangle$ with respect to the solid. Values about six to eight times larger than $L_b = 0.05\mu_B$ are not uncommon. On the other hand, comparison with the atomic value $L_a = 2\mu_B$ shows that the largest part of the quenching of L takes place already at the smallest clusters, as soon as full rotational symmetry is lost. For example, for Ni_3 (triangle) we obtain $\langle L_z \rangle = 0.47\mu_B$, and for Ni_4 (rhombus) $\langle L_z \rangle = 0.35\mu_B$. Concerning the size dependence, one observes that $\langle L_\delta \rangle$ decreases with increasing N showing some oscillations as bulk-like quenching is approached (see Figure 2.6). It should be however noted that an important enhancement of $\langle L_\delta \rangle$, about 100%, is still present even for the largest considered sizes (e.g. $\langle L_\delta \rangle / L_b = 1.8$ for $N = 165$). In smaller clusters, $\langle L_\delta \rangle / L_b$ ranges from $\langle L_\delta \rangle / L_b \simeq 3$ for $N = 50$–80, to $\langle L_\delta \rangle / L_b \simeq 5$ for $N = 10$–20. One concludes that orbital magnetism is the source of an important contribution to the size-dependent magnetic properties of Ni clusters.

In Figure 2.6, results are also given for icosahedral clusters having $N \leq 55$ atoms. $\langle L_\delta \rangle$ decreases with increasing N in a similar way as for the fcc geometries. Nevertheless, the fivefold symmetric structures yield in general significantly larger $\langle L_\delta \rangle$ than the cubic symmetric ones. The results shown in Figure 2.6 for $N \geq 13$ concern mainly highly symmetric structures with nearly spherical shape and closed NN shells. Clusters with lower symmetry usually present a richer size dependence, particularly when the proportion of surface atoms is large. This is illustrated in the inset, where the optimized geometries derived from *ab initio* calculations for $N \leq 13$ are considered.[83] Structural changes may also affect the convergence to bulk-like quenching for large N, particularly for clusters on surfaces, which shape can be experimentally tuned at least to some extent.[3,5,6,24] This problem has been investigated by performing calculations for two-layer-thick coin-like Ni_N, which are indicated by the triangles in Figure 2.6 ($N = 144$ and 158). For these 2D islands, $\langle L_\delta \rangle$ is much larger than for 3D fcc clusters of comparable size. The calculated enhancement with respect to L_b is here about a factor of 3. Similar trends are found for other TMs, in qualitative agreement with recent experiments on Fe and Co particles.[3,5,6,24] Although structure and shape usually have a strong influence on the quantitative values of the orbital moments, the main trends are not affected significantly by the precise geometry.

From the solid-state perspective, the enhancement of $\langle L_\delta \rangle$ can be qualitatively understood as the result of the following major contributions, which are related to the changes in the local environment of the atoms. First, the reduction of local coordination number with decreasing N causes an increase of the local spin polarizations $S_\delta(i)$, which induces larger orbital moments by

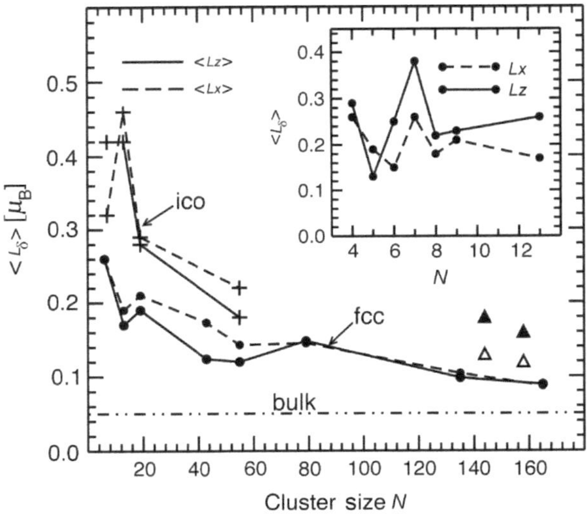

FIGURE 2.6 Average orbital magnetic moment per atom $\langle L_\delta \rangle$ of Ni_N clusters with fcc-like (dots) and icosahedral-like structures (crosses). The magnetization direction δ is a principal C_n symmetry axis ($\delta = z$) or an NN bond perpendicular to z ($\delta = x$). Full (open) triangles refer to coin-like bilayer clusters with perpendicular (in-plane) magnetization. Results for small clusters are given in the inset. *After Ref. 25.*

means of the spin–orbit interactions. Second, the orbital dependence of intra-atomic Coulomb interaction favours the occupation of high-m states, thereby amplifying the enhancement of $\langle L_\delta \rangle$. Although this contribution is quantitatively important, typically 30–40% of the value of $\langle L_\delta \rangle$, it does affect significantly the trends as a function of size and structure. Third, the presence of degeneracies in the single-particle spectrum allows a more effective spin–orbit mixing, which enhances $L_\delta(i)$ even in situations where $S_\delta(i)$ is saturated. Notice that the changes in $S_\delta(i)$ and $L_\delta(i)$ involve different energy scales. Therefore, the occupations of the different m orbitals may vary without altering the spin polarizations (compare icosahedral and fcc clusters in Figure 2.6). Finally, one also observes more subtle effects related to the details of the electronic structure and its dependence on cluster geometry, for example, the presence of high-symmetry axes or changes in bond length.

It is interesting to compare our results with the average magnetizations per atom $\overline{\mu}_N$ derived from SG experiments.[11,14] Despite some quantitative differences among the various calculations, one observes that the predicted ground-state spin moments are systematically smaller than the experimental $\overline{\mu}_N$, by about $(0.3–0.6)\mu_B$ for $N \leq 13$. The orbital moments have been determined by using the structures reported in Ref. 83. As shown in the inset of Figure 2.6, one obtains $\langle L_\delta \rangle \simeq (0.2–0.4)\mu_B$ for $N \leq 13$. Therefore, $\langle L_\delta \rangle$ is an important contribution to the cluster magnetization (20–40%). In fact, taking into account the enhanced orbital moments, which in Ni align parallel to the spin

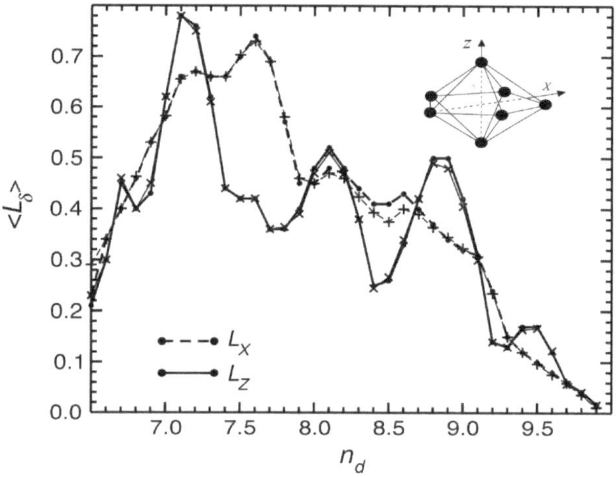

FIGURE 2.7 Average orbital moment per atom $\langle L_\delta \rangle$ in a seven-atom TM cluster as a function of d-band filling n_d. The structure and magnetization directions δ are illustrated. *After Ref. 25.*

moments, removes the largest part of the discrepancy between theory and experiment. The remaining differences, about $(0.1–0.2)\mu_B$, are not far from the estimated experimental uncertainties. These could also be related to correlation effects, which should favour electron localization leading to more atomic-like orbital moments and probably to a further enhancement of $\langle L_\delta \rangle$.

The trends for other TMs and the effect of sp–d charge transfers can be inferred from Figure 2.7, where $\langle L_\delta \rangle$ of a pentagonal bipyramid is shown as a function of d-band filling n_d. $\langle L_\delta \rangle$ increases approximately linearly with increasing number of d holes, as we move from Ni ($n_d \simeq 9$) to Fe ($n_d \simeq 7$). The orbital moment per d hole $\langle L_\delta \rangle/(10 - n_d) \simeq (0.25–0.3)\mu_B$ is quite close to recent experimental results on Fe_N clusters ($N \leq 9$) deposited on Ni/Cu(001).[5] In addition, one observes oscillations and changes of sign of the anisotropy $\Delta L = L_z - L_x$, which are associated with changes in the occupations of molecular states k with different orbital moments $\langle l \rangle_k$. This is consistent with the oscillations of the magnetic anisotropy energy (MAE) as a function of n_d, to be discussed in the following section.[27] One concludes that the size-dependent enhancement of $\langle L_\delta \rangle$ is important for all magnetic TM clusters.

Before closing this section, the properties of Rh clusters deserve to be briefly mentioned, since here the spin moments are smaller and the SO couplings stronger than in 3d TMs. The calculations show that the orbital contributions to the total magnetic moment are also significant in Rh_N. If bulk-like NN distances are assumed, one obtains $\langle L_\delta \rangle = (0.1–0.25)\mu_B$ for $N \leq 19$, which is similar to what is found in Ni_N.[25] Relaxed Rh_N geometries present considerable bond-length contractions, which tend to reduce $\langle L_\delta \rangle = (0.1–0.18)\mu_B$ ($N \leq 19$). This still represents about 15–25% of the total magnetic moment.

Larger Rh_N clusters show a particularly strong size and structural dependence of $\langle L_\delta \rangle$ due to the rapid decrease of the spin magnetization with increasing N and the presence of weak unsaturated spin magnetism. In conclusion, the above discussed results on orbital magnetism in TM clusters demonstrate the remarkable dependence of L as a function of size, geometry, and composition, including the crossover from atomic to bulk-like behaviour. They also establish the importance of orbital contributions to the magnetic moments and their role in the comparison between theory and experiment.

6 MAGNETIC ANISOTROPY OF TRANSITION-METAL CLUSTERS

The magnetic anisotropy is one of the main characteristics of a magnetic material. It determines the low-temperature orientation of the magnetization with respect to the structure of the system and the stability of the magnetization direction in the case of single-domain particles. These properties are crucial for technological applications (e.g. magnetic recording and memory devices) where the magnetization must be pinned to a given direction in space. In past years, an intensive research activity has been devoted to the study of low-dimensional systems, which present novel magneto-anisotropic properties as a result of their reduced dimensionality and symmetry. In this context, the magnetic anisotropy of small 3d TM clusters deserves special attention, not only from a purely theoretical point of view but also because of its implications in cluster-beam SG experiments.[2,15] Indeed, the magnetic anisotropy, and the resulting coupling of the magnetization to rotations and internal vibrations, provides a natural mechanism for describing the phenomenon of spin relaxation within *isolated* clusters in an external magnetic field.[2,15] The MAE, defined as the energy involved in rotating the magnetization from a low-energy direction (easy axis) to a high-energy direction (hard axis), is one of the key parameters characterizing the dynamics of rotating clusters in a SG magnet. It determines the blocking temperature, above which a superparamagnetic behaviour holds, and may lead to a resonance-like coupling between the rotational frequency and the Zeeman splittings.[2,15,84] These investigations, as well as from Mössbauer studies on supported Fe NPs,[26] indicate that the MAE of TM clusters is considerably larger than in the corresponding crystals. It is the purpose of this section to discuss the microscopic origin of the magnetic anisotropy of small TM clusters and to show how the MAE depends on relevant external variables such as cluster size, structure, bond length, and d-band filling.[27]

Figure 2.8 shows results for the MAE ΔE, the orbital angular momentum $\langle L_\delta \rangle$ along the magnetization direction $\delta = x, y, z$, and the average perpendicular spin projection $\langle S_z \rangle$ of Fe_4 having a rhombohedral structure (see inset Figure 2.9). The results have been obtained in the framework of the SCTB

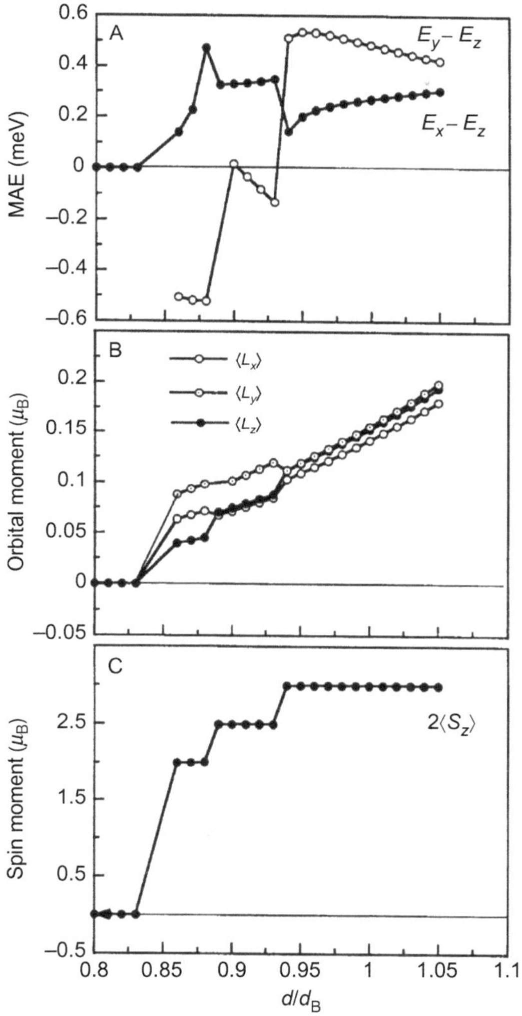

FIGURE 2.8 Magnetic anisotropy energy (MAE), orbital magnetic moment, and spin magnetic moment of Fe_4 with rhombohedral structure as a function of the bond length d (d_B = bulk NN distance). The considered directions of the magnetization, x, y, and z, are illustrated in the inset of Figure 2.9. *After Ref. 27.*

theory outlined in Section **2.2** and are representative of a much larger number of studied sizes and structures. The MAE is given as a function of the bond length d, from which the role of cluster relaxation and the possible coupling of the magnetization direction to vibrations and distortions can be inferred. The variations of d/d_B (d_B = bulk bond length) correspond to a uniform

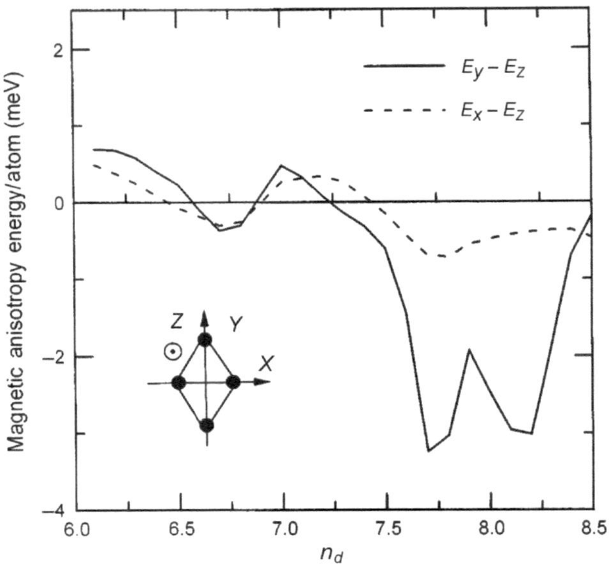

FIGURE 2.9 Magnetic anisotropy energy (MAE) of a rhombohedral 4-atom transition-metal cluster as a function of the d-band filling n_d. Results are given for the off-plane (full line) and in-plane (dashed line) MAEs. The considered directions of the magnetization, x, y, and z, are illustrated in the inset. *After Ref. 27.*

relaxation for the given cluster geometry and quantify how ΔE, $\langle L_\delta \rangle$, and $\langle S_z \rangle$ depend on the local environment of the atoms. For all studied clusters, the modulus of the spin magnetization $\left| \left\langle \vec{S} \right\rangle \right|$ depends very weakly on the direction of the magnetization (e.g. $|\langle S_z \rangle - \langle S_x \rangle| \sim 10^{-3} - 10^{-4}$). Since the changes in $\left| \left\langle \vec{S} \right\rangle \right|$ for the different orientations would be indistinguishable in the scale of the plots, only the results for $\langle S_z \rangle$ are shown.

From Figure 2.8A and C, it is clear that the changes in the MAE as a function of d/d_B are related to the changes in $\langle S_z \rangle$ and in the electronic spectrum. For large values of d/d_B, the spin magnetic moments are saturated: $\langle S_z \rangle \simeq (10 - n_d)/2 = 3/2$. As d/d_B decreases, discrete modifications $\Delta \langle S_z \rangle$ in the spin polarization occur and non-saturated spin magnetizations are obtained. As discussed in Ref. 18, $\Delta \langle S_z \rangle = 2n_e/N$, where n_e is the number of electrons which flip their spin. For constant values of $\langle S_z \rangle$, that is, for d/d_B between two spin flips, the MAE and $\langle L_\delta \rangle$ vary continuously since the electronic spectrum and the local magnetic moments are continuous functions of d/d_B. This is not the case when a spin-flip occurs, since here a strong and discontinuous redistribution of the spin-polarized electronic density takes place. This causes important changes in the energy-level structure around the Fermi energy ϵ_F, which modify the details of the SO mixing between these states.

The resulting changes in the electronic energy depend, of course, on the explicit form of H_{SO} and therefore on the direction of the magnetization (see Figure 2.8). Consequently, very significant and discontinuous variations of the MAE are observed, which may even lead to a change of sign of the MAE as the spin magnetization $\langle S_z \rangle$ decreases. The importance of determining the spin-polarized charge distribution *self-consistently* is clearly demonstrated.

Under certain assumptions (second-order perturbation in the SO coupling and neglect of interactions between unlike spins), $\Delta E = E_\delta - E_\gamma$ is approximately proportional to the anisotropy of the orbital moments, that is, to the difference of the projections of the orbital angular momenta $L_\gamma - L_\delta$.[46] Our self-consistent calculations show that this relation is not valid in general (see Figure 2.8). However, it often works qualitatively well when the difference in $L_\gamma - L_\delta$ is large. Notice, furthermore, that the rhombohedral Fe_4 cluster presents a remarkable in-plane anisotropy $E_x - E_y$ which in the range $0.83 \leq d/d_B \leq 0.93$ is even larger than the usually considered off-plane anisotropy $E_z - E_\delta$ ($\delta = x$, y). A similar situation is found for other clusters and band fillings, as it will be discussed below. In these cases, simple uniaxial anisotropy models are not applicable. The results of ΔE as a function of cluster size N and bond length d defy easy generalizations and simple general rules. Nevertheless, one may certainly conclude that the MAE of small TM clusters is usually much larger than in the corresponding crystalline solids and thin films.

The trends for other TMs and the effects of possible sp–d charge transfers can be inferred from Figure 2.9, where the MAE of a four-atom rhombohedral cluster is given as a function of the d-band filling n_d. In the considered range of n_d, the self-consistent calculations yield saturated magnetic moments. For values of n_d close to that of Fe ($6.5 \leq n_d \leq 7.5$), the MAE is relatively small ($\Delta E \sim 0.1$ meV) and the easy magnetization direction alternates between the xy plane and the z-axis. The situation changes, however, for values of n_d close to that of Co ($7.5 \leq n_d \leq 8.5$). Here, a remarkably large in-plane anisotropy is found ($E_x - E_y \simeq 2-3$ meV) favouring the orientation of the magnetization along the y-axis.

To close this section, it is interesting to compare the magnetic anisotropy of clusters with that of extended low-dimensional systems in order to put apart the effects of finite size from those due to the reduction of coordination number. For example, one expects that one-dimensional structures should show an important enhancement of the MAE. In Ref. 85, finite and infinite Co chains deposited on Pd(110), as well as free-standing chains, have been considered. The monoatomic free-standing chains show an in-line easy axis with remarkably large MAE per Co atom ($\Delta E_{xz} < 0$ with $|\Delta E_{xz}| = 6.2$–20 meV). In the case of deposited chains, the interaction with the substrate changes the easy axis from in-line to off-plane, thereby reducing by an order of magnitude the absolute value of the MAE. These results have been obtained assuming that the Co–Co NN distance is the same as in the Pd substrate ($d_{Pd} = 2.75$ Å). A bond-length contraction to typical Co–Co NN distances ($d_{Co} = 2.5$ Å) would yield an important reduction of $\langle L_\delta \rangle$ and ΔE_{xz}, even if the interactions with the substrate are neglected (e.g. $\Delta E_{xz} \simeq 7$ meV for $d_{Co} = 2.5$ Å and $N \geq 10$).[28] It should also be noted that the largest part of the enhancement of the MAE, which is characteristic of the monoatomic chain, disappears in the bichain. In fact, $\Delta E_{yz} = -0.22$ meV for long bichains, which is similar to the values found in thin films. This suggests a quite rapid crossover to the 2D behaviour.[86] One concludes that the remarkably large anisotropy energies of monoatomic wires are intrinsically related to the 1D character, which is also consistent with the reduction of ΔE_{yz} found upon deposition. The interactions between magnetic clusters and metallic substrates often lead to redistributions of the spin-polarized density and to changes in the electronic structure, which have a strong influence on the spin–orbit energies. This offers uncountable possibilities of optimizing the magneto-anisotropic properties by suitable choices of the substrate, the deposited magnetic material, and their morphology.

7 TAILORING THE MAGNETIC ANISOTROPY OF CORH NANOALLOYS

Magnetic nanoalloys are particularly interesting in view of manipulating and optimizing the spin and orbital moments, magnetic order, and MAE by varying the composition and distribution of the elements within the NPs. These are very important issues from both fundamental and technological perspectives. It is therefore important to understand the microscopic origin of the MAE in nanoalloys in order to control the possibilities of systematic material optimization that it offers. Early experimental studies on CoRh NPs have determined the concentration dependence of the average magnetic moments,[61] while theory revealed interesting changes in the MAE for small sizes ($N < 100$ atoms).[87,88] More recently, experiment and theory have been combined to determine the MAE of CoRh NPs as a function of Co concentration. A non-monotonous dependence of the MAE has been identified, which illustrates the possibilities of nanoalloy design.[73]

In this work, the CoRh NPs have been synthesized in solution by decomposition of organometallic precursors in the presence of a stabilizing polyvinylpyrrolidone polymer.[61] Samples with Co concentration $x = 0.76$, 0.49, and 0.25, and particle diameter ϕ in the range 1.6 nm $\leq \phi \leq$ 2.5 nm were obtained. The wide-angle X-ray scattering patterns of $Co_{0.49}Rh_{0.51}$ and $Co_{0.76}Rh_{0.24}$ NPs correspond to a compact structure having an NN distance $d_{NN} \simeq 0.269$ nm and $d_{NN} \simeq 0.263$ nm, respectively.[80] The $Co_{0.25}Rh_{0.75}$ NPs show a bulk-like fcc structure with $d_{NN} \simeq 0.269$ nm. The contrast observed in high-resolution transmission-electron microscopy on larger isolated particles indicates that the NPs are all bimetallic, with close-packed structures, most probably composed of an Rh-rich inner core and a Co-rich outer shell.

Electronic calculations have been performed in order to quantify the role of chemical order in clusters having $N \simeq 500$ atoms, Co concentrations $x \simeq 0.25$, 0.5, and 0.75, and different distributions of the Co and Rh atoms.[73] Following the SCTB theory described in Section **2.2**, the MAE $\Delta E_{\delta\gamma} = E_\delta - E_\gamma$ is derived in a non-perturbative way as the difference between the electronic energies E_δ. Accurate self-consistent calculations have been performed for each orientation δ of the spin magnetization $\vec{S}$. In this way, the redistributions of the spin- and orbital-polarized density (SOPD), and the SO interactions $\hat{H}_{SO}$ at the origin of magnetic anisotropy, are treated at the same electronic level.[27,87] This is important particularly for weakly ferromagnetic systems, such as Rh-based alloy clusters, in which rotating $\vec{S}$ often yields significant changes in the SOPD.

The magnetization profiles of 2 nm fcc-like $Co_{0.5}Rh_{0.5}$ NPs (e.g. $N = 489$ atoms) show very contrasting behaviours for different chemical orders.[73] In the case of a fully segregated Co core with an Rh outer shell, the Co magnetic moments at the innermost atoms are similar to the Co bulk moment and decrease as one goes from the centre to the CoRh interface. Only small magnetic moments are induced at the Rh, due to the proximity of the spin-polarized Co. These Rh moments oscillate weakly as one moves away from the interface towards the surface. A significant reduction of the Co moments is observed at the interface, which is not fully compensated by the induced Rh moments. The situation changes qualitatively if one considers an Rh core with Co outer shells. In this case, the Co moments are largest at the surface and decrease only slightly at the CoRh interface. The induced Rh moments are quite important, particularly at the interface, yielding a significant net contribution to the average moment $\overline{\mu}_{\text{CoRh}} = 2.08\mu_B$ ($\mu_S = 1.75\mu_B$ and $\mu_L = 0.33\mu_B$) which is now larger than in bulk alloys of similar concentrations. These contrasting behaviours illustrate, once more, the crucial role of chemical order on the magnetic properties of nanoalloys. In fact, only one of these arrangements (an Rh core with a outer Co shells) is consistent with the measured saturated magnetization ($\overline{\mu}^{\text{expt}}_{\text{CoRh}} = (2.38 \pm 0.05)\mu_B$ for 2 nm $Co_{0.5}$ $Rh_{0.5}$).[61]

The dependence on chemical order can be understood by contrasting the different local atomic environments. In the first case (Co core and Rh shell), all Co atoms have bulk-like coordination with less Co than Rh NNs at the interface. This increases the effective local d-band width at Co atoms and reduces the local Co moments. Moreover, the interface Rh atoms have few Co NNs with weakened moments so that the induced Rh polarization is quite small. Finally, the curvature at the surface of 2 nm particles ($N \simeq 500$) is not large enough to sustain the formation of local Rh moments (Rh_N is magnetic only for $N \leq 30$–50).[33] In contrast, in the second case (Rh core and Co shell), there are several factors that enhance magnetism: (i) The reduction of coordination number at the surface Co atoms increases the local moments, in particular, the orbital ones. (ii) The interface Co atoms, being outside, have more Co than

Rh NNs so that the d-band broadening is smaller. (iii) The Rh atoms at the interface have a majority of strongly magnetic Co NNs, which induce important Rh moments over several interatomic distances. These trends are common to compact CoRh clusters with different surface shapes and sizes ($N > 100$). Since 4d magnetism and the associated MAE can only survive close to the 3d–4d interfaces, the shape and structure of the interface and the possible interactions with the cluster surface are central to the magnetic behaviour of nanoalloys.

Real NPs rarely have perfect interfaces with fully segregated species. It is therefore interesting to quantify the consequences of deviations from the ideal core–shell arrangements. In Ref. 73, representative examples of clusters having an Rh core with Co outer shells have been considered by including intermixing between the atoms at the CoRh interface. One observes that intermixing yields an important enhancement of the spin and orbital moments of the Rh atoms, which have now a Co-rich local environment. At the same time, the Co moments are not much affected. The net result is a significant increase of the average magnetic moment per atom, which improves the agreement between theory and experiment ($\overline{\mu}_{\mathrm{CoRh}} = 2.24\mu_{\mathrm{B}}$ with $\mu_S = 1.90\mu_{\mathrm{B}}$ and $\mu_L = 0.34\mu_{\mathrm{B}}$ for $N = 489$ atoms). Further calculations including surface segregation and random alloy arrangements show that the precise choice of intermixing does not change the main trends in the average moment or in the MAE.[87] The enhancement of μ_S and μ_L due to intermixing is explained by the large spin and orbital polarizability of Rh atoms with increasing number of Co NNs and by the robustness of the nearly saturated Co spin moments.

The role of 4d magnetic contributions in CoRh clusters has been determined experimentally by means of XMCD measurements at the $L_{2,3}$ Rh thresholds. These show that the Rh atoms in $Co_{0.5}Rh_{0.5}$ clusters carry significant magnetic moments. Using the usual sum rules,[22] one derives $\mu_L/\mu_S = 0.066$ for the orbital to spin ratio. This is in good agreement with the theoretical result $\mu_L/\mu_S = 0.076$ for 2 nm NPs having an Rh core and outer Co shells. In contrast, the calculations assuming a Co core with Rh shells yield far too small values ($\mu_L/\mu_S \lesssim 0.01$). Comparison between experimental and theoretical results for the spin and orbital moments indicates therefore that the synthesized NPs have an Rh core with Co outer shell structure. This is consistent with the information derived from independent microscopy studies.[80]

In Figure 2.10A, experimental and theoretical results for the MAE of 2 nm CoRh NPs are shown. The experimental values were obtained by fitting the zero-field-cooled and field-cooled magnetization curves using a standard uniaxial Stoner–Wohlfarth model and a log-normal size distribution as described in Ref. 4. The theoretical results correspond to fcc-like octahedral clusters having an Rh core with a Co outer shell ($N = 489$ atoms). A remarkable non-monotonous behaviour is observed. Starting from pure Co NPs ($x = 1$) and increasing the Rh content, the MAE first increases, reaching a maximum around $x = 0.5$, and then decreases rapidly as x is further reduced. Experiment and theory deliver a quite consistent picture. The microscopic origin of the concentration dependence can be understood by analysing the local moments, in particular, the induced Rh

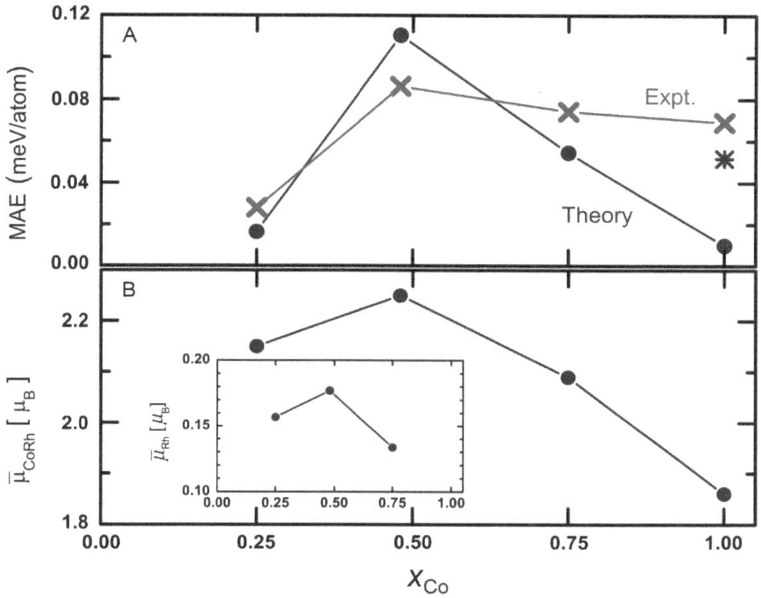

FIGURE 2.10 (A) Anisotropy energy and (B) average magnetic moment per CoRh unit $\bar{\mu}_{CoRh}$ of 2 nm $(Co_xRh_{1-x})_N$ as obtained from experiment (crosses) and theory (dots, $N=489$ atoms). The star for $x=1$ corresponds to a dodecahedral cluster shape. The inset shows the calculated average magnetic moment $\bar{\mu}_{Rh}$ induced at the Rh atoms. *After Ref. 73.*

moments as a function of x. As shown in Figure 2.10B, the magnetic moment per Co atom and the average magnetic moment $\bar{\mu}_{Rh}$ at the Rh atoms (inset figure) increase with increasing Rh content until the Co concentration becomes so low that the overall cluster magnetization breaks down. Moreover, the higher magnetic susceptibility of the Rh clusters, as compared with bulk Rh, explains why the optimal Rh concentration is larger in CoRh NPs than in macroscopic CoRh alloys ($x_{max} \simeq 0.5$ for 2 nm NPs).[89] Furthermore, calculations have been performed in order to confirm the dominant role of the Rh contribution to the MAE by switching off artificially the SO coupling at the Rh atoms. In this case, no enhancement of the calculated MAE, but rather a decrease with increasing Rh content is found, in contrast with experiment. The correlation between induced Rh moments and MAE is found to be a quite general trend, which reflects the microscopic mechanisms controlling the magneto-anisotropic behaviour of 3d–4d nanoalloys. New possibilities for tailoring the magnetic behaviour of nanostructures for specific applications are thereby opened.

8 FINITE TEMPERATURE MAGNETIC PROPERTIES

The theory of cluster magnetism at finite temperatures is not only more challenging methodologically but also more rewarding in view of applications and the comparison with experiment. One would like to understand the nature of the temperature-induced spin fluctuations, the stability of the magnetic order within the NPs, and the temperature dependence of the average magnetization per atom. The variety of ground-state behaviours discussed in the previous sections shows that any finite temperature modellization should be based on an electronic theory, which takes into account the itinerant character of the d-electron states. Moreover, only an explicit treatment of the electronic degrees of freedom would allow one to take into account the effects of spin–orbit interactions, which are responsible for magnetocrystalline anisotropy and spin-reorientation transitions (SRTs). Temperature-driven SRTs have been observed in thin films[90] and are expected to occur also in finite clusters. Characterizing their microscopic origin and the circumstances for their occurrence is therefore of considerable interest. In this section, we discuss three different aspects of TM-cluster magnetism at finite T: the low-temperature spin-fluctuation energies, the short-range magnetic order above the cluster Curie temperature, and the size and structural dependence of the magnetization curves and spin-correlation functions.

8.1 Low-Temperature Spin-Fluctuation Energies in TM Clusters

A first insight on the magnetic behaviour of 3d transition-metal clusters at $T>0$ can be obtained by considering the low-temperature limit of the electronic free energy $F_l(\xi)$ at the different atoms l in the cluster. This corresponds to the fluctuations of a single exchange field ξ at atom l keeping all the rest of the atoms in the ground state ($\xi_l=\xi$ and $\xi_m=\xi_m^0$ for $m\neq l$, see Equation 2.24). The local free-energy difference

$$\Delta F_l(\xi)=F_l(\xi)-F_l\left(\xi_l^0\right), \tag{2.39}$$

with $\xi_l^0=\mu_l^0$ being the local moment of atom l at $T=0$, represents the energy involved in an exchange-field fluctuation at atom l above the Hartree–Fock ground state. $\Delta F_l(\xi)$ determines the probability of the fluctuation $\Delta\xi=\xi-\xi_l^0$ and thus gives a measure of the stability of the ground-state magnetic order within the cluster at finite T. In the following, we discuss representative results for the low-temperature limit of the local free-energies $F_l(\xi)$ and the resulting spin-fluctuation energies $\Delta F_l(\xi)$ in small Fe_N and Ni_N clusters.

In Figure 2.11, results are given for $\Delta F_l(\xi)$ in small Fe clusters.[55] First of all, one observes that $\Delta F_l(\xi)>0$ for all $\xi\neq\mu_l^0$ which indicates, as expected, that in these clusters the FM order is stable at low temperatures. For the smallest clusters, that is, Fe_2 and Fe_4, $F_l(\xi)$ shows two minima located at the exchange fields $\xi^+=\mu_l^0$ and $\xi^-\simeq-\mu_l^0$, where μ_l^0 refers to the ground-state magnetic moment. This double-minimum structure indicates that the dominant magnetic excitations are flips of the magnetic moments keeping their amplitude approximately constant. In fact, at finite T, the probability $P_l(\xi)\propto\exp\{-\beta\Delta F_l(\xi)\}$ has two maxima at ξ^+ and ξ^-. It is thus more probable to find $\xi\simeq-\mu_l^0$ than $\xi\simeq0$. In the case of Fe_6 and Fe_9, $F_l(\xi)$ also shows

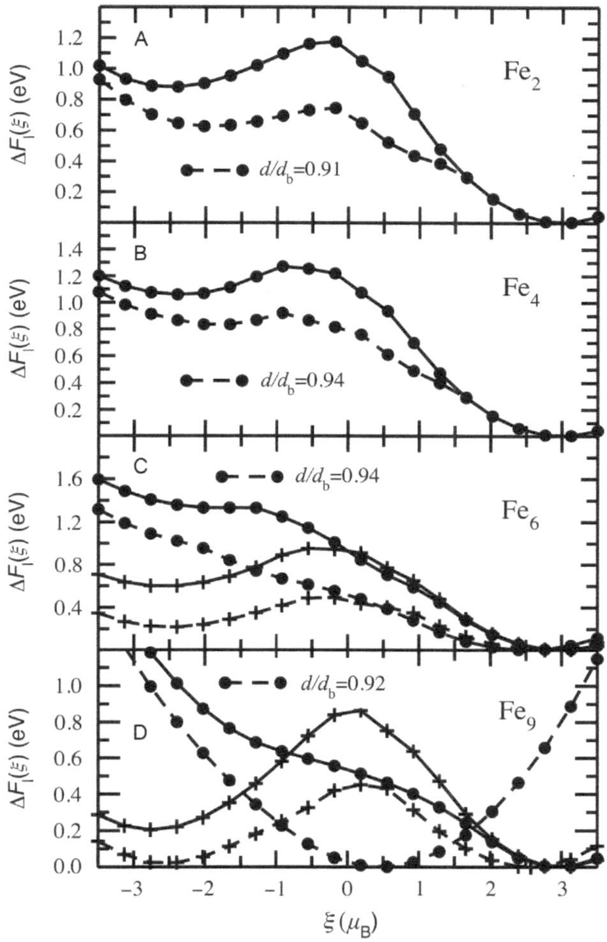

FIGURE 2.11 Local spin-fluctuation energy $\Delta F_l(\xi) = F_l(\xi) - F_l(\xi_l^0)$ as a function of the exchange field ξ at different atoms l of Fe_N clusters: (A) Fe_2, (B) Fe_4 (tetrahedron), (C) Fe_6 (square bipyramid), and (D) Fe_9 (bcc-like). Results are given for bulk NN distances $d/d_b = 1$ (full curves) and for relaxed NN distances (dashed curves). In (C) and (D), the dots refer to the atom $l = 1$ with the largest local coordination number (i.e. the central atom for Fe_9) and the crosses to one of the surface atoms, $l = 2$. *After Ref. 55.*

two minima for the lowest coordinated atoms $l = 2$, which have the largest local magnetic moments μ_l^0 at $T = 0$. In these cases, as in Fe_2 and Fe_4, only small fluctuations $\Delta\xi = \xi - \mu_l^0$ are possible with an excitation energy $\Delta F_l(\xi)$ smaller than the energy $\Delta F_l(\xi^-) = F_l(\xi^-) - F_l(\xi^+)$ required to flip the local moment. In contrast, for the most coordinated atoms which have smaller μ_l^0 (e.g. the central atom in bcc-like Fe_9), a single minimum is found. This implies that the fluctuations of the amplitude of the local moments dominate.

The fact that very small clusters and the atoms having the smallest local coordination numbers z_l show a Heisenberg- or Ising-like behaviour should not be surprising. On the one side, the kinetic energy loss ΔE_K caused by flipping a local magnetic moment ($\xi \simeq \mu_l^0 \rightarrow \xi \simeq -\mu_l^0$) decreases with decreasing z_l, since the perturbation introduced by flipping ξ_l is in general less important when l has a small number of neighbours. The contribution of atom l to E_K is approximately proportional to $\sqrt{z_l}$ (second moment approximation). On the other side, the exchange energy $\Delta E_X = (J/4)\sum_l \mu_l^2$ is basically a local property which is much less affected by the change of sign of ξ. Thus, when z_l is reduced, the local character is enhanced and it becomes energetically more favourable to have $\xi \simeq -\mu_l^0$ rather than a significant fluctuation of ξ around ξ_0 (e.g. $\xi \simeq 0$).

The results for Ni clusters show, as in the case of Fe, that $\Delta F_l(\xi)$ is much more sensitive to cluster size and to changes in the interatomic distances than the magnetic moments at $T=0$.[55] In small Ni_N, the magnetic moments are saturated and are thus nearly independent of cluster structure and local atomic environment. Nevertheless, $\Delta F_l(\xi)$ presents significant quantitative changes as a function of l and N , which let us expect interesting temperature-dependent properties. Moreover, bond-length contractions ($d < d_b$) usually result in a reduction of $\Delta F_l(\xi)$ even if μ_l^0 remains unchanged. Comparing surface and inner atoms, we obtain that the spin-fluctuations energies are in general smaller at the cluster surface, that is, as the local coordination number is smaller. Thus, one expects that in Ni clusters, the surface atoms should trigger the decrease of the average magnetization per atom as T increases.

It is important to keep in mind that the low-temperature limit of $\Delta F_l(\xi)$ cannot be used as a simple means of estimating the temperature $T_C(N)$ above which the FM order is lost within the cluster. The Curie temperature is not the energy necessary to flip a spin at $T=0$ but rather the temperature at which it costs no free-energy to flip one.[51] In other words, collective fluctuations of ξ_m at sites $m \neq l$ modify the effective spin-excitation energies, thereby reducing the stability of the FM order. Moreover, the low-temperature limit of $\Delta F_l(\xi)$, where all but one field ξ_l are kept equal to the $T=0$ value, ignores that the local exchange fields at different sites fluctuate in some correlated fashion, showing some degree of SRMO (see Section **8.2**). The fluctuations of the ensemble of exchange fields modify the magnetic environment at which individual spin fluctuations occur. They should be therefore taken into account in quantitative studies at finite T (see Section **8.3**).

8.2 Short-Range Magnetic Order in Fe_N and Ni_N

The experimental results for the temperature dependence of the magnetization per atom $\overline{\mu}_N(T)$ of Fe_N ($25 \leq N \leq 700$, $100\text{ K} \leq T \leq 1000\text{ K}$) and Ni_N ($40 \leq N \leq 600$, $100\text{ K} \leq T \leq 700\text{ K}$)[9,13] show that for all studied sizes, $\overline{\mu}_N(T)$ decreases with increasing T reaching an approximately constant value above a characteristic size-dependent temperature $T_C(N)$, which corresponds to a magnetically disordered state.[91] The magnetization above $T_C(N)$,

$\overline{\mu}_N(T > T_C)$, was found to be significantly *larger* than $\overline{\mu}_N(T=0)/\sqrt{N}$, the value corresponding to N randomly oriented atomic magnetic moments.[9,13] This is a very interesting, since the results for randomly oriented moments should be independent of the details of the interactions between the local moments, provided that the size of the local moments remains approximately constant. It is the purpose of this section to analyse the high-temperature values of $\overline{\mu}_N$ by taking into account the presence of a certain degree of short-range magnetic order (SRMO) in these clusters above $T_C(N)$.

The degree of SRMO within the cluster can be characterized by the average number of atoms ν involved in an SRMO domain. The magnetization per atom of a N-atom cluster at $T > T_C(N)$ is then approximately given by

$$\overline{\mu}_N(T > T_C) \simeq \overline{\mu}_N(T=0)\sqrt{\nu/N}, \tag{2.40}$$

which represents the average $\sqrt{\langle \mu^2 \rangle}$ of N/ν randomly oriented SRMO domains, each carrying a magnetic moment $\nu\overline{\mu}_N(T=0)$. The disordered local moment picture, that is, the situation without SRMO, corresponds thus to $\nu=1$. The actual value of ν for 3d TMs can be estimated first from known bulk and surface results.[92–94] For bulk Fe, Haines *et al.*[93] have retrieved a range of SRMO near T_C of about 5.4 Å (at least 4 Å), which corresponds to $\nu \simeq 15$ (up to next nearest neighbours). Similar values are obtained in calculations of SRMO in Fe bulk and Fe surfaces.[94] For Fe_N clusters, one finds a particularly high stability of ferromagnetism for $N \simeq 15$. The energy gain associated with the development of the magnetic moments, $\Delta E(N) = E(\mu) - E(\mu=0)$, is quite large around this size: $\Delta E(15) - \Delta E(\text{bulk}) \simeq 0.3$ eV.[18] Furthermore, the energy involved in flipping a single local magnetic moment within Fe_{15} is about 0.1 eV larger than for the bulk.[55] Therefore, $\nu = 15$ seems a reasonable estimation of the degree of SRMO in Fe clusters. For Ni, the SRMO is generally expected to be stronger than for Fe.[92]

The comparison between experiment and the results neglecting SRMO (i.e. with $\nu=1$) is very poor. The root mean square deviation σ amounts to $\sigma = 0.75\mu_B$ for Fe_N and $\sigma = 0.21\mu_B$ for Ni_N. This rules out the disorder-local-moment picture for Fe and Ni clusters, as it is also the case for the bulk and for plane surfaces.[92–94] On the contrary, the results including SRMO are in very good agreement with experiment if we assume $\nu = 15$ for Fe_N and $\nu = 19$–43 for Ni_N: $\sigma(Fe_N) = 0.12\mu_B$ and $\sigma(Ni_N) = 0.06\mu_B$ (see Ref. 94). These values of σ are of the order of the experimental uncertainties (about 7%[9,13]). One concludes that the low- and high-temperature measurements of $\overline{\mu}_N(T)$ of Fe and Ni clusters[9,13] are consistent with each other. The large observed values of $\overline{\mu}_N(T > T_C)$ provide a strong evidence for the existence of SRMO in these clusters above $T_C(N)$, which is in agreement with known surface and bulk properties of itinerant magnetism.[92–94]

Clearly, the existence of SRMO alone cannot explain the complete temperature dependence of $\overline{\mu}_N(T)$. As in the solid, there are many element-specific features to consider, such as the details of the electronic densities of states, the d-band filling, the Coulomb interaction strength, and the lattice structure. This challenging question is addressed in the following section.

8.3 MC Simulations of Itinerant Cluster Magnetism

In Section **2.3**, we have seen that the finite-temperature magnetic properties of an N-atom cluster are obtained by performing a multidimensional integral over N exchange fields. Each calculation of $F'(\xi_1, \ldots, \xi_N)$ involves a self-consistent electronic calculation that is almost as involved as a ground-state one. The integration procedure must be therefore efficient and unbiased. Simple Metropolis MC methods[95] are insufficient in this context, since the energy landscape defined by $F'(\xi_1, \ldots, \xi_N)$ is complex, showing numerous local minima often separated by large barriers. The results shown in Section **8.1** illustrate this behaviour clearly (see Figure 2.11). Several methods have been proposed to overcome this by far not usual situation. One of these is exchange MC also known as parallel-tempering MC.[96–101] In this approach, the ergodicity times are drastically reduced by performing several simulations at different temperatures in a parallel way and allowing for exchange of configurations between the various temperatures.[101] In the exchange MC method, one considers many replicas of the system of interest, each of which is simulated simultaneously and independently at a different temperature following the conventional Metropolis MC algorithm. In addition to the usual local updates of the spin configurations $\vec{\xi}$, one allows the exchange of configurations at nearby temperatures according to a Metropolis criterion by taking into account the energy difference between the configurations. This introduces additional nonlocal Markov steps by which a simulation at low temperature can avoid being trapped in metastable local minima. Further details of the parallel-tempering MC method may be found in Ref. 101. For specific applications in the context of the spin-fluctuation theory of itinerant magnetism, see Ref. 20. In the following, we discuss results for the temperature dependence of the average magnetizations, $\overline{\mu}_N(T)$, local magnetic moments μ_l, and pair correlation functions γ_{lk} for Fe_N clusters having $N \leq 24$ atoms.[20]

In Figure 2.12, results are shown for the temperature dependence of the average magnetization $\overline{\mu}_N$ as given by Equation (2.34). Among the general common features of all curves, one observes the low-temperature saturation of $\overline{\mu}_N$ for small clusters ($\overline{\mu}_N(T \to 0) \simeq 3.0\mu_B$ for $N \leq 6$) and the enhancement of $\overline{\mu}_N(0)$ with respect to the bulk for $N = 15$ and 24. In the other extreme, at very high temperatures $(T4000\text{K})\overline{\mu}_N$ is approximately constant, as expected for a disordered magnetic state. Notice that the high-temperature values are somewhat smaller than $\overline{\mu}_N(0)/\sqrt{N}$, which would be the result predicted by a simple localized Ising-like model.[102] This reflects a moderate reduction of the local magnetic moments μ_l, which can be ascribed to Stoner excitations and the delocalized character of the d-states. The high-temperature limit of

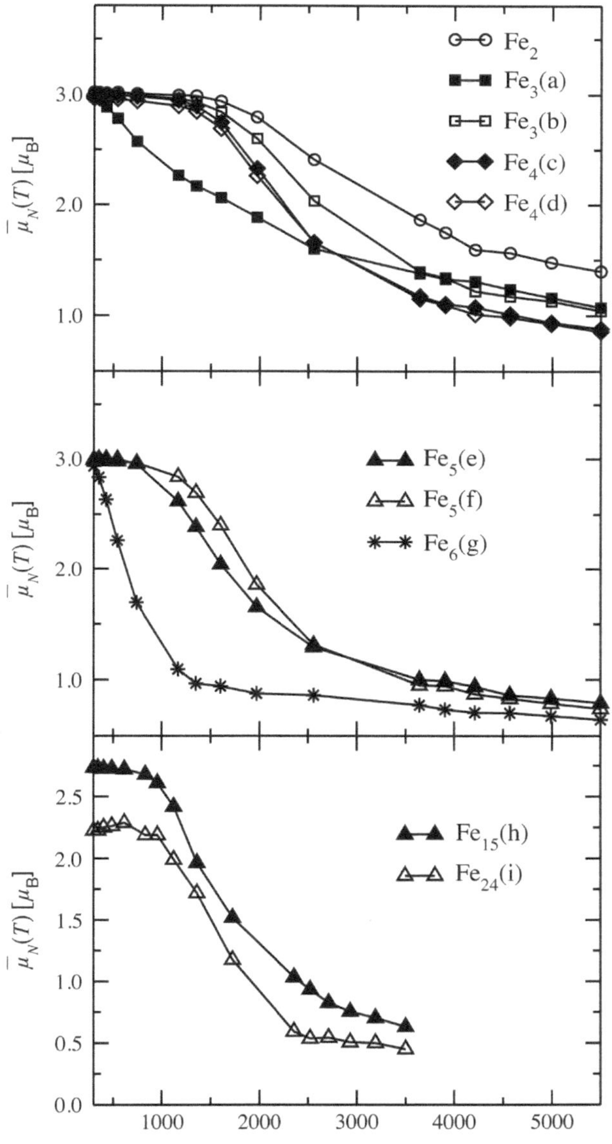

FIGURE 2.12 Temperature dependence of the average magnetic moment per atom $\overline{\mu}_N(T)$ of Fe_N clusters (see Equations 2.34 and 2.35). The labels (a)–(i) refer to different structures. *After Ref. 20.*

$\overline{\mu}_N$, as well as of the local moments μ_l to be discussed below, is essentially independent of the details of the electronic structure and of the cluster geometry. It is mainly a statistical local effect behaviour, which sets in once all the spin correlations vanish.

The temperature dependence linking the low and high T limits is not universal, that is, it depends strongly on the cluster geometry and on the details of the single-particle spectrum. A strong structural dependence of $\overline{\mu}_N(T)$ is found, even in situations where the ground-state moments are saturated and therefore do not depend significantly on structure (e.g. in small clusters). Quantitatively, the differences in $\overline{\mu}_N(T)$ are found to be very important in Fe_3 and to lesser extent in Fe_4 and Fe_5 (see Figure 2.12). At high temperatures, in the disordered local moment regime, the differences in $\overline{\mu}_N$ among different isomers disappear completely, even in cases where the deviations at low and intermediate temperatures are important (e.g. Fe_3 chain and triangle, or Fe_5 bipyramid and trust). This occurs for $T \simeq 2500$–3000 K, when the pair correlation functions essentially vanish.

It is interesting to compare the temperature dependence of $\overline{\mu}_N$ for different cluster sizes, particularly for the most compact and highly symmetric clusters ($N = 2$–5) since the average coordination number increases here almost linearly with N. Remarkably, the results show that $\overline{\mu}_N(T)$ decreases more rapidly with T as N increases (see Figure 2.12). This means that at $T > 0$, the ferromagnetic order becomes comparatively less stable as N and the coordination number z increase. This trend is strictly opposite to the predictions of simple spin models with size-independent exchange couplings J_{ij} (for instance, the Ising or the Heisenberg models). In fact, if one would attempt to derive effective exchange couplings J_{ij} by fitting the electronic calculations, one would conclude that J_{ij} decreases rapidly with N or z, namely, as $1/z$ or faster. Obviously, this surprising behaviour has its origin in the itinerant character of the d-electron states. As z increases the d-band width increases and with it the relative importance of the kinetic energy as compared to the local exchange energy. This contribution appears to be so strong in small Fe clusters, that it overcomes the effects of the perturbation introduced by the fluctuations of the exchange field ξ_l, which affects a larger number of atoms as z increases and should therefore imply a higher excitation energy. These results illustrate the fundamental interplay between spin fluctuations and electronic structure, and the subtle competition between localized and itinerant aspects of TM-cluster magnetism.

The temperature dependence of the magnetic order within the cluster can be analysed in more detail by considering the spin-correlation functions γ_{lk} (see Equation 2.37). Comparing the various local moments $\mu_l = \sqrt{\gamma_{ll}}$ and interatomic γ_{lk} allows us to understand the behaviour of $\overline{\mu}_N = \sqrt{\sum_{lk} \gamma_{lk}}$ from a local perspective. Moreover, γ_{lk} gives a useful insight on the environment dependence of finite-temperature cluster magnetism.

The temperature dependence of γ_{12} for Fe_2 and Fe_3 (triangle) follows the behaviour of the corresponding $\overline{\mu}_N(T)$ curves, as expected for highly symmetric clusters having local moments μ_l which depend weakly on temperature.[20] For large temperatures, γ_{12} decreases monotonically, remaining positive and approaching zero ($\gamma_{12} \ll \gamma_{11}$) at approximately $T \simeq 4000$ K.[20] For $T \gtrsim 4000$ K, the local moments fluctuate in an uncorrelated way and therefore $\overline{\mu}_N \simeq \mu_l / \sqrt{N}$. An interesting temperature dependence of the spin-correlation functions is found in rhombohedral Fe_4, namely, a ferromagnetic coupling between first NNs and antiferromagnetic coupling between second NNs.[20] These antiferromagnetic correlations result in a faster decrease of the average magnetization with increasing temperature.

The characteristic behaviour found in very small low-symmetry clusters, that is, FM correlations for first NNs and AF correlations for second NNs above $T \simeq 1500$, no longer applies to Fe_{15}. In this case, all the correlation functions are positive. Remarkably, the spin-correlation functions involving the central atom $i = 1$ (e.g. γ_{12} and $\gamma_{1,10}$) show an unusual temperature dependence. They start from rather small values at $T = 0$ and then *increase* with increasing T, as if they were driven by the still strong FM correlations between all the other atoms in the cluster. A change of trend and a decrease of γ_{12} and $\gamma_{1,10}$ are only observed when the fluctuations are so important that the correlations between the atoms $i = 2$–15 start to decrease. This effect is most probably due to changes in the local electronic structure with temperature. It suggests that low-energy states with higher spin are occupied at $T > 0$. A similar behaviour has been found in other clusters showing non-saturated ground-state moments, including fully correlated exact diagonalization studies of Hubbard clusters.

9 CONCLUSION

In this chapter, state-of-the-art first principles and model theories of the magnetic properties of nanostructures have been presented and a number of representative applications to pure and alloy TM clusters have been reviewed. Although the presentation did not aim to be exhaustive, it is expected to have covered the fundamental aspects of cluster magnetism and enough examples of problems of current interest, so as to illustrate the achievements and challenges in the field. Among the subjects which deserve more detailed and systematic investigations, one should mention the dependence of the magnetic properties on the distribution of the different components within the nanoalloys, and the correlation between magnetic behaviour and local chemical environment. This is particularly relevant for the comparison between theory and experiment, since the morphology of the NPs can be tuned, at least to some extent, by changing the synthesis method, the deposition conditions, or by subsequent annealing. The strong sensitivity of TM magnetism to the specific local atomic environments, several examples of which have been given in the previous sections, lets us expect a wide variety of interesting behaviours. The combination of experimental and theoretical structural and magnetic studies should ultimately provide a reliable physical picture of the actual cluster morphology and magnetic behaviour.

Besides the analysis and interpretation of specific experiments and their implications in material science, the future theoretical research on magnetic clusters is confronted with a number of interesting methodological and conceptual issues. For instance, it seems important to introduce a larger flexibility in the self-consistent magnetic calculations by allowing for non-collinear spin polarizations $\vec{m}\left(\vec{r}\right)$. On the one side, the diversity of local and chemical environments found in NPs often leads to magnetic frustrations, such as competing FM and AF couplings between NNs, which are likely to lead to non-collinear spin arrangements. On the other side, the SO interactions are known to introduce hybridizations between up and down spin states, which imply deviations from collinearity. Moreover, the magnetization direction yielding the lowest local energy $E_\delta(i)$ can be different for different atoms in the cluster, particularly when 4d and 5d elements are involved. This could also lead to some degree of non-collinearity particularly among the orbital moments. Other interesting phenomena which deserve further study are the SRTs, that is, the changes in the direction of the easy axis, as a function of size, composition, and chemical order. In fact, such qualitative modifications of the magneto-anisotropic behaviour have already been observed at the interfaces of thin CoPt and CoPd films.[103] They could therefore be induced in TM clusters by manipulating the interfaces between different elements, for example, by means of annealing or by coating magnetic clusters with 4d or 5d elements. Furthermore, detailed investigations of the MAE surface, for all directions of the magnetization, are central to understanding the mechanisms of magnetization reversal and the associated dynamical processes in NPs.

An even broader range of challenging research directions opens in the context of finite-temperature properties. A subject which deserves special attention is the role of temperature-induced fluctuations of the cluster structure, which could coexist with the spin fluctuations of electronic origin considered in the chapter. Previous exact diagonalization studies on the single-band Hubbard model have shown that the isomerization energies of magnetic clusters are often comparable to the spin-excitation energies. Even if the stability of ground-state structures is expected to be higher in realistic *d*-band model calculations, the contributions of structural fluctuations to the temperature dependence cannot be excluded *a priori*. The effect could be significant, particularly in weak unsaturated ferromagnets like Fe_N, whose magnetic moments are known to be very sensitive to structure already for $T=0$.[18] As shown in this chapter, the stability of ferromagnetism at finite T depends on the cluster geometry. Therefore, the population of low-energy isomers can modify $\overline{\mu}_N(T)$, even if the $T = 0$ moments of the excited isomers are nearly the same as for the optimal geometry. More detailed investigations taking into account electronic spin excitations and structural rearrangements on the same footing are certainly worthwhile.

The functional-integral theory of spin fluctuations and the local approach to the cluster electronic structure reviewed in this chapter are well suited for investigating extended nanostructures with reduced symmetry. It would be interesting to quantify how the finite-temperature magnetic properties of free clusters are modified by the interaction with macroscopic supports such as matrices, surfaces, and thin films. From a more methodological perspective, one would like to restore the spin-rotational symmetry of the effective Hamiltonian $\hat{H}'$ by allowing for three-dimensional fluctuations of vector exchange fields $\vec{\xi}_l$ at each atom l. In this way, non-collinear magnetic orders and transversal fluctuations of the magnetic moments could be quantified. These are likely to affect the magnetization curves and probably reduce the calculated values of the cluster Curie temperature T_C (N). Last but not least, the effects of spin–orbit interactions should be investigated at finite temperatures, since they are responsible for magnetic anisotropy. Indeed, the temperature dependence to the MAE is a subject of central importance for applications, which remains essentially unexplored. Temperature-driven SRTs, which have already been observed in thin films,[90] are expected to take place in small clusters. Theory should be able to identify them and to characterize the microscopic mechanisms behind such remarkable effects. Experiments on clusters deposited of surfaces (e.g. using STM manipulation and STM spectroscopy) would allow us to quantify the validity of the predictions and motivate further theoretical improvements.

ACKNOWLEDGEMENTS

It is pleasure to thank L. Díaz-Sánchez, V. Dupuis, R. Garibay-Alonso, J.H. Mokkath, M. Muñoz-Navia, and M. Respaud for their helpful discussions and useful comments. This work has been financed in part by CONACyT-Mexico, the Deutsche Forschungsgemeinschaft, and the DAD-CONACyT exchange programme PROALMEX.

ABBREVIATIONS

1D	one dimensional
DFT	density-functional theory
FM	ferromagnetic
GGA	generalized gradient approximation
LDA	local density approximation
MAE	magnetic anisotropy energy
NN	nearest neighbour
NP	nanoparticle

SCTB	self-consistent tight-binding
SG	Stern–Gerlach
SO	spin–orbit
SOPD	spin- and orbital-polarized density
SRT	spin-reorientation transition
TM	transition metal
VASP	Vienna *ab initio* simulation package
WS	Wigner–Seitz
XC	exchange and correlation
XMCD	X-ray magnetic circular dichroism

REFERENCES

1. Cox DM, Trevor DJ, Whetten RL, Rohlfing EA, Kaldor A. Magnetic behavior of free-iron and iron oxide clusters. *Phys Rev B* 1985;**32**:7290–8.
2. de Heer WA, Milani P, Châtelain A. Spin relaxation in small free iron clusters. *Phys Rev Lett* 1990;**65**:488–91.
3. Binns C. Nanoclusters deposited on surfaces. *Surf Sci Rep* 2001;**44**:1–49.
4. Respaud M, Broto JM, Rakoto H, Fert AR, Thomas L, Barbara B, et al. Surface effects on the magnetic properties of ultrafine cobalt particles. *Phys Rev B* 1998;**57**:2925–35.
5. Lau JT, Föhlisch A, Nietubyc R, Reif M, Wurth W. Size-dependent magnetism of deposited small iron clusters studied by X-ray magnetic circular dichroism. *Phys Rev Lett* 2002;**89**:057201-1–4.
6. Gambardella P, Rusponi S, Veronese M, Dhesi SS, Grazioli C, Dallmeyer A, et al. Giant magnetic anisotropy of single cobalt atoms and nanoparticles. *Science* 2003;**300**:1130–3.
7. Yin S, Moro R, Xu X, de Heer WA. Magnetic enhancement in cobalt-manganese alloy clusters. *Phys Rev Lett* 2007;**98**:113401-1–4.
8. Gruner ME, Rollmann G, Entel P, Farle M. Multiply twinned morphologies of FePt and CoPt nanoparticles. *Phys Rev Lett* 2008;**100**:087203-1–4.
9. Billas IML, Châtelain A, de Heer WA. Magnetism from the atom to the bulk in iron, cobalt, and nickel clusters. *Science* 1994;**265**:1682–4.
10. Gerion D, Hirt A, Billas IML, Châtelain A, de Heer WA. Experimental specific heat of iron, cobalt, and nickel clusters studied in a molecular beam. *Phys Rev B* 2000;**62**:7491–501.
11. Knickelbein MB. Nickel clusters: the influence of adsorbed CO on magnetic moments. *J Chem Phys* 2001;**115**:1983. Knickelbein MB. Magnetic moments of small bimetallic clusters: Co_nMn_m. *Phys Rev B* 2007;**75**:014401-1–6.
12. Xu X, Yin S, Moro R, de Heer WA. Magnetic moments and adiabatic magnetization of free cobalt clusters. *Phys Rev Lett* 2005;**95**:237209-1–4.
13. Billas IML, Becker JA, Châtelain A, de Heer WA. Magnetic moments of iron clusters with 25 to 700 atoms and their dependence on temperature. *Phys Rev Lett* 1993;**71**:4067–70.
14. Apsel SE, Emmert JW, Deng J, Bloomfield LA. Surface-enhanced magnetism in nickel clusters. *Phys Rev Lett* 1996;**76**:1441–4.
15. Bucher JP, Douglass DC, Bloomfield LA. Magnetic properties of free cobalt clusters. *Phys Rev Lett* 1991;**66**:3052–5. Douglass DC, Cox AJ, Bucher JP, Bloomfield LA. Magnetic properties of free cobalt and gadolinium clusters. *Phys Rev B* 1993;**47**:12874–89. Douglass DC, Bucher JP, Bloomfield LA. Magnetic studies of free nonferromagnetic clusters. *Phys Rev B* 1992;**45**:6341–4.

16. Knickelbein MB. Experimental observation of superparamagnetism in manganese clusters. *Phys Rev Lett* 2001;**86**:5255–7.
17. Lee K, Callaway J, Dhar S. Electronic structure of small iron clusters. *Phys Rev B* 1984;**30**:1724–30. Lee K, Callaway J, Kwong K, Tang R, Ziegler A. Electronic structure of small clusters of nickel and iron. *Phys Rev B* 1985;**31**:1796–803. Lee K, Callaway J. Electronic structure and magnetism of small V and Cr clusters. *Phys Rev B* 1993;**48**:15358–64.
18. Pastor GM, Dorantes-Dávila J, Bennemann KH. The magnetic properties of small Fen-clusters. *Physica B* 1988;**149**:22–5. Pastor GM, Dorantes-Dávila J, Bennemann KH. Size and structural dependence of the magnetic properties of small 3*d*-transition-metal clusters. *Phys Rev B* 1989;**40**:7642–54.
19. In a finite system it is not possible to define a Curie temperature T_C (N) univocally due to the absence of true phase transitions. As a result, the peaks in the specific heat and in the magnetic susceptibility as a function of T do not coincide. In the present context, T_C (N) should be understood as a measure of the stability of the magnetic order within the clusters or as the characteristic temperature defining the crossover from the low-temperature ordered FM state to the high-temperature disordered state. Experimentally, this corresponds approximately to the inflection point in $\overline{\mu}_N(T)$.
20. Garibay-Alonso R, Dorantes-Dávila J, Pastor GM. Electronic spin-fluctuation theory of finite-temperature cluster magnetism: size and environment dependence in Fe_N. *Phys Rev B* 2009;**79**:134401-1–12.
21. Erskine JL, Stern EA. Calculation of the M23 magneto-optical absorption spectrum of ferromagnetic nickel. *Phys Rev B* 1975;**12**:5016–24. van der Laan G, et al. Experimental proof of magnetic x-ray dichroism. *Phys Rev B* 1986;**34**:6529–31. Schütz G, Wagner W, Wilhelm W, Kienle P. Absorption of circularly polarized x rays in iron. *Phys Rev Lett* 1987;**58**:737–40. Thole BT, Carra Paolo, Sette F, van der Laan G. X-ray circular dichroism as a probe of orbital magnetization. *Phys Rev Lett* 1992;**68**:1943–6. Carra P, Thole BT, Altarelli M, Wang X. X-ray circular dichroism and local magnetic fields. *Phys Rev Lett* 1993;**70**:694. van der Laan G. Angular momentum sum rules for x-ray absorption. *Phys Rev B* 1998;**57**:112–5.
22. Thole BT, Carra P, Sette F, van der Laan G. X-ray circular dichroism as a probe of orbital magnetization. *Phys Rev Lett* 1992;**68**:1943–6.
23. Edmonds KW, Binns C, Baker SH, Thornton SC, Norris C, Goedkoop JB, et al. Doubling of the orbital magnetic moment in nanoscale Fe clusters. *Phys Rev B* 1999;**60**:472–6. Ohresser P, Ghiringhelli G, Tjernberg O, Brookes N, Finazzi M. Magnetism of nanostructures studied by x-ray magnetic circular dichroism: Fe on Cu(111). *Phys Rev B* 2000;**62**:5803–9. Edmonds K, Binns C, Baker SH, Maher MJ, Thornton SC, Tjernberg O, Brookes NB. Size dependence of the magnetic moments of exposed nanoscale iron particles. *J Magn Magn Mater* 2001;**231**:113–9. Fauth K, Ballentine GE, Praetorius C, Kleibert A, Wilken N, Voitkans A, Meiwes-Broer K-H. Magnetic properties of Fe nanoclusters on Cu(111) studied with X-ray magnetic circular dichroism. *Phys Status Solidi B* 2010;**247**:1170–9. Jamet M, Dupuis V, Melinon P, Guiraud G, Perez A, Wernsdorfer W, Traverse A, Baguenard B. Structure and magnetism of well defined cobalt nanoparticles embedded in a niobium matrix. *Phys Rev B* 2000;**62**:493–9.
24. Koide T, Miyauchi H, Okamoto J, Shidara T, Fujimori A, Fukutani H, et al. Direct determination of interfacial magnetic moments with a magnetic phase transition in Co nanoclusters on Au(111). *Phys Rev Lett* 2001;**87**:257201-1–4.
25. Guirado-López RA, Dorantes-Dávila J, Pastor GM. Orbital magnetism in transition-metal clusters: from Hunds rules to bulk quenching. *Phys Rev Lett* 2003;**90**:226402-1–4.

26. Bødker F, Mørup S, Linderoth S. Surface effects in metallic iron nanoparticles. *Phys Rev Lett* 1994;**72**:282–5. Jamet M, Wernsdorfer W, Thirion C, Mailly D, Dupuis V, Mélinon P, Pérez A. Magnetic anisotropy of a single cobalt nanocluster. *Phys Rev Lett* 2001;**86**:4676–9. Luis F, Torres JM, García LM, Bartolomé J, Stankiewicz J, Petroff F, Fettar F, Maurice J-L, Vaurés A. Enhancement of the magnetic anisotropy of nanometer-sized Co clusters: influence of the surface and of interparticle interactions. *Phys Rev B* 2002;**65**:094409-1–10. Gilmore K, Idzerda YU, Klem MT, Allen M, Douglas T, Young M. Surface contribution to the anisotropy energy of spherical magnetite particles. *J Appl Phys* 2005;**97**:10B301–1–3. Yanes R, Chubykalo-Fesenko O, Kachkachi H, Garanin DA, Evans R, Chantrell RW. Effective anisotropies and energy barriers of magnetic nanoparticles with Néel surface anisotropy. *Phys Rev B* 2007;**76**:064416-1–13. Berger L, Labaye Y, Tamine M, Coey JMD. Ferromagnetic nanoparticles with strong surface anisotropy: spin structures and magnetization processes. *Phys Rev B* 2008;**77**:104431-1–10. Tournus F, Blanc N, Tamion A, Hillenkamp M, Dupuis V. Dispersion of magnetic anisotropy in size-selected CoPt clusters. *Phys Rev B* 2010;**81**:220405-1–4.
27. Pastor GM, Dorantes-Dávila J, Pick S, Dreyssé H. Magnetic anisotropy of 3*d* transition-metal clusters. *Phys Rev Lett* 1995;**75**:326–9.
28. Dorantes-Dávila J, Pastor GM. Magnetic anisotropy of one-dimensional nanostructures of transition metals. *Phys Rev Lett* 1998;**81**:208–11.
29. Galicia R. Estudio del Rh_{13} por medio de orbitales moleculares. *Rev Mex Fis* 1985;**32**:51–65.
30. Dorantes-Dávila J, Dreysse H, Pastor GM. Exchange interaction and local environment effects on the magnetic properties of Fe_N clusters. *Phys Rev B* 1992;**46**:10432–6.
31. Reddy BV, Khanna SN, Dunlap BI. Giant magnetic moments in 4*d* clusters. *Phys Rev Lett* 1993;**70**:3323–6.
32. Villaseñor-González P, Dorantes-Dávila J, Dreyssé H, Pastor GM. Size and structural dependence of the magnetic properties of rhodium clusters. *Phys Rev B* 1997;**55**:15084–91.
33. Cox AJ, Louderback JG, Bloomfield LA. Experimental observation of magnetism in rhodium clusters. *Phys Rev Lett* 1993;**71**:923–6. Cox AJ, Louderback JG, Apsel SE, Bloomfield LA. Magnetism in 4*d*-transition metal clusters. *Phys Rev B* 1994;**49**:12295–8.
34. Hohenberg P, Kohn W. Inhomogeneous electron gas. *Phys Rev* 1964;**136**:B864–B871.
35. Kohn W, Sham LJ. Self-consistent equations including exchange and correlation effects. *Phys Rev* 1965;**140**:A1133–8.
36. von Barth U, Hedin L. A local exchange-correlation potential for the spin polarized case. *J Phys C Solid State Phys* 1972;**5**:1629–42. Kübler J, Höck K-H, Sticht J. Local spin-density functional theory of noncollinear magnetism (invited). *J Appl Phys* 1988;**63**:3482. Kübler J, Höck K-H, Sticht J, Williams AR. Density functional theory of non-collinear magnetism. *J Phys F Met Phys* 1988;**18**:469–83.
37. Schwarz K, Mohn P. Itinerant metamagnetism in YCo2. *J Phys F Met Phys* 1984;**14**: L129.
38. Bader RFW. *Atoms in molecules, a quantum theory*. Oxford: Oxford University Press; 1990.
39. Alvarado P, Dorantes-Dávila J, Pastor GM. Magnetic properties of 3*d* transition-metal nanostructures: Cr and V clusters embedded in bulk Fe. *Phys Rev B* 1998;**58**:12216–23.
40. Nicolas G, Dorantes-Davila J, Pastor GM. Orbital polarization effects on the magnetic anisotropy and orbital magnetism of clusters, films, and surfaces: a comparative study within tight-binding theory. *Phys Rev B* 2006;**74**:014415-1–10.
41. Slater JC. *Quantum theory of atomic structure*, vols. I and II. New York: McGraw-Hill Book Co., Inc; 1960.

42. Dorantes-Dávila J, Vega A, Pastor GM. Self-consistent theory of overlap interactions in the tight-binding method. *Phys Rev B* 1993;**47**:12995–8. Dorantes-Dávila J, Pastor GM. Alternative local approach to nonorthogonal tight-binding theory: environment dependence of the interaction parameters in an orthogonal basis. *Phys Rev B* 1995;**51**:16627–34.
43. Harrison WA. *Electronic structure and the properties of solids*. San Francisco: Freeman; 1980.
44. van Schilfgaarde M, Harrison WA. Theory of the multicenter bond. *Phys Rev B* 1986;**33**:2653–9.
45. Pastor GM, Dorantes-Dávila J, Bennemann KH. A theory for the size and structural dependence of the ionization and cohesive energy of transition-metal clusters. *Chem Phys Lett* 1988;**148**:459–64.
46. Bruno P. Dederichs PH, Grünberg P, Zinn W, editors. *Magnetismus von Festkörpern und Grenzflächen*. Jülich: KFA; 1993 [chapter 24].
47. Haydock R. Ehreinreich H, Seitz F, Turnbull D, editors. *Solid state physics*. vol. 35. New York: Academic; 1980. p. 215, $\rho_{i\alpha\sigma}(E) = -1/\pi\ \mathrm{Im}\{G_{i\alpha\sigma}\}$, where $G_{i\alpha\sigma}$ refers to the local Green's function.
48. Robles R, Martínez E, Stoeffler D, Vega A. Noncollinear magnetism of thin Cr films deposited on a stepped Fe (001) surface. *Phys Rev B* 2003;**68**:094413-1–6. Ojeda Miguel A, Dorantes-Dávila J, Pastor GM. Noncollinear cluster magnetism in the framework of the Hubbard model. *Phys Rev B* 1999;**60**:6121–30.
49. Ruiz-Díaz P, Garibay-Alonso R, Dorantes-Dávila J, Pastor GM. Noncollinear magnetism in transition-metal nanostructures: exchange interaction and local environment effects in free and deposited clusters. *Phys Rev B* 2011;**84**:024431-1–11.
50. Bansmann J, Baker S, Binns C, Blackman J, Buecher J-P, Dorantes-Dávila J, et al. Magnetic and structural properties of isolated and assembled clusters. *Surf Sci Rep* 2005;**56**:189–275.
51. Hubbard J. The magnetism of iron. *Phys Rev B* 1979;**19**:2626. Hubbard J. The magnetism of iron II. *Phys Rev B* 1979;**20**:4584. Hasegawa H. Single-site spin fluctuation theory of itinerant-electron systems with narrow bands. *J Phys Soc Jpn* 1980;**49**:178–88. Hasegawa H. Single-site spin fluctuation theory of itinerant-electron systems with narrow bands. II. Iron and nickel. *J Phys Soc Jpn* 1980;**49**:963–71.
52. Kakehashi Y. Magnetism of Fe, Co and Ni in the CPA and full static approximation. *J Phys Soc Jpn* 1981;**50**:2251.
53. Moriya T, editor. *Electron correlation and magnetism in narrow-band systems. Springer series in solid state sciences*, vol. 29. Heidelberg: Springer; 1981. Vollhardt D, et al. *Advances in solid state physics*, vol. 38. Wiesbaden: Vieweg; 1999, p. 383.
54. Pastor GM. *Theory of the electronic properties of metal clusters*. PhD Thesis, Berlin: Freie Universität; 1989, unpublished.
55. Pastor GM, Dorantes-Dávila J, Bennemann KH. Spin-fluctuation energies in transition-metal clusters. *Phys Rev B* 2004;**70**:064420.
56. Stratonovich RL. On a method of calculating quantum distribution functions. *Sov Phys Dokl* 1958;**2**:416. Hubbard J. Calculation of partition functions. *Phys Rev Lett* 1959;**3**:77.
57. Landau David P, Binder Kurt. *A guide to Monte Carlo simulations in statistical physics*. 2nd ed. Cambridge: Cambridge University Press; 2005.
58. Lee K, Callaway J, Dhar S. Electronic structure of small iron clusters. *Phys Rev B* 1984;**30**:1724–30. Lee K, Callaway J, Kwong K, Tang R, Ziegler A. Electronic structure of small clusters of nickel and iron. *Phys Rev B* 1985;**31**:1796–803.
59. Hund F. *Linienspektren und periodisches System der Elemente*. Berlin: Springer; 1927, p.124. Condon EU, Shortley GH. *The theory of atomic spectra*. Cambridge: Cambridge University Press; 1957, p. 209.

60. See, for instance, Nanoalloys: from theory to applications, Johnston RL, Ferrando R, editors. *Faraday Discuss* 2008;**138**:1.
61. Zitoun D, Respaud M, Fromen MC, Casanove MJ, Lecante P, Amiens C, et al. Magnetic enhancement in nanoscale CoRh particles. *Phys Rev Lett* 2002;**89**:037203-1–4.
62. Dennler S, Ricardo-Chávez JL, Morillo J, Pastor GM. Density functional calculations on small bimetallic magnetic clusters. *Eur Phys J D* 2003;**24**:237–40.
63. Efremenko I, Sheintuch M. DFT study of small bimetallic palladium–copper clusters. *Chem Phys Lett* 2005;**401**:232–40.
64. Ganguly S, Kabir M, Datta S, Sanyal B, Mookerjee A. Magnetism in small bimetallic Mn-Co clusters. *Phys Rev B* 2008;**78**:014402-1–10.
65. Entel P, Gruner ME. Large-scale ab initio simulations of binary transition metal clusters for storage media materials. *J Phys Condens Matter* 2009;**21**:064228.
66. Andriotis AN, Mpourmpakis G, Froudakis GE, Menon M. Magnetic enhancement and magnetic reduction in binary clusters of transition metal atoms. *J Chem Phys* 2004;**120**:11901–4.
67. Rollmann G, Sahoo S, Hucht A, Entel P. Magnetism and chemical ordering in binary transition metal clusters. *Phys Rev B* 2008;**78**:134404-1–5.
68. Antoniak C, Lindner J, Spasova M, Sudfeld D, Acet M, Farle M, et al. Enhanced orbital magnetism in $Fe_{50}Pt_{50}$ nanoparticles. *Phys Rev Lett* 2006;**97**:117201-1–4.
69. Knickelbein B. Magnetic moments of small bimetallic clusters: Co_nMn_m. *Phys Rev B* 2007;**75**:014401-1–6.
70. Wang RM, Dmitrieva O, Farle M, Dumpich G, Ye HQ, Poppa H, et al. Layer resolved structural relaxation at the surface of magnetic FePt icosahedral nanoparticles. *Phys Rev Lett* 2008;**100**:017205-1–4.
71. Gruner ME, Rollmann G, Entel P, Farle M. Farle multiply twinned morphologies of FePt and CoPt nanoparticles. *Phys Rev Lett* 2008;**100**:087203-1–4.
72. Dorantes Dávila J, Pastor GM. In: Calvo F, editor. *Nanoalloys: from fundamentals to emergent applications.* Amsterdam: Elsevier; 2013 [chapter 8; p. 257].
73. Muñoz-Navia M, Dorantes-Dávila J, Zitoun D, Amiens C, Jaouen N, Rogalev A, et al. Tailoring the magnetic anisotropy in CoRh nanoalloys. *Appl Phys Lett* 2009;**95**:233107-1–3.
74. Kresse G, Furthmüller. Efficient iterative schemes for *ab initio* total-energy calculations using a plane-wave basis set. *J Phys Rev B* 1996;**54**:11169. Kresse G, Joubert D. From ultrasoft pseudopotentials to the projector augmented-wave method. *Phys Rev B* 1999;**59**:1758–75.
75. Perdew JP, Chevary JA, Vosko SH, Jackson KA, Pederson MR, Singh DJ, et al. Atoms, molecules, solids, and surfaces: applications of the generalized gradient approximation for exchange and correlation. *Phys Rev B* 1992;**46**:6671–87.
76. Blöchl PE. Projector augmented-wave method. *Phys Rev B* 1994;**50**:17953–79.
77. Mokkath J, Pastor GM. First-principles study of the structural, magnetic, and electronic properties of small Fe-Rh alloy clusters. *Phys Rev B* 2012;**85**:054407-1–7.
78. Mokkath J, Pastor GM. Interplay between chemical and magnetic order in Fe-Rh clusters. *J Phys Chem C* 2012;**116**:17228.
79. Díaz-Sánchez LE, Dorantes-Dávila J, Pastor GM. Local and chemical environment dependence of the magnetic properties of CoRh core-shell nanoparticles. *Phys Rev B* 2013;**88**: 134423.
80. Fromen MC, Morillo J, Casanove M-J, Lecante P. Structure and chemical order in CoRh nanoparticles. *Europhys Lett* 2006;**73**:885–91.
81. Blanc N, Díaz-Sánchez LE, Ramos AY, Tournus F, Tolentino HCN, de Santis M, et al. Element-specific quantitative determination of the local atomic order in CoPt alloy nanoparticles: experiment and theory. *Phys Rev B* 2013;**87**:155412.

82. Nonas B, Cabria I, Zeller R, Dederichs PH, Huhne T, Ebert H. Strongly enhanced orbital moments and anisotropies of adatoms on the Ag(001) surface. *Phys Rev Lett* 2001;**86**:2146.
83. Castro Miguel Castro M, Jamorski Christine, Salahub Dennis R. Structure, bonding, and magnetism of small Fen, Con, and Nin clusters, $n \leq 5$. *Chem Phys Lett* 1997;**271**:133. Reddy BV, Nayak SK, Khanna SN, Rao BK, Jena P. Physics of nickel clusters. 2. Electronic structure and magnetic properties. *J Phys Chem A* 1998;**102**:1748.
84. Khanna SN, Linderoth S. Magnetic behavior of clusters of ferromagnetic transition metals. *Phys Rev Lett* 1991;**67**:742.
85. Félix-Medina R, Dorantes-Dávila J, Pastor GM. Spin moments, orbital moments, and magnetic anisotropy of finite-length Co wires deposited on Pd (110). *New J Phys* 2002;**4**:1.1–1.14.
86. Dorantes-Dávila J, Pastor GM. Magnetic reorientation transitions along the crossover from one-dimensional to two-dimensional transition-metal nanostructures. *Phys Rev B* 2005;**72**:085427.
87. Muñoz-Navia M, Dorantes-Dávila J, Zitoun D, Amiens C, Chaudret B, Casanove M-J, et al. Magnetic properties of Co_N Rh_M nanoparticles: experiment and theory. *Faraday Discuss* 2008;**138**:181–92.
88. Muñoz-Navia M, Dorantes-Dávila J, Respaud M, Pastor GM. Theoretical study of the magnetic moments and anisotropy energy of CoRh nanoparticles. *Eur J Phys D* 2009;**52**:171–4.
89. Harp GR, Parkin SSP, O'Brien WL, Tonner BP. Induced Rh magnetic moments in Fe-Rh and Co-Rh alloys using x-ray magnetic circular dichroism. *Phys Rev B* 1995;**51**:12037–40.
90. Pappas DP, Kämper K-P, Hobster H. Reversible transition between perpendicular and in-plane magnetization in ultrathin films. *Phys Rev Lett* 1990;**64**:3179. Allenpasch R, Bischof A. Magnetization direction switching in Fe/Cu(100) epitaxial films: temperature and thickness dependence. *Phys Rev Lett* 1992;**69**:3385. Qiu ZQ, Pearson J, Bader SD. Asymmetry of the spin reorientation transition in ultrathin Fe films and wedges grown on Ag(100). *Phys Rev Lett* 1993;**70**:1006. Arnold CS, Pappas DP, Popov AP. Second- and first-order phase transitions in the magnetic reorientation of ultrathin Fe on Gd. *Phys Rev Lett* 1999;**83**:3305. Kukunin A, Prokop J, Elmers HJ. Temperature-driven spin reorientation transition in Fe/Mo(110) nanostructures. *Phys Rev B* 2007;**76**:134414.
91. Binder K, Rauch R, Wildpaner V. Monte Carlo calculation of the magnetization of superparamagnetic particles. *J Phys Chem Solids* 1970;**31**:391.
92. Korenman V, Prange RE. Local-band-theory analysis of spin-polarized, angle-resolved photoemission spectroscopy. *Phys Rev Lett* 1984;**53**:186.
93. Haines EM, Clauberg R, Feder R. Short-range magnetic order near the Curie temperature iron from spin-resolved photoemission. *Phys Rev Lett* 1985;**54**:932.
94. Pastor GM, Dorantes-Dávila J. Short range magnetic order in Fe_N and Ni_N clusters. *Phys Rev B* 1995;**52**:13799.
95. See for example Newman MEJ, Barkema GT. *Monte Carlo methods in statistical physics.* Oxford: Oxford University Press; 1999.
96. Swendsen RH, Wang JS. Nonuniversal critical dynamics in Monte Carlo simulations. *Phys Rev Lett* 1987;**58**:86.
97. Wolff U. Collective Monte Carlo updating for spin systems. *Phys Rev Lett* 1989;**62**:361.
98. Berg BA, Neuhaus T. Multicanonical algorithms for first order phase transitions. *Phys Lett B* 1991;**267**:249.
99. Berg BA, Neuhaus T. Multicanonical ensemble: a new approach to simulate first-order phase transitions. *Phys Rev Lett* 1992;**68**:9.
100. Marinari E, Parisi G. Simulated tempering: a new Monte Carlo scheme. *Europhys Lett* 1992;**19**:451.

101. Hukushima K, Nemoto K. Exchange Monte Carlo method and application to spin glass simulations. *J Phys Soc Jpn* 1996;**65**:1604.
102. As in an N-step random walk, the root mean square average of the total magnetic moment per atom of a cluster having N uncorrelated local moments of size μ_0 is $\mu_0/\sqrt{N}$.
103. Dorantes-Dávila J, Dreyssé H, Pastor GM. Magnetic anisotropy of transition-metal interfaces from a local perspective: reorientation transitions and spin-canted phases in Pd capped Co films on Pd(111). *Phys Rev Lett* 2003;**91**:197206. Valvidares SM, Dorantes-Dávila J, Isern H, Ferrer S, Pastor GM. Interface-driven manipulation of the magnetic anisotropy of ultrathin Co films on Pt(111): substrate deposition of hydrogen and model calculations. *Phys Rev B* 2010;**81**:024415.

Chapter 3

Novel Methods for the Synthesis of Magnetic Nanoparticles

Sarah S. Staniland*, Andrea Rawlings*, Jonathan Bramble*, Juan Tolosa†, Orla Wilson‡, Joaquín C. García-Martínez† and Chris Binns§

*Department of Chemistry, University of Sheffield, Sheffield, United Kingdom
†Pharmaceutical and Organic Chemistry. Facultad de Farmacia, Edificio Polivalente, Universidad de Castilla-La Mancha, Albacete, Spain
‡Department of Materials Science and Engineering, Johns Hopkins University, Baltimore, Maryland, USA
§Department of Physics and Astronomy, University of Leicester, Leicester, United Kingdom

1 INTRODUCTION

As described throughout this volume, the applications of magnetic nanoparticles (MNPs) are diverse and a number of synthesis methods have been developed, each having its own advantages or special abilities. The requirements for the materials, sizes and morphology of nanoparticles vary widely depending on the application. For example, in the case of fundamental experimental research in nanomagnetism (Chapter 2), extremely monodisperse nanoparticles (preferably with a specific number of atoms) are required combined with a flexible choice of materials used, including elements, alloys and oxides. For material applications (Chapter 4), the nanoparticles are required to be encapsulated in a thin film and the quantities are expressed in terms of the film thickness and area. For medical applications (Chapter 6), the nanoparticles are required to be in the form of a hydrosol in modest quantities (ml, mg) for a single therapeutic or imaging treatment. On the other hand for environmental applications (Chapter 7) large quantities of nanoparticle hydrosol are required.

The synthesis of MNPs that are monodisperse (or at least have a narrow size distribution) can be delineated into three basic classes, that is, chemical, biological and physical. The chemical and physical routes in turn encompass a number of different fabrication methods. In this chapter, we will discuss all

Frontiers of Nanoscience, Vol. 6. http://dx.doi.org/10.1016/B978-0-08-098353-0.00003-8

three production classes with a special emphasis on new and emerging methods that enable increased flexibility in terms of materials that can be used and structures that can be formed. The simplest generic chemical methods is the precipitation of the MNPs out of solution. This is the case for both the reduction of metal and alloy nanoparticles and the oxidation of metals oxides from their soluble salts. This is the most universal and evolved method and can produce large quantities of nanoparticle suspensions. Given the extensive range of materials and the multiple variations of conditions, a full review of this method is beyond the scope of this chapter. Instead, in section 2 we demonstrate this approach with the example of iron oxide, as this material is directly relevant to the next section. Section 3 describes a biological route, which can produce naturally biocompatible magnetite nanoparticles. An emerging method using dendrimer templates, which can make monometallic and bimetallic nanoparticles, is presented in Section 4. Finally, Section 5 describes gas-phase methods that can produce both solid films and hydrosols of elemental, alloy or core–shell nanoparticles. In addition, the production of nanoparticles suited to environmental and medical applications are discussed in Chapters 6 and 7, respectively.

2 PRODUCTION OF IRON OXIDE NANOPARTICLES BY CO-PRECIPITATION, PARTIAL OXIDATION OF FERROUS HYDROXIDE AND THERMAL DECOMPOSITION

A facile method to produce magnetic iron oxide nanoparticles is the room temperature co-precipitation (RTCP) of mixed valence iron salts by titration of base in an inert atmosphere.[1,2] This method produces particles around 15 nm in diameter. The size, shape, crystallinity and polydispersity are all dependent on the conditions of the reaction. The temperature, concentration of starting materials, Fe^{2+}/Fe^{3+} ratio, the type of iron salts used, and the rate of base addition are all important reaction parameters to produce high-quality particles. It is generally difficult to achieve the low polydispersity seen in other nanoparticle synthesis methods.[3] As the blocking temperature of the magnetite nanoparticles depends on the size (see Chapter 1, Section 4), polydispersity widens the blocking temperature range and for most applications this is not ideal.

Magnetite nanoparticles can become oxidised in solution, converting them to another magnetic iron oxide, maghemite. Small magnetite nanoparticles (<5 nm diameter) are also unstable and quickly dissolve. Surface coatings can be applied to magnetite nanoparticles to protect them. It has been shown that various surfactants, polymers, metals and silicon can be used to protect the particles and prevent aggregation.[4,5]

The RTCP of ferrous and ferric ions of iron oxides from solution is complex[6,7] and proceeds through a number of intermediate iron oxides. Some of the intermediate steps in this process have only recently been studied.[8,9] The method demonstrated by Ruby *et al.*[8] shows how careful measurements of pH during a slow titration can be used to determine the proportion of different iron oxides in the solution at the end of the reaction and how it depends on the starting Fe^{2+}/Fe^{3+} ratio. From detailed study, they were able to propose a model for the formation of green rust intermediates. Some magnetite synthesis schemes start out with an excess of Fe^{2+} to ensure that if there is any oxidation during the reaction there is sufficient Fe^{2+} to complete the formation of magnetite. This usually results in an excess of unstable green rust particles which must be removed from the solution prior to analysis or use.

Additives present in starting solutions can subtly change the processes involved in the formation of these intermediates and can greatly change the particle shape, size and polydispersity. When comparing the results of additives on magnetite nanoparticle synthesis, the morphology of the particles before and after cleaning should be carefully observed by TEM to ensure that any positive effect on synthesis products are not simply due to the difference in non-magnetite intermediates.

Magnetite forms at a particular equivalent point during a titration; at a ratio of OH to total Fe concentration. We can compare the position of this point in the presence or absence of additives. Some additives can move the position of this point such that no magnetite is formed or force the production of more green rust. By carefully controlling the starting Fe^{2+}/Fe^{3+} ratio and the additive used, we can create more magnetite in the final product.

Partial oxidation of ferrous hydroxide (POFH) is an alternative magnetite synthesis method to RTCP that forms larger particles compared to co-precipitation. In this method, ferrous hydroxide forms in aqueous solution from a ferrous salt and is then partially oxidised with nitrate at 90 °C. The mixture of ferrous and ferric hydroxides forms magnetite. Again, the reaction conditions are critical to control the size, shape and iron oxide formed.[10,11]

Thermal decomposition methods have been developed recently that offer high levels of control over both size and shape of magnetite nanoparticles. Sun *et al.* have shown that monodisperse particles could be created via thermal decomposition of iron acetylacetonate $Fe(acac)_3$, in a high-boiling point solvent phenyl ether in the presence of surfactants.[12] Small 4 nm nanoparticles were created but could be enlarged to 16 nm through a second reaction. Ho *et al.* have shown that cubic magnetite nanoparticles could be produced by controlling the concentration of iron.[13] Here, $Fe(acac)_3$ was added slowly to boiling benzyl ether with a number of surfactant additives. Initially, small particle nucleate in solution, but the surfactant binds preferentially to the [100] faces such that when the iron concentration in solution increases, the [111] faces grow out to form a cube. Figure 3.1 shows some images of particles created by these three methods.

The thermal decomposition methods produce high-quality particles with much higher levels of control over size and shape, compared to RTCP. However, the particles must be transferred into an aqueous phase for a number of applications.[14] Also the conditions and materials required are not environmentally friendly and are only suitable in small-scale experiments in

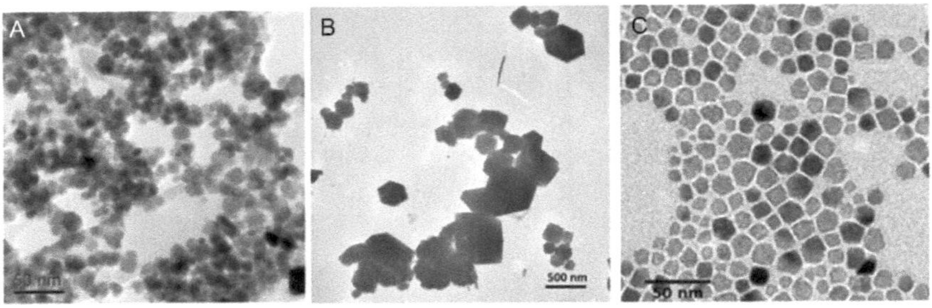

FIGURE 3.1 (A) Magnetite nanoparticles formed using the co-precipitation method at room temperature (RTCP). (B) Magnetite nanoparticles formed by the partial oxidation of ferrous hydroxide (POFH). (C) Magnetite nanoparticles created by a hot-injections polyol process.

well-equipped laboratories. Routes to the synthesis of nanoparticles with similar levels of control over size and shape, in environmentally friendly and ambient conditions, should lead to lower costs and a wider range of potential applications. To achieve this goal, we look for inspiration from nature, where we find an organism capable of producing such particles in mild conditions.

3 SYNTHESIS OF MNPs BY MTB

3.1 Introduction to Magnetotactic Bacteria

Magnetotactic bacteria (MTB) were first identified as early as 1963 by Bellini[15] but were rediscovered and more widely reported by Blakemore in 1975.[16] They can be classified as prokaryotic, gram-negative using flagellae to move and are classified as micro-aerobes, and as such can be found at the oxic–anoxic transition zone (OATZ) in aquatic environments where they find their optimum oxygen concentration to survive.[16] MTB biomineralise magnetite[17] or greigite[18,19] MNPs within internal vesicles called magnetosomes.[20,21] Each magnetosome consists of a lipid membrane containing biomineralisation proteins surrounding a MNP of precise morphology dictated by the bacterial strain[22–24] (Fig. 3.2). The nanoparticles are typically 35–120 nm in length[25] (Fig. 3.2). For magnetite, this is the size range of single domain nanoparticles (see Chapter 1, Section 3), so they have the maximum magnetic moment possible.[26] There are typically between 20 and 40 magnetosomes per cell and are arranged in long chains, this acts as a compass needle inside the bacteria.[27] There is sufficient torque from the applied geomagnetic field to orientate the bacteria. This response is known as magnetotaxis.[28,29] When the bacteria move they move along the geomagnetic field lines, but as the geomagnetic field is tilted with respect the surface of the earth (magnetic dip) the bacteria can transverse the vertically layered OATZ. It is suggested that magnetotaxis along with chemotaxis are used to seek optimal conditions in their environment.

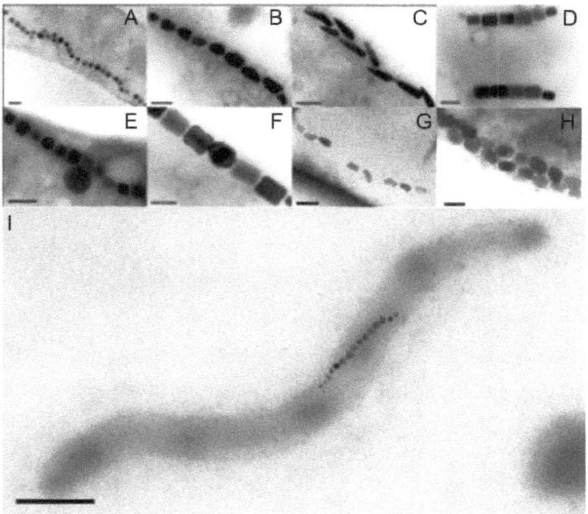

FIGURE 3.2 TEM images of (A–H) different magnetosome morphologies. (A–G) Magnetite particles and (H) greigite particles. (A and E) Show cubo-octahedral magnetosomes from (A) *M. gryphiswaldense* MSR-1 and (E) *M. Magnetotacticum* MS-1. (B) Bullet-shaped magnetosomes*, (C) tooth-shaped magnetosomes*. (D and F) Are elongated magnetosomes from (D) *M. coccus* MC-1* and (F) *M. vibros* MV-1*. (G and H) Irregular bullet-shaped magnetosomes from (G) *D. Magneticus* RS-1† and (H) uncharacterised greigite-producing bacterium*. Scale bars are 100 nm. (I) The magnetic bacterium *M. Magnetotacticum* MS-1 whole cell scale bar is 500 nm. *Some images are reproduced with the kind permission from D. Schuler (*) and T. Matsunaga (†) with images (A–H) reprinted with permission from Wiley.*

3.2 Microbiology and Diversity

MTB represent a diverse and phylogenetically unrelated collection of bacteria, grouped simply by their ability to synthesise the magnetic mineral magnetite (or in some cases greigite). To understand the molecular basis of the biomineralisation process, a detailed understanding of the genes and proteins which control the mineral formation is required. Difficulties in culturing these bacteria due to their fastidious, microaerobic growth conditions initially hampered approaches to conduct detailed analysis of their magnetite biomineralisation capabilities. The first species which proved amenable to culturing and fruitful for investigation was *Magnetospirillum magnetotacticum* MS-1[30] (Fig. 3.2 I), which was later followed by *Magnetospirillum magneticum* AMB-1,[31] and *Magnetospirillum gryphiswaldense* MSR-1.[32] Recently these were joined by *Magnetovibrio blakemorei* MV-1[33] and *Magntococcus marinus* MC-1 amongst others[34,35] (Fig. 3.2 A, F and D respectively). All the species mentioned so far are alpha-proteobacteria of the order *Rhodospirillales* and *Magnetococcales* and are the most studied but this does not well represent the diversity of MTB present in nature with examples such as the delta-proteobacterium *Desulfovibrio magneticum* RS-1 (Fig. 3.2 G),[36,37] strain BW-2 which is a gammaproteobacter,[38] and even magnetotactic multicellular bacteria such as *Candidatus Magnetoglobus multicellularis*,[39] and no doubt other species and classes will be cultured and identified successfully in the future. The MTB mentioned are all considered to be Gram negative. This term refers to the cell membrane arrangement, with Gram negative cells having both an inner and an outer cell membrane separated by a region termed the periplasmic space.

3.3 Magnetosome Membrane

The magnetic crystals are surrounded by a lipid membrane, which defines the perimeter of the biomineralisation nanoreactor, termed the magnetosome. In MTB, which were cultured in the absence of iron, empty magnetosomes[40] or those featuring vestigial nanoparticles were observed, giving rise to the idea that the magnetosome is formed prior to crystal nucleation and awaits sufficient iron for nanoparticle formation to occur.

Lipid analysis has demonstrated high similarities in composition between the magnetosome membrane (MM) and the inner cell membrane indicating that the former may derive from the cytoplasmic membrane via a "budding" process.[24] This is supported by a recent cryo electron tomographic study of MTB showing that the magnetosomes are attached to the inner membrane, although whether this is the case once the particle has fully matured is not clear.[41] Having a lipid vesicle as the site of magnetite formation makes sense for the bacteria, as it allows conditions favourable to magnetite synthesis to be strictly controlled by regulating the flow of molecules across the membrane and the confines of the magnetosome may act to limit the size of the nanoparticles produced. Interestingly, it appears that cells produce many more empty magnetosomes than will eventually yield magnetite nanoparticles, pointing towards an as yet undiscovered activation mechanism, which turns a dormant magnetosome into a functioning magnetite nanoreactor.

3.4 The Magnetosome Genes

One of the key steps towards understanding the biological basis of cellular processes at the molecular level is to sequence the genome of the organism. This was previously a lengthy, complex and expensive process, but with the advent of more advanced sequencing techniques the cost and difficulty of sequencing a new organism is much reduced, resulting in new genomes being deposited at an ever increasing rate.[42] The genes identified within a sequenced genome can be used to search for similar sequences in genome databases of both related and unrelated organisms.

The majority of the genes found in the MTB are identified, unsurprisingly, as having roles in essential cellular processes and metabolism and are well conserved in their non-magnetic counterparts.[43] However, what became apparent from the sequences was a unique region of the genomes of these bacteria, which are not present in non-magnetic species.[44,45] This approximately 100 kb region of the genome is termed the magnetosome island (MAI) and encodes around 100 proteins. The majority of the Mam (magnetosome associated) and Mms (MM specific) proteins are encoded within the MAI. Flanking this region are repeat sequences and transposase genes commonly associated with genome insertions. These sections are thought to facilitate efficient transfer of the genetic material sandwiched between them to different organisms in a process termed horizontal gene transfer. This represents a plausible mechanism by which the magnetite biomineralisation genes have been widely distributed throughout many different bacterial species, although recent analysis of the evolution and divergence of the different species places a new complexion on this theory by suggesting that most MTB may have originated from a common magnetic ancestor and that non-magnetic bacteria may have simply lost these genes over time.[46]

Genes that encode proteins with similar functions, involved in similar processes, or co-dependent on one another, are often found together in the genome in regions called clusters. Such genes are typically co-transcribed, meaning they are copied into mRNA in one long strand, and the rate of production of the mRNA, the transcription rate, is determined by a promoter which regulates when and how many copies, and therefore proteins, are produced. Major clusters in the magnetosome island are the mamAB cluster encoding 11 genes for iron transport and protein sorting, and which on its own is capable of producing a minimal magnetosome,[47] the mamGDFC cluster, the mamXYZ cluster and the mms6 gene cluster.

3.5 Magnetosome Proteins

Alone, an empty lipid vesicle does not spontaneously produce magnetite nanoparticles and certainly not with the consistency and homogeneity that the MTB do. The difference being that the bacteria produce an abundance of specialised proteins that control every aspect of magnetite biosynthesis. Initially, a proteomics approach was used to identify the proteins specific to the MM and a unique selection of proteins was uncovered.[24,48] MTB were lysed and the magnetosomes magnetically separated from the lysate. The MM was disrupted with detergents and the resulting protein mixture analysed by sodium dodecyl sulphate polyacrylamide gel electrophoresis (SDS-PAGE), which separates the proteins according to molecular weight, enabling the proteins to be identified through mass spectrometry analysis. Some proteins were found to be only loosely associated with the MM, other were tightly bound to the magnetite crystal.[49] It was interesting to note that many of the proteins present in the MM were not found in the inner membrane which suggests that such proteins are targeted specifically to the magnetosome, perhaps through a network of protein:protein and protein:magnetite interactions.

The identified proteins were given designations to reflect their affinity for the magnetosome with the principle ones termed Mms (MM specific) and Mam (magnetosome associated membrane) proteins, with letters and numbers used to describe their order of discovery or molecular weight, respectively. Based on sequence homology many of the proteins belong to well-characterised families, which give tantalising clues to the likely role of these proteins in the bacteria. However, a number of the proteins are distinct to MTB with no homologues or similarity to known proteins to give indications of function. As magnetite nanoparticles are unique to these organisms, it is probable that some of these unique proteins may interact directly with the nanoparticle and act to control the particle nucleation, maturation and direct the particle to assume specific morphologies.

3.5.1 *Iron Transport Proteins*

The magnetic crystals formed in the magnetosome require high concentrations of iron and a controlled ratio of ferrous to ferric ions of 1:2. The bacteria have therefore developed specialised iron transporters, which are capable of transporting iron into the magnetosome against the concentration gradient from a region of low iron concentration in the cytoplasm to the highly concentrated magnetosome interior. MamB and MamM for instance, are members of the cation diffusion facilitator (CDF) family of membrane transporter proteins, which may be a likely transporter of iron ions across the MM to fuel the growth of the magnetite nanoparticles.[50] Most CDF proteins transport zinc ions but a well-characterised CDF protein, FieF in *E. coli* has a ferrous iron transport capability[51] to which MamB and MamM both share critical amino acid similarities. These proteins are antiporters that transport metal ions in one direction and protons in the other. The formation of magnetite produces protons as a by-product, which could act to drive the transport of iron ions across the MM. The MTB also contain members of the FeoB family of membrane transporters which transports ferrous iron across the membrane.[52,53]

3.5.2 *Magnetosome Chain Proteins*

A striking observation for MTB is the alignment of the individual magnetosomes of the cell into chain-like structures. One identified protein present in the magnetosome, MamK, has significant homology to the sequence of an actin-like protein MreB. Well characterised in bacteria, MreB is a cyto-skeletal protein in which many monomers self-assemble into long fibres which aid shape determination of the cell amongst other things.[54] Through careful genetic experiments in AMB-1 the gene encoding MamK was deleted, producing a phenotype where the magnetosomes coalesced together, losing the ability to form ordered chains. Cryo electron tomography of the bacteria has demonstrated long fibrous structures traversing the cell to which the magnetosomes appear to be attached like beads on a string. The string in this analogy is thought to be the MamK fibres; anchoring the magnetosomes through the action of a partner protein MamJ,[55] which traverses the MM and binds tightly to both MamK and the MM. Having the magnetosomes held in such a way is thought to improve the ability of the organism to sense and react to external magnetic fields.

3.5.3 Magnetite Interacting Proteins

After stripping of the MM from the nanoparticles, a small number of proteins remain strongly attached to the crystal. In the AMB-1 strain, these were found to be Mms5, Mms6, Mms7 and Mms13 (number based on the apparent molecular weight of the proteins in SDS-PAGE experiments).[49] All of these proteins are predicted to be integrated to the MM due to regions of their sequences that are strongly hydrophobic and most have an overall isoelectric point (pI) of >8 except one, Mms6, which has a pI of approximately 4, giving it a net negative charge at neutral pH.

Mms6 features an N-terminal region, which is dominated by hydrophobic residues and is the part thought likely to be located within the lipidic MM, and followed by a highly conserved glycine–leucine repeat motif, which is also present in other Mms proteins. This motif is typically observed in silk proteins and is involved in the bundling of silk into fibres.[56] The C-terminal region is characterised by a high proportion of amino acids with acidic side chains.[49] This part of the protein is thought to interact tightly with the positively charged iron ions of the magnetite nanoparticle[49] and is able to bind ferric ions through the glutamic acid residues present in this part of the protein. Mms6 has been introduced into *E. coli* cells using molecular cloning techniques to generate larger quantities of protein for *in vitro* analysis to determine its function.[49,57] Unusually for a predicted membrane protein it has been shown to assemble into water soluble oligomers when purified from *E. coli* possibly via interactions through the glycine–leucine repeat sections.[58] The hydrophobic glycine–leucine repeats could fold up into helical structures with the bulky leucine residues packing into holes created by the small glycine side chains. This assembly motif could allow hetero-oligomers to form *in vivo* between different Mms proteins that share similar repeat regions as a way of packing similar proteins together. The pure Mms6 oligomers feature a ferric ion binding capability and are oriented so that the C-terminal acidic regions are together, thereby creating a proposed iron/magnetite binding surface. Purified Mms6 can be added to synthetic magnetite precipitation reactions to promote particle homogeneity, suggesting a role for Mms6 as a regulator of particle size and shape *in vivo*.[57] This is confirmed by an *mms6* gene deletion from the bacteria which results in a subtle deterioration in particle quality,[59] again pointing to a role for this protein in particle morphology and size regulation.

Recent genetic studies have uncovered a new protein, MmsF which has been described as the master morphological regulator for magnetite biomineralisation *in vivo*.[60] The MmsF protein is encoded by a gene residing in the same gene cluster as *mms6*. A *ΔmmsF* mutant displays a phenotype with much smaller, misshapen particles compared to the wild-type cells. When the *mms6* gene cluster (including the *mmsF* gene) is deleted ($\Delta mms6_{cl}$) then a similar but only slightly more severe phenotype is observed. By re-introducing the *mmsF* gene alone back into the $\Delta mms6_{cl}$ strain, the cells are rescued and a near normal magnetosome is observed.[60] These experiments indicate a critical role for MmsF in controlling the growth, size and shape of the formed magnetite crystals.

3.6 Magnetosome Formation Mechanism

A recent proposed mechanism of magnetosome formation has three key stages, which are shown schematically in Figure 3.3. In stage 1, the inner cell membrane is distorted through the action of MamY, which has a high similarity to the lipid membrane deformation proteins of the dynamin and BAR families and acts to constrict the membrane to initiate vesicle formation.[61] A second protein Mms16, a GTPase, is thought to hydrolyse GTP to help to form a vesicle-like structure.[62] MamK is present as a filament, spanning the long axis of the cell. Under sufficient iron conditions the process proceeds to stage two where the concentration of iron within the vesicle compartment rapidly increases as specific iron transporters such as the MamB, MamM and FeoB proteins mentioned earlier bring about iron accumulation into the interior cavity across the MM. Much protein sorting occurs possibly through interaction with MamB as many magnetosome specific proteins are recruited

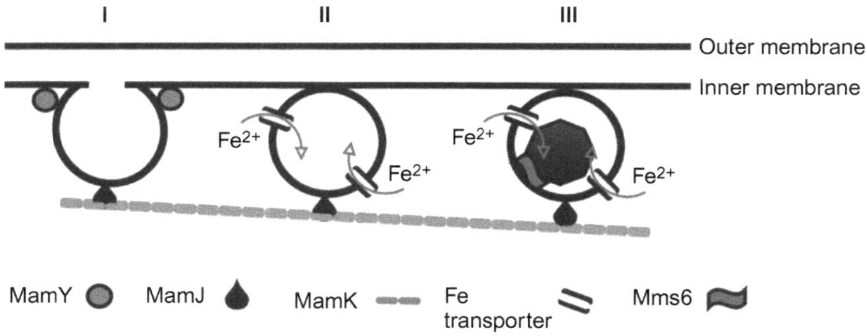

FIGURE 3.3 Proposed magnetosome formation scheme. Some of the key proteins involved in this process are depicted. Stage I sees the formation of a vesicle from the inner cell membrane through the action of the MamY protein. The formed vesicle is attached to the MamK filament via MamJ. In Stage II, iron is imported to the magnetosome interior through dedicated ferrous transporters such as MamB and MamM. In Stage III, the iron is nucleated to form a magnetite crystal. Proteins such as Mms6 act to regulate this process.

to the MM. MamJ anchors the magnetosomes in place by interacting with the MamK fibre. Once a critical iron concentration is reached, stage three commences where proteins such as Mms6 begin to nucleate the iron into magnetite. Iron continues to be loaded into the magnetosome allowing the crystal to grow and mature into the correct mineral and with species specific morphology. A variety of proteins are present in the magnetosome during nucleation and crystal growth which include MamE, P and T,[63] which are c-like cytochrome redox proteins, and MamX, a protein important in controlling crystal growth.[64] It is likely that many of the array of different proteins involved in stage III may act as magnetite morphological regulators, thereby providing a pool of candidate proteins for *in vitro* magnetite mineralisation control.

3.7 Exploiting Magnetosome Synthesis

3.7.1 *Enhanced Magnetosome Biosynthesis* In Vivo

The precise MNPs formed by the bacteria feature high crystallinity and are extremely homogeneous, making them suitable for a range of therapeutic and diagnostic biomedical applications.[65] Additionally, it must be remembered magnetosomes have an intrinsic biological lipid coating onto which proteins and drug targets can easily be attached, making these materials very attractive for biomedical applications.

Magnetosomes, due to their liposomal coating disperse exceptionally well in aqueous solution, meaning they are unlikely to clump or aggregate during use. Similar quality synthetically produced particles need to be coated post-production in a lipid to bring about similar properties, a step which is often non-trivial.

Iron oxide MNPs can be used as carriers for drugs, DNA, proteins or other relevant molecules to facilitate magnetically targeted treatments (see Chapter 6, Section 7). For magnetosomes, this is fairly easy as the majority of the lipids in the membrane are amine terminated which allow easy decoration with biotin.[66] Streptavidin robustly binds to the biotin molecules and due to the remaining unoccupied binding sites on the streptavidin can pick up any biotin functionalised molecule for a particular application. Another approach to functionalisation is to make genetically engineered MTB to produce biotinylated proteins anchored to the magnetite nanoparticles which can then be tagged with streptavidin linked molecules as before.[67] These methods are in addition to direct covalent attachment to the lipid membrane. The MNP can then be magnetically guided to a particular site for localised drug treatment or gene therapy.

In one study doxorubicin (DOX), a well-known anti-cancer agent, was attached to magnetosomes and applied to mice bearing a tumour.[68] The attached magnetosome allows the drug to be guided to and accumulated at the site of the tumour giving a high-tumour kill rate with reduced side effects. Such targeted treatments allow gene or drug delivery to be carefully directed to the best location for a successful outcome and often means lower doses can be used.

Magnetosomes also display superior properties when utilised as contrast agents in magnetic resonance imaging (MRI—see Chapter 6, Section 4). The high magnetisation of magnetosomes over synthetic nanoparticles gives improved contrast allowing a more detailed picture of disease states to be observed.[69] Similarly magnetosomes show exceptional properties with respect to alternating magnetic field heat loss for hyperthermic applications (see Chapter 6, Section 6). Magnetosomes have the largest specific heating power compared to comparable synthetic MNPs, making them well suited to this promising treatment area.[70,71] Although it is not clear why this is the case it has been proposed it is due to the high degree of crystallinity within magnetosomes or conversely due to the subtly different, slightly reduced form of magnetite the magnetosomes produce, which cannot be emulated synthetically.

Whilst magnetosomes offer highly desirable qualities for biomedical applications, they are not currently used commercially as there are two main drawbacks of the biosynthetic method. The first is the slow growth rate and low growth density of the cells giving low yields of magnetosomes. Several research groups have developed improved fermenter methods with improved yields[72] achieving a growth rate of 0.32 g/L/day, with a final cell density of 1.42 giving a magnetosome yield of 39 mg/g dry cells by optimising media nutrient and oxygen tension.[72] Later Sun *et al.* improved this further simply by increasing the air aeration/stirring rate with growth rate. This resulted in a final cell density of 7.24 ($>5\times$ increase), with cells (dry weight) of 2.17 g/L at a growth rate of 0.868 g/L/day, giving a magnetosome yield of 41.7 mg/L or 16.7 mg/L/day, all >2.5 times the higher than previous result.[73] While these improvements go some way to meeting commercialisation goals, biomineralisation in MTB, even in an improved fermenter system, is still an inefficient method of producing MNPs when compared to other commercial methods. While this presents a real barrier to the successful commercialisation of natural magnetosomes, it must be weighed against the high value of these particles, offering superior quality and unique functionalisation, especially when only small quantities are required for biomedical applications.

The second drawback of the biosynthesis of magnetosomes within the cells is the lack of flexibility this methodology offers. Magnetosomes are almost exclusively pure magnetite of highly specific sizes and shapes due to the high degree of control exhorted by the cell. Indeed, this precision is one of their assets. However, this means applications for natural unaltered magnetosomes are limited to a small size range of pure magnetite. This is beginning to be addressed in several ways: In our research, we have successfully doped magnetite magnetosomes with cobalt *in vivo*, simply by adding cobalt ions into the growth media.[74] There is a narrow iron and cobalt concentration window where the bacteria are able to take up cobalt ions into the magnetosome and incorporate them into the growing mineral; too little cobalt and the cell avoids cobalt uptake resulting in no doping; too much and the cobalt is toxic to the cells. The maximum doping with cobalt (before cell inhibition) is approximately 3% doping (metal content) of magnetosomes with the addition of 40 μM Co^{2+} ions to the media.[75] The amount of doping is also slightly dependent on the bacterial strain.[74] We have explored a full range of rational candidate transition metals as potential magnetosome dopants, but the only other first row transition metals we have found to successfully dope magnetosomes are manganese and copper.[75] The cells can tolerate very high levels of manganese (over 1 mM) but will only dope up to 1.2% metal into the magnetosomes while conversely, copper is very toxic to the cell (cell death at 30 μM) but can achieve very high levels of doping of 16% (in a 20-μM Cu^{2+} ion concentration), showing the toxicity may be due to the cells poor defence against penetration of this metal. The doped magnetosomes display various magnetic properties, differing from the natural magnetosomes. The cobalt-doped magnetosomes have been most extensively studied due to the anisotropic character cobalt doping gives to magnetite by increasing its coercivity. Magnetosomes show superior hyperthermic heat loss over nanomagnetite of the same size with increased heating effect on the surrounding solution. Unaltered magnetosomes show heating from 25 to 52 °C in 15 min. The addition of cobalt into magnetosomes increases the coercivity, and with an increase in temperature of 34 °C from 25 to 59 °C in 15 min.[70] This superior heating could again be coupled with the highly functional MM to tether therapeutics such as anti-cancer drugs. We have shown doping by this method is tunable, with approximately 1% doping in 20 μM media going up to 3% doping in 40 μM media. The ability to adjust the composition of the nanoparticles in this way is a useful biosynthetic tool, which allows their properties to be fine-tuned to suit a variety of biomedical and nanotechnology applications, for example in this case, highly specific heating effects.

3.7.2 *Enhanced MNP Biokleptic Synthesis* In Vitro

3.7.2.1 In Solution

An alternative approach to increasing the flexibility and range of applications that magnetosomes can be used for is to take components of the biosynthesis (key biomineralisation proteins) to make nanomagnetite without need to use the whole bacterial cell. In this biomediated method, we use key magnetosome synthesis protein as additives to mediate an *in vitro* chemical precipitation of magnetite to control the crystallisation. Ultimately this simplifies the system by removing all the cellular machinery not dedicated to magnetosome synthesis. This approach allows us to increase the level of flexibility as removing the need for the cell also removes any problems caused by toxicity, limiting doping to low levels with the MTB. It also offers a new level of control, allowing a choice of which proteins to add. As discussed above, previous research to investigate the function of Mms6 showed that the purified recombinant protein can control MNP size and shape in an *in vitro* chemical precipitation. This exploratory assay to investigate the action of Mms6 has since been used to develop a revolutionary biokleptic[76] synthetic method for the production of improved MNPs. The addition of the protein to this inexpensive and environmentally friendly room temperature mineralisation process significantly increases the control over MNP size and quality. We have developed this further to produce magnetite MNPs containing various levels of cobalt to increase the coercivity of the particles in a controlled manner.[77]

Currently our group is using a suite of candidate Mms proteins, such as Mms6, MmsF and others to control *in vitro* precipitation of magnetite. These proteins are responsible for controlling the crystallisation of cubo-octahedral MNPs within the magnetosomes. However, there are many strains of MTB that synthesise magnetosomes with morphologies other than cubo-octahedral, from elongated hexagonal prisms to bullet-shaped particles (Figure 3.2). We are therefore also investigating proteins that may control the crystallisation of different morphologies to see if these will control the formation of different shaped MNPs in a biokleptic *in vitro* precipitation. Further to this, the research is being extended to use proteins and peptides to control the formation of MNP for different magnetic materials other than magnetite such as doped magnetite, ferrites and even magnetic alloys such as CoPt.[78]

3.7.2.2 Surfaces

The beauty of the biokleptic synthetic method is that it can be applied across a range of situations and in combination with a variety of other syntheses. We can use the Mms proteins to template the formation of MNPs in solution, but can go a step further by immobilising the protein onto a patterned surface before the formation of the MNPs. The protein is selectively orientated onto a patterned substrate to ensure that it is able to remain functional, and thus template the formation of uniform magnetite MNPs in patterned arrays on surfaces. This has been achieved by binding the N-terminus of Mms6 to a carboxylic acid micrometre patterned self-assembled monolayer. Protein is prevented from attaching to the background by passivating these areas with PEG self-assembled monolayers to produce patterned areas of Mms6 that have their active C-terminus exposed. The precipitation reaction was then performed with aqueous reactants at relatively low temperatures ($\approx$80 °C) to produce magnetite MNP arrays (Figure 3.4).[79] The Mms6 protein is now dual functional, as it not only controls the size distribution and morphology of the MNPs, but it also anchors these uniform MNPs onto the patterned arrays. The magnetism of the Mms6 templated particles appears to be exchange coupled to form multi-particle zones of magnetic attraction and repulsion within the micrometre patterns (Figure 3.4). Interestingly, we see that different pattern morphologies on the surface produce differently shaped zones of magnetic

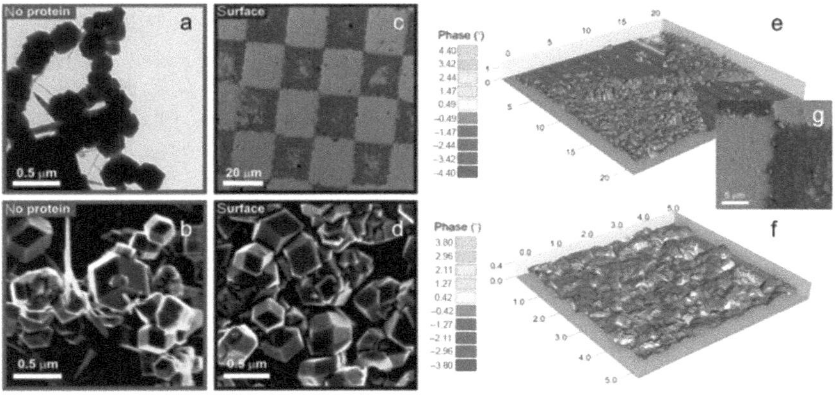

FIGURE 3.4 TEM (A), SEM (B–D and G) and MFM (E and F) images of the Mms6 biotemplated magnetite MNP 20 μm^2 square patterned array: (A and B) Shows the control sample of precipitated MNPs, note irregular size and shapes. (C and D) Shows the magnetite MNPs on the patterned arrays and close up on the surface, respectively. (E and F) Composite of tapping mode AFM topography and MFM phase shift. (E) Image 20 μm^2 at 200 nm lift height, (F) image 5 μm^2 and 50 nm lift height. MNPs show preferred orientation of magnetisation (E) and multi-particle zones of attraction and repulsion (F) (dark areas are attractive and light repulsive). (G) Insert SEM image of representative AFM/MFM scanned region. *Images are reprinted with permission from Wiley.*

interactions. For example, the MNPs on the striped patterns show longer, thinner zones of interaction than are seen on the square patterns. Such magnetic patterned arrays could be used in applications such as data storage. We are now working on reducing the size of the pattern islands of biotemplated MNPs to achieve single particle nano-arrays which could be used in high-density data storage to form bit-patterned media. Unfortunately, magnetite has a low coercivity, so is not likely to be suitable for use in data storage applications, as recorded information would be lost easily. However, ferrites of cobalt, strontium and barium have much larger coercivities, so could be ideal for this purpose. We have already shown that Mms6 is able to template cobalt-doped ferrite in solution, so this biokleptic synthetic method could also be used to form high-coercivity materials on patterned surfaces. Further to this, the peptides used to control the formation of high-coercivity CoPt MNPs can be patterned to produce arrays on CoPt in a similar manner.[78]

The biokleptic protein/peptide mediated synthetic methods offer strong promise for the future of MNP synthesis. Biotemplating of minerals is flexible, high yield and can be adapted to produce a range of industrially relevant MNPs. The precise control of biomolecules in room temperature mild mineralisation systems offer "greener" routes to the production of high-quality materials for use in technology.

3.8 Future Perspectives for Biological Synthesis

There have been great strides made in understanding the biomineralisation of magnetite in MTB since they were first discovered over 40 years ago. The difficulties in culturing the MTB (mentioned earlier) mean that biologically synthesised nanoparticles will always have a high cost, limiting them to high value, low-quantity applications such as advanced anti-cancer therapies, where other innovative treatments are often equally expensive. Future advances in this field are focused on transferring the magnetite generating genes from these bacteria into alternative host species such as *E. coli* which are well characterised and easy to cultivate in large amounts. This would not only reduce costs, making magnetosomes a viable treatment option, but due to the ease of genetic manipulation of such strains, the magnetosomes could be readily adapted to contain specialised proteins or molecules for specific applications. For example, a magnetosome developed in a strain of *E. coli* with an expanded genetic code[80] could allow incorporation of unnatural amino acids on key magnetosome proteins, perhaps with a photoactive cross-linking group. This could make labelling magnetosomes with drugs, DNA, cells or proteins as simple as turning on a light. Furthermore, such materials could be produced cheaply and in high yields. The first key steps towards realising this goal has very recently been achieved by the Schüler group by inserting a minimal number of magnetosome island gene clusters into *Rhodospirillum rubrum* and found, for the first time, this foreign host could express these proteins to induce the production of magnetosomes.[80a] This is a very important first step. However, *Rhodospirillum rubrum* is genetically similar the *magnetospirillum* but less so to *E. coli*. Currently biotechnology is based on the model organism *E. coli* and being able to replicate this bioengineering in *E. coli* will have substantial benefits to achieve the goals outlined above.

The knowledge gained by studying the genes and proteins which are able to produce these high quality, uniform MNPs with such high fidelity will undoubtedly pave the way for future innovations. The effects that proteins such as Mms6 have to exert control over the formation of magnetite nanoparticles *in vitro* shows the potential to harness the effects of morphology and size controlling proteins in synthetic reactions. Only a limited number of such proteins have been investigated to date but we envisage that this number will increase over the coming years to build up a "tool box" of proteins and peptides that can control specific features of MNPs, such as their composition, crystal phase, morphology and size. This would allow different MNPs to be tailored and synthesised to the requirements of specific applications simply by selecting and adding the correct protein or peptide additive to an appropriate aqueous mineralisation solution.

These proteins can also be used in completely different contexts as demonstrated by the results of surface patterning of biomineralisation proteins. By reducing the scale of the surface patterns, this technique could lead to new data storage methods.

However, MNPs produced using biokleptic protein mediated precipitations are not yet as precise as magnetosomes. Therefore we are attempting to further this approach by using not just Mms proteins, but aim to produce biomimetic artificial magnetosomes with all the precision of natural magnetosomes, but with the higher yields, flexibility and tailoring ability of the synthetic system. Such bio-coated MNPs are destined for a range of high-value biomedical targeted applications in diagnostics, therapies and theranostics.

MTB are still little known outside of the niche biomineralisation community, but the unique properties of the biogenic nanoparticles they synthesise makes them, and the synthetic mineralisation they have inspired, perfectly poised to become serious contenders in the fields of nano-medicine and nanotechnological material synthesis.

4 DENDRIMER ENCAPSULATED NANOPARTICLE OR DENs

The use of polymers, more specifically dendrimers, as templates for the preparation of small metal nanoparticles is a promising synthetic route to particles with sizes less than 3 nm, and high control of the size, and distributions close to the monodispersity[81,82] These compounds are called *Dendrimer Encapsulated Nanoparticle* or *DENs*. Dendrimers are three-dimensional (3D), highly branched polymers with an arboreal structure. They can be prepared in a controlled fashion following an iterative sequence of reactions involving convergent and divergent routes.[83–85] These synthetic methodologies allow an accurate knowledge of the dendrimers structure and composition and are the basis of their theoretical monodispersity. The globular shape of a dendrimer is defined by three components: a central core, an interior dendritic structure (the branches, also called generations) and a periphery with functional surface groups, responsible for most of the properties of the macromolecule, such as solubility. The number of functional groups on the surface of the macromolecule increases exponentially with each generation and causes a gradient of density from the core to the periphery (Figure 3.5 and Table 3.1). Therefore, the branching also generates a series of internal voids in the molecule suitable for host–guest interactions.[86,87] Within this discipline, polyamidoamine (PAMAM) dendrimers are the most common class of commercially available dendrimers. They have been widely employed, in everything from materials science to biotechnology applications.[88–90] The backbone of this type of dendrimers presents amide and tertiary amine functionalities suitable for interaction with metal ions. The extensive use of PAMAM dendrimers is probably due to their tailored commercial availability.[91]

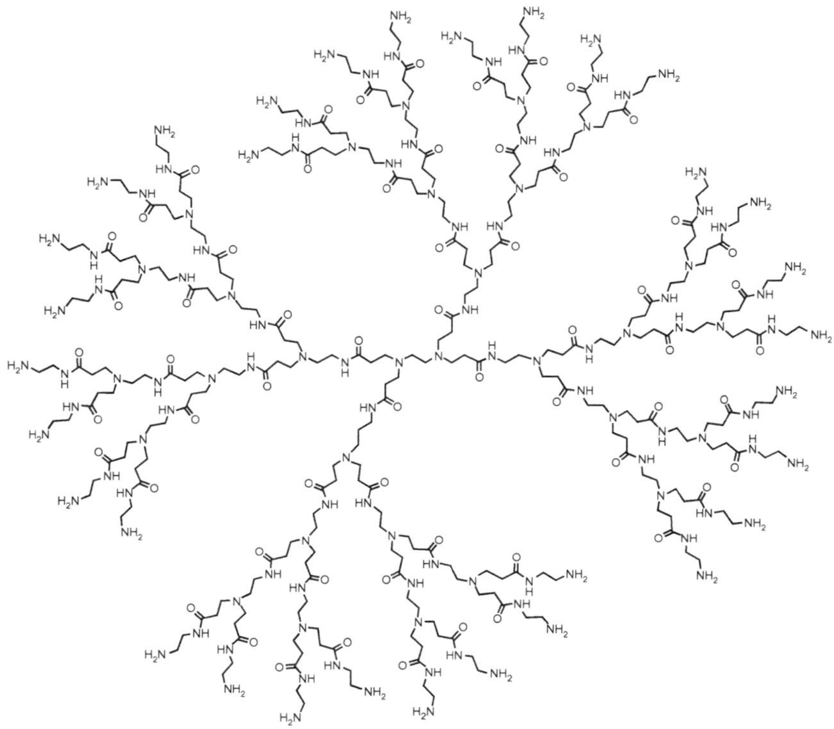

FIGURE 3.5 G3-PAMAM dendrimer (32 amine groups in the surface of the macromolecule).

TABLE 3.1 Evolution of the Number of Peripheral Groups, Molecular Weight and Size of a PAMAM Dendrimer while Growing

Generation	Peripheral Groups	Molecular Weight	Diameter (nm)
0	4	517	1.5
2	16	3256	2.9
4	64	14,215	4.5
6	256	58,048	6.7
8	1024	233,383	9.7
10	4096	934,720	13.5

4.1 Synthesis and Characterisation of Dendrimer Encapsulated MNPs

PAMAM dendrimers have been used as templates for metal nanoparticles in order to control not only their sizes but also the stability and solubility of the nanoparticles. There are several reasons why PAMAM dendrimers are suitable for hosting metal nanoparticles: (1) their uniform structure and composition allows well-defined replicas, (2) nanoparticles are stabilised by encapsulation within the dendrimer since the dendrimer template prevents aggregation, (3) the nanoparticles are imprisoned by steric effects, therefore the surface of the nanoparticle remains unpassivated and available for catalysis or molecular binding interaction, (4) the branches of the dendrimers may be used as a selective gate to control the access of molecules, (5) the peripheral groups of the dendrimer can be selected at will in order to tune physical properties such as solubility, or as ties to other polymers o biomolecules. All these five characteristics are due to the structural and chemical properties of the dendrimers. In the case of DENs, both template and replica together show a greater functionality than the individual components, analogous to a nanocomposite, the dendrimer, plays a role more important than a simple template.

In 1998, Prof. Crooks *et al.* described the preparation of Cu nanoparticles using PAMAM dendrimers as templates for the first time and shortly afterwards many different metal nanoparticles encapsulated in dendrimers were reported including metallic (Cu, Au, Ag, Pd and Pt), bimetallic (PdPt, AuAg, PdAu, PdRh and PtAu), and semiconducting (CdS) DENs.[81] Briefly, this methodology employs PAMAM dendrimers although other dendrimers such as polypropyleneimine (PPI) can be used as well. The preparation consists of two steps: (1) the loading of the metal ions into the interior of the dendrimer by coordination with the internal amine groups and (2) the chemical reduction to yield the zerovalent metal atoms that collapse to form the nanoparticle imprisoned within the dendrimer as shown in Figure 3.6.

Experimentally, DENs can be easily obtained by mixing two solutions containing PAMAM dendrimer and the salt source of the metal ions. The stochiometric relationship between both components is the key for controlling the final size of the nanoparticle and independent of the other factors such as functionalisation of the dendrimer. The only important issue to consider in selecting the dendrimer is ensuring the dendrimer generation contains enough tertiary amines to coordinate all the metal ions. A related issue is that the size of the dendrimer needs to be sufficient to host the resulting particle in its interior. For example, if a solution of fourth generation PAMAM dendrimer containing 64 tertiary amines with a concentration of 2.0 μM is mixed with a solution of 110 μM metal ion salt, the nanoparticle prepared is expected to have 55 atoms. In the

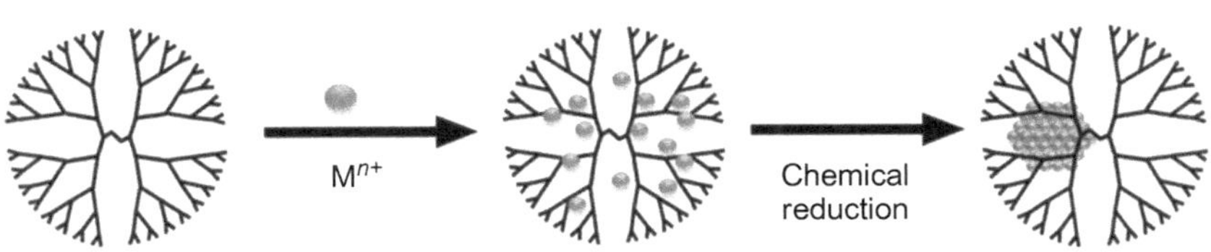

FIGURE 3.6 Synthesis of a metal nanoparticle by loading metal ions into the interior of a dendrimer followed by chemical reduction to yield the zerovalent metal atoms.

case of the preparation of 147 atoms nanoparticles, fourth generation PAMAM dendrimer cannot be used as a template because each dendrimer only has 64 tertiary amine—enough for a 55-atom particle but insufficient for any ratio greater than 1:64. For a 147-atom nanoparticle, a sixth-generation PAMAM dendrimer can be used since each has 254 tertiary amines, enough to coordinate all 147 ions.[92] Bearing this in mind, the nanoparticle size is independent of the generation dendrimer and a 55 atom particle has the same size if the template is generation fourth, fifth, sixth or eighth.[81] It is worth noting that when the quantity of metal ions exceeds the number of internal amines, or the size of the dendrimer is too small, chemical reduction will induce the formation of stable larger metal nanoparticle formed outside the dendrimers but stabilised by multiple dendrimers. These materials are called *dendrimer stabilised nanoparticles* (DSN)[93,94] and their size and polydispersity are more difficult to control. The chemical reduction of metal ions to zerovalent atoms within the dendrimer typically occurs with an excess of sodium borohydride but other reducing agents such as hydrazine, ascorbic acid, hydrogen, etc. can be used. The formation of the nanoparticle is easy to follow because reduction and the formation of zerovalent particles are accompanied by a colour change.

Following a similar strategy, bimetallic (or polymetallic) nanoparticles can be prepared either as alloys or as core–shell structures (see Figure 3.7). If ions from different metals are loaded within the dendrimer simultaneously and subsequently reduced, the resulting DEN will be an alloy of both metals. Once again, the composition of the alloy as well as the particle size are only dependent on the stochiometric relationship of the metal ions before reduction and the amount of metal loaded into the dendrimer. This methodology has been used to synthesise PdPt,[95] AuPd,[96] AuAg,[97] PtCu[98] or PdRh[99] alloys with different ratios between metals. Interestingly, a monometallic DEN can itself be

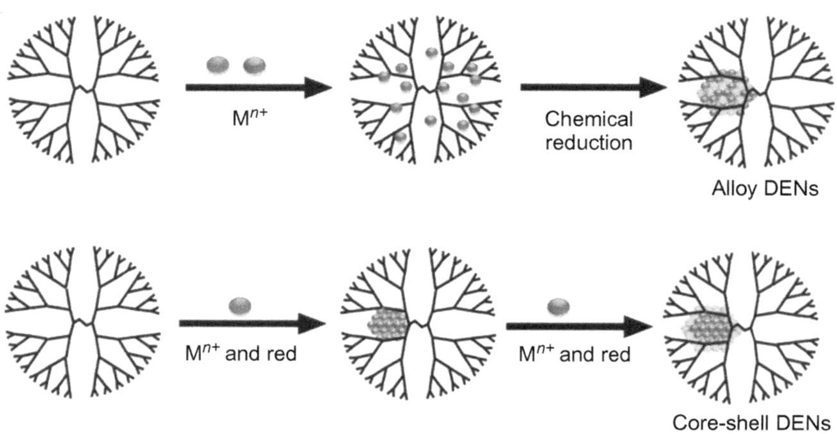

FIGURE 3.7 (Top) Synthesis of bimetallic alloy nanoparticles. (Bottom) Synthesis of core–shell nanoparticles.

used as a template for the formation of a core–shell DEN, with the original DEN constituting the core. The second (shell) metal must also coordinate with the tertiary amines and is subsequently reduced onto the surface of the already encapsulated DEN resulting in the formation of a core–shell nanoparticle (see Figure 3.7). This second reduction is carried out using a weaker reducing agent than $NaBH_4$, in fact a catalytic surface is needed for the reduction to occur on. The already-formed monometallic DEN acts as this catalytic surface, thus ensuring the second reduction causes formation of the shell. This sequential synthesis of the core and shell enables a fine control over the size and composition of the nanoparticle independently. Core–shell bimetallic nanoparticles combining metals like Pd, Pt, Au or Ag have been described.[81] In theory, a third metal could be added in an onion-like structure and a layer by layer nanoparticle could potentially be prepared. As far as we know, no reports of these three metal combinations have been described in the literature to date.

This methodology has some interesting features: (1) the particle sizes obtained are lower than 3 nm and the distribution of sizes is close to monodisperse, (2) high control of the particle size is made possible by fine-tuning the molar ratio between metal and dendrimer, (3) high control of the structure of the bimetallic nanoparticle from alloys to perfectly controlled and well-defined core–shell arrangements.

In the synthesis of nanoparticles with magnetic properties, some special variations are required in order to address the distinctive characteristics of these nanoparticles, for instance their ease of oxidation. As already mentioned, the synthesis of DENs is normally carried out in water because PAMAM dendrimers are soluble in water. Commercially available PAMAM dendrimers whose peripheries have been modified with dodecyl functional groups[91] pave the way for preparing DENs in non-polar organic solvents and in the absence of water or oxygen.[100] Particular attention must be paid to the use of dry and degassed solvents in order to avoid these issues. Among the solvents typically used to prepare magnetic DENs are toluene and THF, previously dried and degasseded following standard methods. The synthesis is carried out under an inert atmosphere. Since the solvents are different (non-polar) for the preparation of MNPs, further consideration has to be given to the reducing agent. Specifically, $NaBH_4$ cannot be used as it will not be soluble in dry THF or toluene. One option is to use a borohydride derivate such as $NaEt_3BH$, which will provide perfectly dry conditions in the preparations of these particles.

The characterisation of magnetic DENs, relies on some of the same techniques as are used for metal DENs although the very small size of magnetic DENs tends to limit the ease of obtaining and interpreting data from routinely-used techniques such as UV–vis, TEM and even HRTEM. In addition to these techniques, X-Ray powder diffraction (XRD), Energy Dispersive X-Ray (EDS or EDX), X-Ray photoelectron spectroscopy (XPS), X-Ray Absorption spectroscopy (XAS) provide information about the composition and the crystal structure of the metal nanoparticle. On the other hand, Fourier Transform Infrared

(FT-IR) and Nuclear Magnetic Resonance spectroscopy (NMR) are used to follow the interaction between the metal and the dendrimer prior to the reduction step.

4.2 Monometallic Dendrimer Encapsulated MNPs

Nickel, cobalt and iron encapsulated nanoparticles (Ni_{55}, Ni_{147}, Co_{55}, Co_{147}, Fe_{55} and Fe_{147}) were prepared and their magnetic properties preliminary evaluated by Crooks *et al.*[101,102]. As previously described, appropriate stoichiometric ratios of $Ni(ClO_4)_2$, $CoCl_2$ or $FeCl_3$ were added to solutions of sixth-generation dodecyl terminated PAMAM dendrimers. A 10-fold excess of sodium triethylborohydride (in toluene or THF) was added to reduce the dendrimer-metal complexes to the analogous zerovalent DENs. Upon reduction, the solutions change colour, which is characteristic of metal nanoparticle formation. The solution of Ni, Co and Fe DENs were stable indefinitely if kept under nitrogen; however, when exposed to air, the colour faded and a precipitation appeared because of the oxidation by atmospheric oxygen. UV–vis spectroscopy (Figure 3.8) can be used to monitor metal–dendrimer complex formation but, occasionally, in organic solvents, the metal ligand

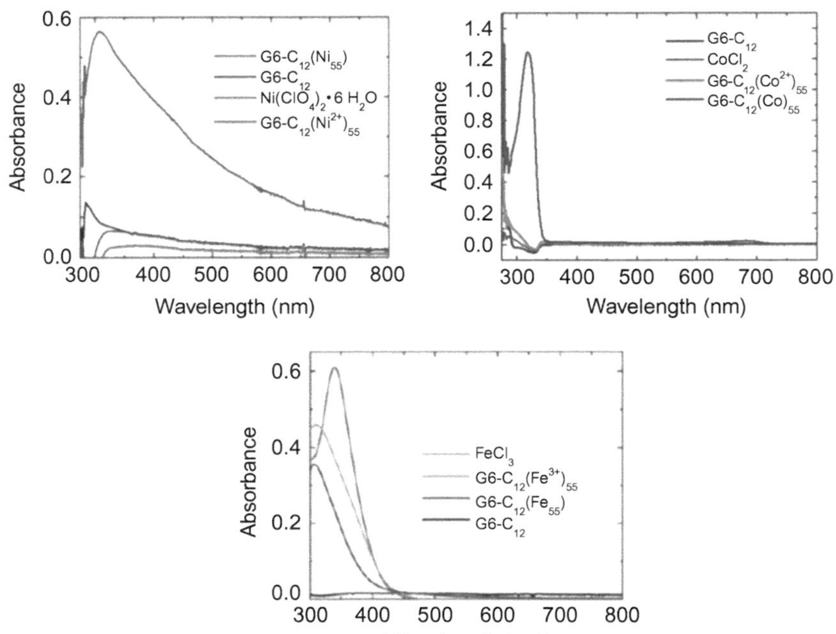

FIGURE 3.8 UV–vis spectra corresponding to the preparation of $G6\text{-}C_{12}(Ni_{55})$, $G6\text{-}C_{12}(Co_{55})$ and $G6\text{-}C_{12}(Fe_{55})$. *Reprinted with permission from Refs. 101 and 102. Copyright 2006 American Chemical Society and 2007 Royal Society of Chemistry.*

charge transfer band overlaps with the absorbance of the organic solvent. Ni or Co complexation cannot be tracked by UV–vis but Fe shows a 35-nm hypsochromic shift in λ_{max} upon formation of the Fe^{3+} dendrimer complex. Following reduction, the spectra of the nanoparticles exhibit an increase in the absorbance towards lower wavelength consistent with previous spectroscopic results for other types of metal nanoparticles in this size range.[81,82]

The sizes of these nanoparticles correspond to an average diameter of 0.8 ± 0.2 nm for Ni_{55}, 1.2 ± 0.3 nm for Ni_{147}, 0.9 ± 0.2 nm for Co_{55}, 1.2 ± 0.2 nm for Co_{147}, 0.9 ± 0.2 nm for Fe_{55}, and 1.1 ± 0.2 nm for Fe_{147} (Figure 3.9). All experimental results were comparable to the calculated diameters for each nanoparticle.[103] The agreements between the measured and calculated diameters, as well as, the high degree of monodispersity are a direct consequence of the template-based synthesis. Attempts to evaluate the

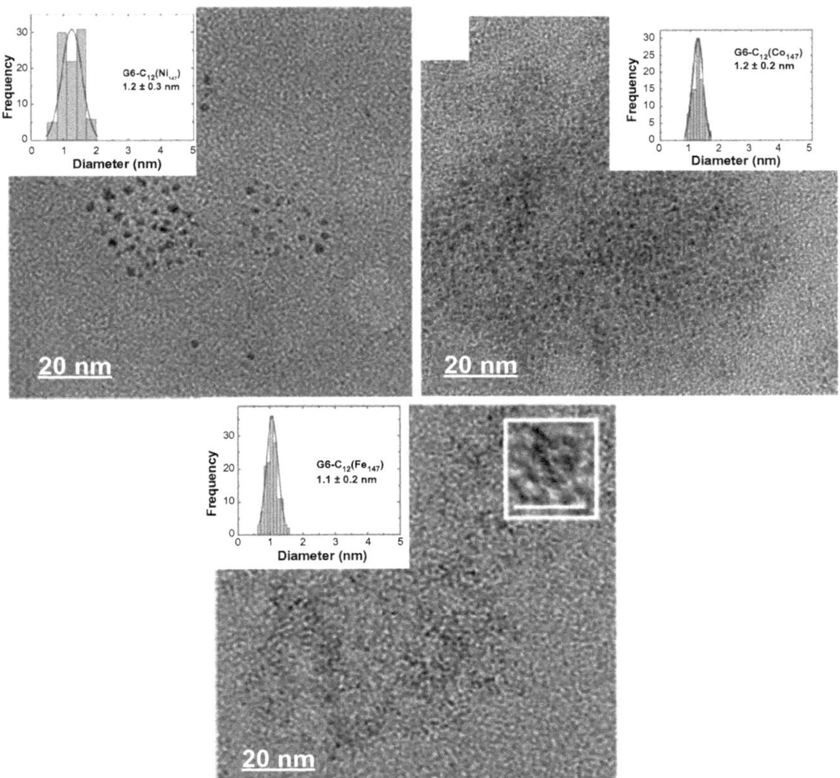

FIGURE 3.9 TEM micrographs and particle size histograms for G6-$C_{12}(Ni_{147})$, G6-$C_{12}(Co_{147})$ and G6-$C_{12}(Fe_{147})$. *Reprinted with permission from Refs. 101 and 102. Copyright 2006 American Chemical Society and 2007 Royal Society of Chemistry.*

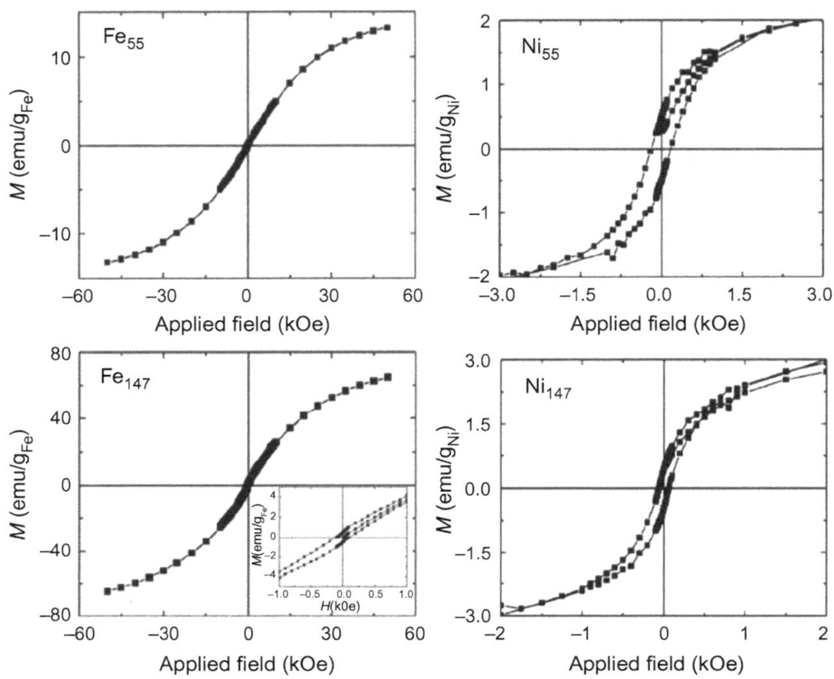

FIGURE 3.10 Magnetic moment (M) of G6-$C_{12}(Ni_{55})$, G6-$C_{12}(Ni_{147})$, G6-$C_{12}(Fe_{55})$ and G6-$C_{12}(Fe_{147})$ DENs as a function of the applied magnetic field. All analyses were carried out at 5 K. *Reprinted with permission from Refs. 101 and 102. Copyright 2006 American Chemical Society and 2007 Royal Society of Chemistry.*

oxidation states of the Ni and Fe DENs using XPS were not successful, presumably due to the small size of the nanoparticles.

The magnetic properties of Ni (Ni_{55} and Ni_{147}) and Fe_{147} DENs were studied using SQUID magnetometry (Figure 3.10) at temperatures ranging from 5 to 300 K with a magnetic field (H) strength of 500 Oe. The effect of thermal energy on the magnetic properties of the DENs (the blocking temperature T_b—see Chapter 1, Section 4) becomes apparent at 200 K for both sizes of Ni particles and at 6 K for Fe_{147}. That is, at $T > T_b$ no remanent magnetisation is observed, however, at $T < T_b$ Ni_{55}, Ni_{147} and Fe_{147} DENs show hysteresis with magnetic saturation values (M_s) of 3.40 emu/gNi for Ni_{55}, 3.95 emu/gNi for Ni_{147} and 70.0 emu/gFe for Fe_{147}. These values are significantly smaller than the bulk value of 55 emu/gNi and 220 emu/gFe at 300 K.[104] The M–H loop for the Fe_{55} DEN shows a hysteresis-free magnetism and complete saturation of the material was not observed over the range of magnetic fields studied. The absence of hysteresis and a blocking temperature indicate that the Fe_{55} DENs are superparamagnetic down to the lowest temperature used due their small particle size.[105]

Ni, Co or Fe DENs or DSNs have been prepared with other dendrimers for applications other than magnetic-based ones, such as catalysis,[106] imaging of cancer cells[107] or synthesis of single-walled carbon nanotubes.[108] Wieckowski *et al.*[109] describe the oxygen reduction catalytic properties of a series of Ni and Co DENs prepared by Crooks' procedure, among others. Although magnetic properties of Co DENs have not been described yet, Atwater *et al.*[110] have described the superparamagnetic behaviour of PAMAM dendrimers loaded with cobalt ions. Even though nanoparticles are not formed, the complex among generation fifth dendrimers and 38 cobalt atoms exhibit ferromagnetic properties below 50 K ($T_b = 50$ K) and an M_s of 159.6 emu/gCo.

A further alternate fabrication involves the use of poly(propylene) imine dendrimers (PPI or DAB-Am) instead of PAMAM dedrimers. McCarley *et al.*[111] described Ni DSNs using DAB-Am_{32} as a template with the resulting particle size in the range 1.9–2.7 nm (Figure 3.11). An extensive study of XPS rules out the formation of metal borides (boron-nickel alloys) during the reduction step because of the presence of the borohydride. Well known Fréchet-type dendrons can also be used if the focal-point on the dendrimer

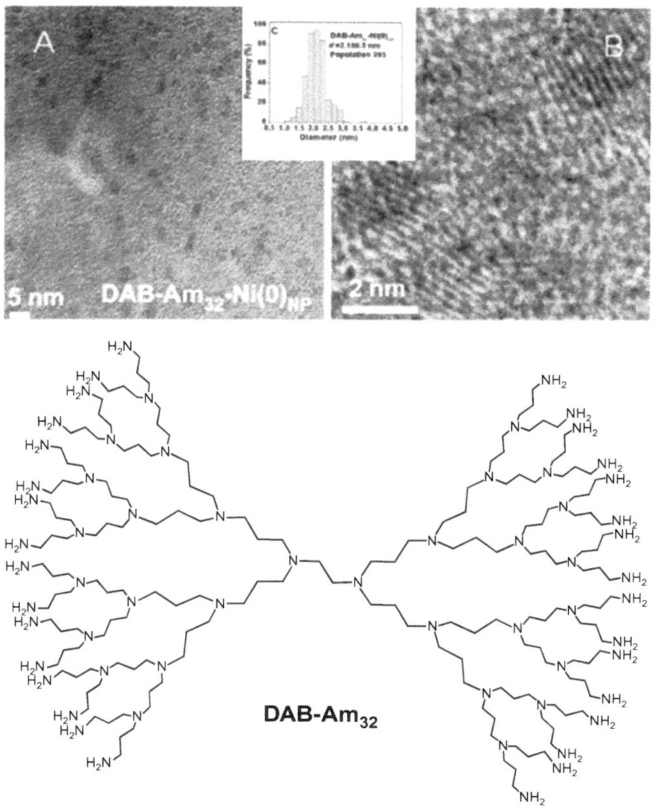

FIGURE 3.11 TEM micrographs, HRTEM and particle size histograms for Ni DSNs using DAB-Am_{32} (dendrimer illustrated on bottom). *Reprinted with permission from Ref. 111. Copyright 2010 American Chemical Society.*

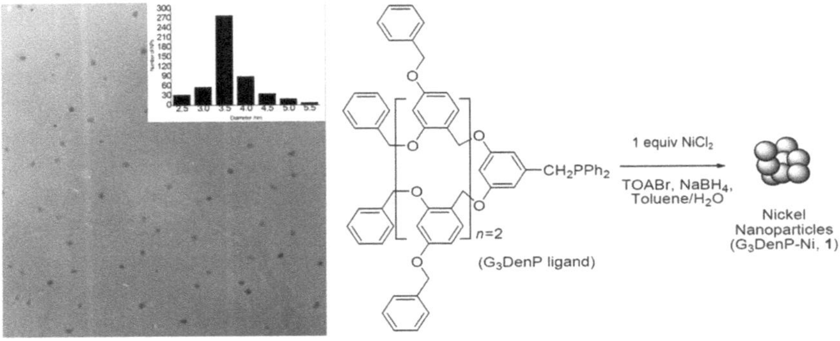

FIGURE 3.12 TEM micrographs and particle size histograms for Ni nanoparticles using phosphine Fréchet-type dendrons. On right the scheme reaction for the synthesis of the Ni nanoparticles. *Reprinted with permission from Refs. 112,113. Copyright 2006 American Chemical Society and 2007 Royal Society of Chemistry.*

is suitably modified (Figure 3.12). Wu *et al.* use phosphine dendrimers to stabilise 3.5 nm Ni nanoparticles and describe their application in catalysed C–C coupling reactions.[112,113]

4.3 Bimetallic Dendrimer Encapsulated MNPs

One immediate concern is that this kind of nanoparticle is very difficult to handle due to its ease of oxidation. Bimetallic nanoparticles have the potential to overcome this issue as the additional component can be used to tune the property in question compared to the monometallic DEN.[114] For example, although nickel and gold are essentially immiscible at temperatures below 500 K[115] bimetallic NiAu nanoparticles have been described by Chandler *et al.*[116,117]. They take advantage of a miscibility gap that enables the preparation of metastable nanometer scale bimetallic particles.[118,119] The nanoparticles were prepared following the method described above (*vide supra*)—essentially combining a solution of both salts containing the metal ions ($NiNO_3$ and $HAuCl_4$) with a fifth generation PAMAM dendrimer solution in water. This method was further improved by anchoring and chemically modifying the periphery of the dendrimer resulting in a more hydrophobic moiety. This modification enabled preparation in toluene instead of water. Afterwards, the bimetallic particles were fully passivated and could be extracted from the dendrimer template with thiols[92,120] in order to prevent oxidation and improve stability. The bimetallic nanoparticles described have different Au/Ni composition, but because the total metal:dendrimer ratio was held constant at 147:1 the particle size are ca. 2.2 ± 0.5 nm in all cases (Figure 3.13 left image). Discussion about the morphology of these bimetallic particles is particularly interesting. Authors claim the formation of core–shell structure instead of an alloy which would be the expected product here. This

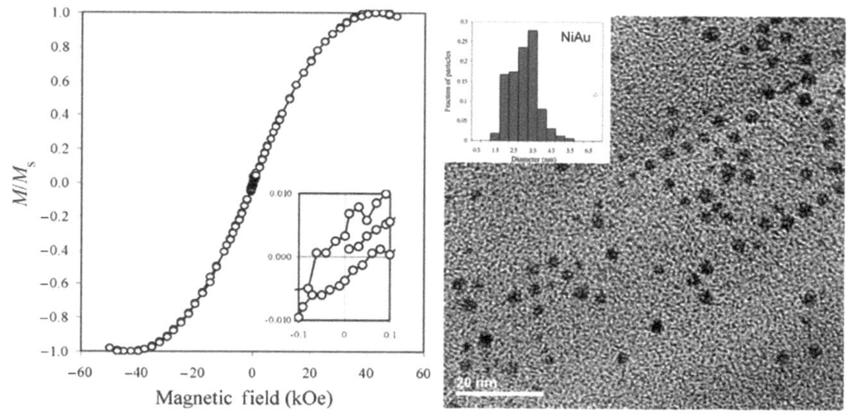

FIGURE 3.13 Magnetic moment (M) as a function of the applied magnetic field, TEM micrographs and particle size histograms for bimetallic NiAu DENs. *Reprinted with permission from Refs. 116,117. Copyright 2008 American Chemical Society.*

is explained by the fact that Au has a lower work function than Ni, making it thermodynamically more stable on the particle surface. Furthermore, since thiol extraction selectively binds through the Au atoms, these are more likely to be at the surface, with the Ni atoms embedded in the core.[97,121]

At 5 K, the saturation magnetisation of NiAu DENs are slightly higher than the corresponding monometallic Ni DENs showing a value of 10 emu/gNi. In this very preliminary study, authors attribute this enhancement to a Au shell effect. Specifically, the Au shell prevents quenching of the magnetism of the surface atoms due to the presence of organic diamagnetic material. Although, the magnetic properties here refer to the thiol passivated nanoparticles, the authors take advantages of the dendrimer template method to prepare the very small Ni alloys.

FePt and CoPt DSNs have been analogously described by Street *et al.*[122,123]. In a typical procedure, two solutions of $FeSO_4$ and H_2PtCl_6 or $CoCl_2$ and H_2PtCl_6 were mixed together and added to a generation four or generation six hydroxyl-terminated PAMAM dendrimer in water. The stoichiometric ratio between Fe:Pt and Co:Pt was 1:1 and after 2 h the mixture was reduced with $NaBH_4$. Bimetallic DENs sizes determined by HRTEM were 3.0 ± 0.4 nm for FePt and 3.5 ± 0.4 nm for CoPt (Figure 3.14). Subsequently, alloys were annealed at 700 ºC and it was observed that the FePt particle size increased to 10.7 ± 1.7 nm and the CoPt size to 7.9 ± 1 nm.

The magnetic properties of these FePt and CoPt DENs (Figure 3.15) show superparamagnetism at room temperature consistent with the low magnetocrystalline anisotropy of the fcc structure. However, the thermal annealing transforms the chemically disordered fcc structure to the chemically ordered fct structure transforming FePt and CoPt to blocked nanoparticles. The change

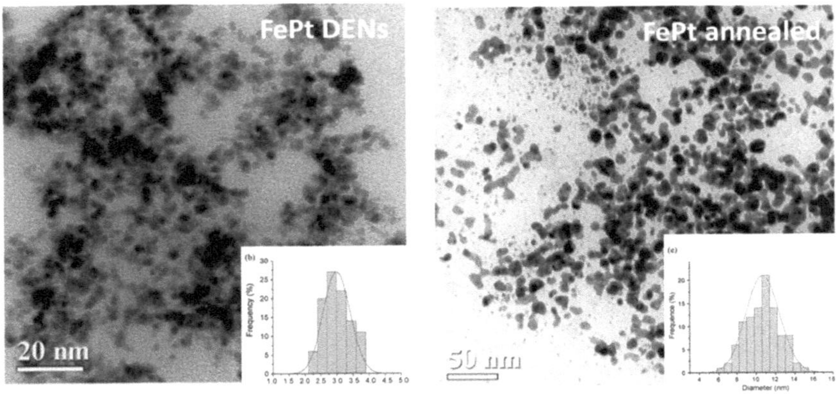

FIGURE 3.14 TEM micrographs and particle size histograms for bimetallic FePt nanoparticles, either DENs as-made and after annealing at 700 °C. *Reprinted with permission from Ref. 122. Copyright 2009 Elsevier.*

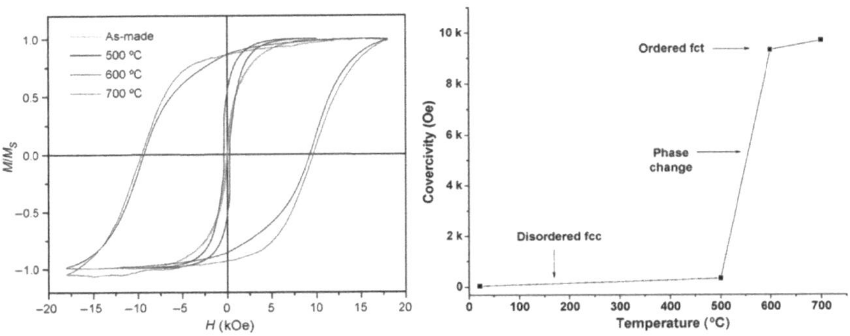

FIGURE 3.15 Magnetic moment (M) as a function of the applied magnetic field and room temperature of coercivity values as a function of annealing temperatures for bimetallic FePt nanoparticles. *Reprinted with permission from Refs. 122. Copyright 2009 Elsevier.*

in the magnetic properties accompanying the phase change at 500 K is seen most clearly in the sharp onset of the coercivity.

Similarly to monometallic DENs, many bimetallic DENs have been described for applications other than magnetism. For example, NiSn DENs have been described in the development of high-energy-density anode material for Li ion batteries.[124] Bimetallic NiSn DENs are prepared in different Ni:Sn ratios by the standard co-complexation procedure for alloy described previously. The sizes of the particles range from 1.2 ± 0.3 nm for monometallic Ni_{147} DENs to 1.9 ± 0.3 nm for monometallic Sn_{147} DENs, and the alloys $Ni_{110}Sn_{37}$, $Ni_{73.5}Sn_{73.5}$ and $Ni_{37}Sn_{110}$ DENs falling in between these two extremes with sizes of 1.1 ± 0.3 nm, 1.2 ± 0.3 nm and 1.3 ± 0.3 nm, respectively.

A further example of bimetallic magnetic DENs are NiFe DENs described by Kuhn *et al.* as active hydrogenation silica-supported catalysts.[125] These bimetallic DENs were prepared by co-complexation of both metal ions with the dendrimer in precise ratios to obtain alloys with compositions of $Fe_{30}Ni_{10}$, $Fe_{20}Ni_{20}$ and $Fe_{10}Ni_{30}$. It is worth mentioning that metal loading is performed over different time intervals because of the different complexation kinetics between the different metal ions and the dendrimer's internal tertiary amine groups. The resulting sizes of all the alloys are approximately 1.1 ± 0.3 nm. Recently, Stevenson *et al.* described a redox-displacement method to prepare Ru_5Ni_{30} and $Ru_{10}Ni_{30}$ DENs and their subsequent application as a catalyst.[126] Nanoparticle sizes are 1.0 ± 0.1 nm as expected for nanoparticles inside the dendrimer template.

4.4 Summary of Synthesis of Dendrimer Encapsulated MNPs

Utilising dendrimers as templates for the synthesis of nanoparticles provides us with a facile route to tight control over the size and, in the case of bimetallic DENs, compostition of the resultant particles. In the case of magnetic DENs, the dendrimer also provides us with the ability to synthesise these easily oxidisable materials, without compromising the control over size and composition. Here, we have summarised the most important strategies, issues and tips for preparation of easily oxidisable small nanoparticles, both monometallic and bimetallic, using dendrimer templates. Preliminary studies of magnetic properties for monometallic DENs demonstrated that, at room temperature, they displayed a superparamagnetic behaviour and became ferromagnetic at low temperature. Bimetallic DENs enable us to study even more complex systems.

5 GAS-PHASE SYNTHESIS OF NANOPARTICLES AND NANOPARTICLE MATERIALS

In this section, the synthesis of nanoparticles in the gas-phase will be presented with a focus on beam sources that can operate at ultra-high vacuum (UHV) enabling the production of pristine particles of reactive magnetic metals like Fe and Co. The extensions of the technique to produce alloy and core–shell nanoparticles will also be presented as well as methods to produce nanomaterials by depositing the nanoparticles in conjunction with various matrices, including solids and liquids.

5.1 Cluster Beam Sources

Sources producing free beams of metal nanoparticles with a controlled size over a wide range were first reported more than 30 years ago.[127,128] The technology continues to develop rapidly, however, as beam intensities are pushed ever higher and new capabilities are introduced such as UHV-compatibility,[129,130] the production of binary metal clusters,[131] new mass-selection techniques, etc. The generic synthesis method of most gas-phase nanoparticle sources is the condensation of a supersaturated metal vapour generated by ejecting a metal plume into a flow of a cooled inert gas. The metal vapour can be generated by laser ablation,[128] sputtering,[132] an arc source that may be pulsed[133] or continuous,[134] or thermally.[127] Monte–Carlo simulations[135] show that the dominant cooling mechanism for the growing clusters is by evaporation of atoms. This is approximately balanced by the heating due to the adsorption of new atoms (latent heat) and the inert gas provides the extra cooling required to allow the clusters to grow. The simulation also revealed that the initial formation of dimers is a critical bottleneck to the clustering process so that methods that produce an initial metal vapour rich in dimers and larger clusters tend to produce larger particles. The growth of clusters is complete after a dwell time in the source of the order of 10 ms after which the cluster size is described by a log-normal distribution[136]:

$$F(n) = \frac{1}{\sqrt{2\pi}\ln\sigma}\exp\left[-\left(\frac{\ln n - \ln\bar{n}}{\sqrt{2}\ln\sigma}\right)^2\right] \tag{3.1}$$

where n is the number of atoms per cluster and σ is the variance. The region in which the metal vapour initially mixes with the rare gas is at a pressure of a few mbar to a few bar and the mixture is expanded through an aperture with a diameter of a few μm to a few mm. The free jet expansion accelerates the clusters and in the limit of a high pressure differential between the two regions separated by the aperture, the clusters can acquire the full thermal distribution of the bath gas atom velocities. In weaker expansions, there is a velocity slip between the carrier gas and the metal clusters. Figure 3.16 shows the design of some of the basic types of gas-phase cluster beam sources. For a comprehensive description of these and other types, the reader is referred to existing reviews on cluster sources.[136,138,139]

A new type of gas-phase cluster source that has recently made an appearance is based on the formation of a beam of superfluid He droplets at 0.4 K that are then passed through a vapour of the material required in the nanoparticles. Most materials pass into the centre of the droplet and if the vapour pressure is raised sufficiently the material aggregates at the centre of the droplet to form a cluster. Originally the method was used to entrap molecules within the He matrix and study chemical reactions at very low temperature.[140] The technique has since been used to synthesise a variety of metal nanoparticles[141–147] and has recently been extended to include the production of MNPs.[148] It is the most flexible of the gas-phase synthesis methods and can easily be adapted to produce core–shell nanoparticles as described in Section 5.2 below.

Gas-phase methods, offer great flexibility and are able to produce nanoparticle beams of any metallic element and combined with in flight mass

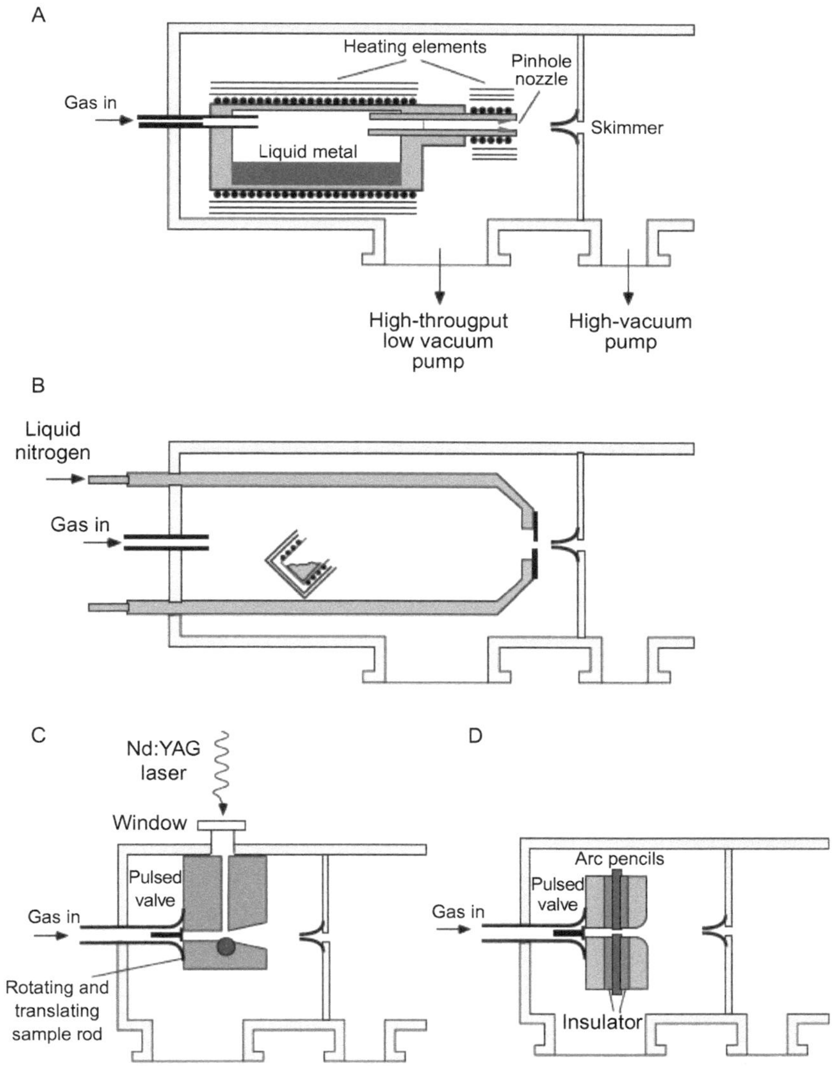

FIGURE 3.16 Schematics of sources using supersaturated metal vapours in rare gases to produce nanoparticles. (A) Seeded supersonic nozzle source. (B) Thermal gas aggregation source. (C) Laser ablation source. (D) Pulsed arc cluster ion source. *Reproduced with permission from Refs. 137.*

spectrometry[137] can provide very tight size control including selecting specific numbers of atoms. The capability of working in UHV also facilitates working with pure reactive metals such as the magnetic transition elements, without converting them to oxide. As discussed in the next section, it is relatively straightforward to extend the synthesis to multi-element nanoparticles with tight control over the stoichiometry. This great flexibility has made gas-phase production the method of choice for carrying out fundamental studies of the development of the properties of nanoparticles with size and composition despite the expense and complexity of the vacuum equipment required. Increasingly, attention is being focused on the manufacture of nanomaterials using this type of source, especially as very high-flux rates can now be achieved.[138] This aspect is described in Sections 5.3 and 5.4.

5.2 Production of Bimetallic Nanoparticles with Various Structures

All gas-phase sources start with a vapour of the material to be formed into nanoparticles so when it is required to produce binary particles the source needs to mix vapours of different materials. By varying the conditions under which the nanoparticles are seeded and grow, it is possible to vary the structure from uniform alloys through core–shell and nano-onion structures to Janus particles, that is, particles that present regions of different elements at the surface.

This was demonstrated in a recent paper that used a sputter nanoparticle source with a bimetallic Cu–Mo target (Figure 3.17) and varied the sputter gas (Ar) flow rate and discharge current to form various structures.[149] Changing the gas flow rate changes the rate of production of the vapours and in

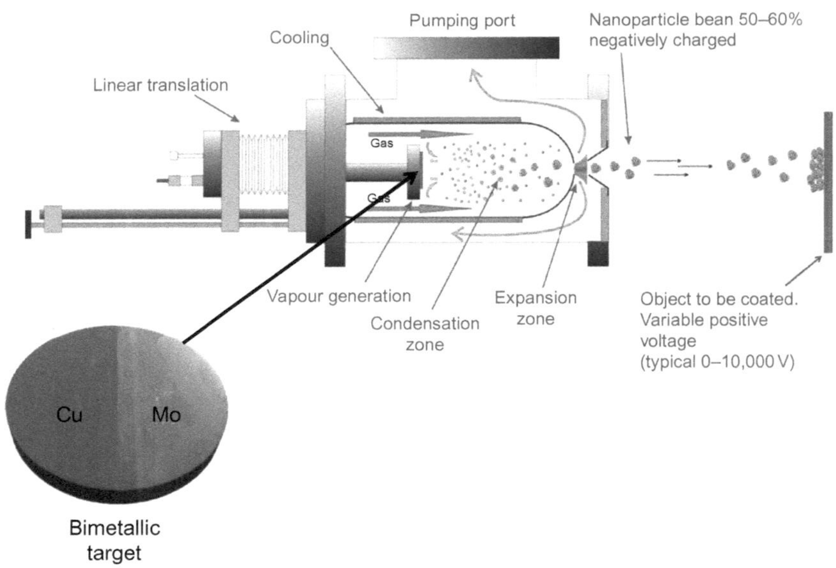

FIGURE 3.17 Sputter source with a bimetallic target used to produce a variety of heterostructures by varying the sputtering conditions (gas flow rate and discharge current). *Reproduced from Refs. 149, supplementary information.*

addition the rate of cooling of the growing clusters while changing the sputtering current alters the energy imparted to the vapours.

A variety of sources employing binary targets has been used to produce bimetallic nanoparticles including FeCo, FePt, CoSm, CoPt, CoAg and AlPb.[150–152] In some cases, the naturally occurring segregation of elements within the nanoparticle has been exploited to produce core–shell structures as described above. In pulsed sources, for example, those exploiting laser ablation, the separation of elements can be encouraged by focusing sequential pulses on separate targets.[131]

If one is interested specifically in producing core–shell structures, an alternative approach is to produce elemental nanoparticles, which are formed into a beam in high vacuum as normal and then deposit a shell of a different material onto the core nanoparticles as they fly through the source. A system demonstrated recently that uses thermal evaporation to coat the shell layer is illustrated in Figure 3.18.[153] The nanoparticles were produced in a thermal gas aggregation source and after being formed into a beam in high vacuum were passed through a hot tube containing the shell material whose temperature was adjusted to control the shell thickness. The system was demonstrated by synthesising core–shell particles containing an Fe core and a Cr shell (Fe@Cr) whose thickness was varied and it was shown that a minimum of 2 Cr monolayers was required in the shell for the nanoparticles to show exchange bias (EB) in their magnetisation loop.[153]

One problem with coating the core nanoparticle thermally is that the particle also heats up to the oven temperature so that surface is at the same temperature as the evaporant. This means that the vapour pressure of the shell material must be higher than that of the core material otherwise the nanoparticle will evaporate within the oven faster than it is coated. A solution is to sputter coat the shell material as demonstrated by the formation of FeCo nanoparticles with a AuAg shell (FeCo@AuAg).[154]

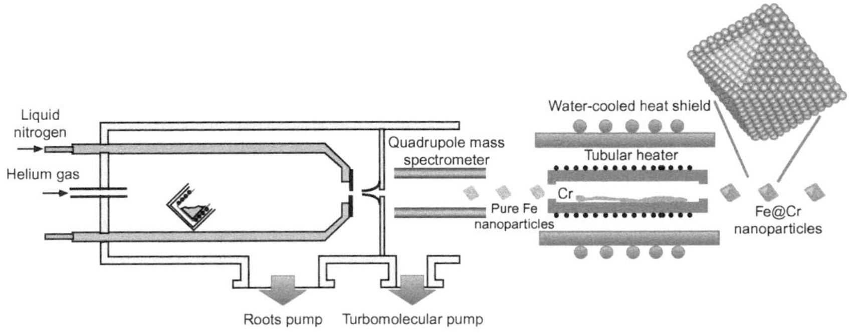

FIGURE 3.18 System for producing core–shell nanoparticles. Cluster beams formed by a thermal gas aggregation source are coated in the high-vacuum stage by passing them through a tubular evaporator filled with the shell material. The system was demonstrated by producing nanoparticles with an Fe core and a variable thickness Cr shell. *Reproduced from Ref. 153.*

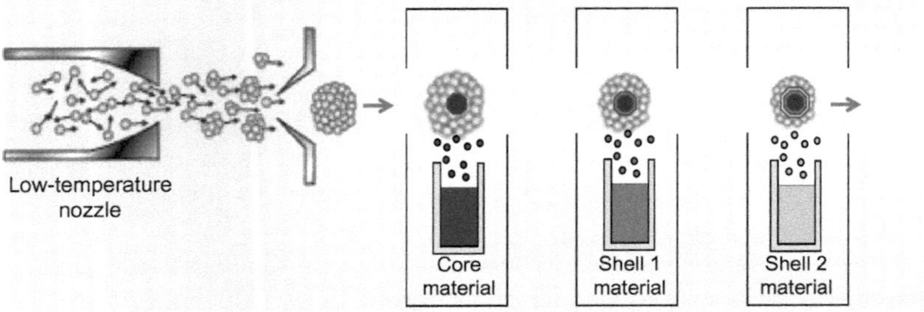

FIGURE 3.19 Formation of nano-onions by passing superfluid He droplets through cells containing vapour of different materials.

The most flexible method for forming core–shell nanoparticles is the He droplet technique introduced at the end of Section 5.1 and coating a shell is achieved simply by passing the He droplet containing the core nanoparticle through a second vapour source as illustrated in Figure 3.19. The process can be continued to further vapour cells downstream to form 'nano-onions' as long as the initial He droplet is sufficiently large. Since each metal atom entering the He droplet and bonding to a pre-existing atom will typically evaporate 1000 He atoms through the heat of condensation, the total number of metal atoms that can be accommodated is limited by the number of He atoms in the initial droplet. The method has recently been demonstrated by producing Ag@Au and Ni@Au core–shell nanoparticles.[148]

5.3 Production of Nanoparticle Assemblies in Solid Matrices By Cluster Deposition

An advantage of gas-phase synthesis is that it lends itself to a simple method of embedding the nanoparticles within a solid matrix by co-depositing the preformed nanoparticles and a molecular beam of the matrix material onto a common substrate as illustrated in Figure 3.20. This is a method for producing nanogranular materials in which there is independent control over grain size and volume fraction and free choice of materials in the grains and matrix. In addition, the grains can be alloy or core–shell type materials as described above and it is also possible to maintain a granular structure with miscible grain and matrix metals. This flexibility in the chemistry and structure of granular materials is not possible with any other synthesis method.

The volume fraction can be varied in the range from the dilute limit (isolated nanoparticles) to pure clusters with no matrix. This degree of control is very useful in fundamental research on the magnetic behaviour of nanostructured films as it is possible to use well-characterised uniform nanoparticles and isolate the

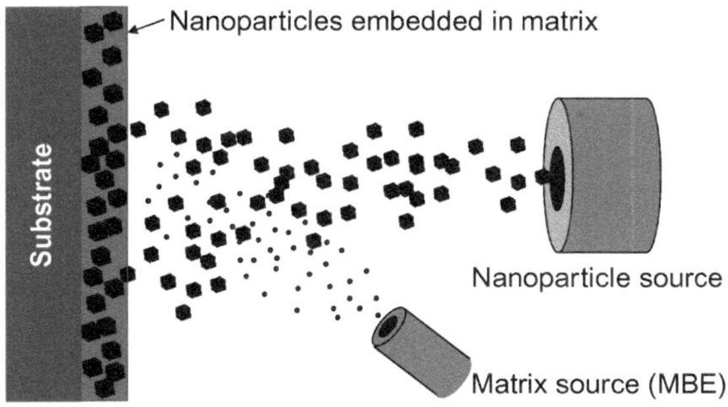

FIGURE 3.20 Formation of nanostructured materials by depositing pre-formed nanoparticles in conjunction with a molecular beam of a matrix material onto a common substrate.

changes in behaviour caused by interaction as the density of nanoparticles is increased. This was demonstrated in a detailed magnetometry study of 2 nm diameter Fe and Co nanoparticles embedded in Ag as a function of volume fraction.[155,156] These showed an evolution from perfect superparamagnetism (see Chapter 1, Section 5) through so-called interacting superparamagnetism,[157] where the nanoparticles start to form clusters, to a strong exchange coupling between all nanoparticles and the formation of a correlated super-spin glass.[158]

The very high proportion of surface atoms in a nanoparticle results in a strong influence of the epitaxy with the matrix material on the nanoparticle atomic structure and it has been discovered that a suitable choice of matrix material enables the structure to be controlled. For example, while 2 nm diameter Fe nanoparticles embedded in Ag maintain the bulk bcc structure,[159] embedding the same particles in Cu at low volume fractions causes them to adopt the fcc structure of the matrix.[160] It was found that the fcc Fe nanoparticles at low volume fraction in Cu adopt a low-moment magnetic state ($\sim$1 μ_B/atom) but increasing the volume fraction to above the percolation threshold ($\sim$25%) causes the nanoparticles to switch to a bcc atomic structure with the bulk Fe magnetic moment of 2.2 μ_B/atom.[160] The method has also been used to prepare Co nanoparticles (normally fcc) in the bcc phase by embedding them in bcc Fe matrices.[161] Recently it has been shown that it is even possible to control the lattice constant in embedded Fe nanoparticles by embedding them in $Cu_{1-x}Au_x$ alloy matrices whose lattice constant can be varied continuously by changing the Cu:Au ratio.[162] Thus it was possible to produce embedded fcc Fe nanoparticles and stretch their lattice constant thereby causing them to switch to the theoretically predicted high-spin state of 2.5 μ_B/atom.[162]

In addition to fundamental research, the great flexibility of cluster deposition can be exploited to produce high-performance magnetic materials. Applications of nanostructured materials in general are discussed in Chapter 4 but magnetic materials produced by the process in Figure 3.20 for technology can be divided into classes distinguished by the nanoparticle/matrix material combinations. They include:

(a) MNPs in non-magnetic matrices, which can be used as magnetic sensors exploiting giant magnetoresistance[150,163,164]
(b) MNPs in antiferromagnetic matrices for films required to produce an EB in the hysteresis loop[165–167]
(c) MNPs in ferromagnetic matrices to synthesise magnetic materials with a saturation magnetisation exceeding the Slater–Pauling limit of 2.45 μ_B/atom.[150]

Note that this last application is possible due to the possibility of producing MNPs in a magnetic matrix composed of a miscible materials without forming a uniform alloy, which is only possible using the cluster deposition method.

5.4 Production of Nanoparticle Assemblies in Liquid Matrices By Cluster Deposition

In many cases and especially for medical applications (see Chapter 6), MNPs need to be dispersed in a liquid. The synthesis methods described in Sections 2–4 of this chapter naturally produce hydrosols containing the nanoparticles but with gas-phase production the starting form of the nanoparticles is in a beam in high vacuum or UHV. For most applications, it is desirable to maintain a metallic magnetic core without converting it to oxide in order to maximise the magnetic moments of the nanoparticles and most liquids, including water, have a vapour pressure at room temperature that is incompatible with UHV operation. A simple solution to this, as demonstrated recently,[168] is based on the method shown in Figure 3.20 but uses a molecular beam of water as the matrix material and maintains the substrate temperature at 77 K. Thus the water molecules hitting the substrate stick with a probability close to unity and form an ice matrix at 77 K, whose vapour pressure is in the range 10^{-14}–10^{-12} mbar depending on whether the ice is crystalline or amorphous. Meanwhile the nanoparticles incident on the substrate are embedded in the ice matrix and UHV conditions are maintained throughout the source. The method is valuable despite its complexity compared to synthesis by wet chemistry because of the enormous flexibility of nanoparticle design offered by gas-phase production outlined above. It has been used to prepare suspensions of Fe@Fe oxide and Fe@Au core–shell nanoparticles.[168]

REFERENCES

1. Schwertmann U, Cornell RM. *Iron oxides in the laboratory: preparation and characterization*. Weinheim: Wiley; 2000.
2. Massart R. Preparation of aqueous magnetic liquids in alkaline and acidic media. *IEEE Trans Magn* 1981;**17**:1247–8.
3. Willard MA, Kurihara LK, Carpenter EE, Calvin S, Harris VG. Chemically prepared magnetic nanoparticles. *Int Mater Rev* 2004;**49**:125.

4. Lu A-H, Salabas EL, Schüth F. Magnetic nanoparticles: synthesis, protection, functionalization, and application. *Angew Chem Int Ed* 2007;**46**:1222–44.
5. Thakur M, De K, Giri S, Si S, Kotal A, Mandal TK. Interparticle interaction and size effect in polymer coated magnetite nanoparticles. *J Phys Condens Matter* 2006;**18**:9093–104.
6. Jolivet J-P. *Metal oxide chemistry and synthesis: from solution to solid state*. Chichester: Wiley; 2000.
7. Jolivet J-P, Chanéac C, Tronc E. Iron oxide chemistry. From molecular clusters to extended solid networks. *Chem Comm* 2004;481–7.
8. Ruby C, Gehin A, Abdelmoula M, Genin JMR, Jolivet JP. Coprecipitation of Fe(II) and Fe(III) cations in sulphated aqueous medium and formation of hydroxysulphate green rust. *Solid State Sci* 2003;**5**:1055–62.
9. Bocher F, Gehin A, Ruby C, Ghanbaja J, Abdelmoula M, Génin JM. Coprecipitation of Fe(II-III) hydroxycarbonate green rust stabilised by phosphate adsorption. *Solid State Sci* 2004;**6**:117–24.
10. Regazzoni AL, Urrutia GA, Blesa MA, Maroto AJC. Some observations on the composition and morphology of synthetic magnetites obtained by different routes. *J Inorg Nucl Chem* 1981;**43**:1489–93.
11. Matijević E, Sugimoto T. Formation of uniform spherical magnetite particles by crystallization from ferrous hydroxide gels. *J Colloid Interface Sci* 1980;**74**:227–43.
12. Sun S, Zeng H. Size-controlled synthesis of magnetite nanoparticles. *J Am Chem Soc* 2002;**124**:8204–5.
13. Ho Chien-Hsin, Tsai Chih-Pin, Chung Chia-Chi, Tsai Chun-Ying, Chen Fu-Rong, Lin Hong-Ji, et al. Shape-controlled growth and shape-dependent cation site occupancy of monodisperse Fe_3O_4 nanoparticles. *Chem Mater* 2011;**23**:1753–60.
14. Mériguet G, Dubois E, Perzynski R. Liquid-liquid phase-transfer of magnetic nanoparticles in organic solvents. *J Colloid Interface Sci* 2003;**267**:78–85.
15. Frankel RB. The discovery of magnetotactic/magnetosensitive bacteria. *Chin J Oceanol Limnol* 2009;**27**:1–2.
16. Blakemore R. Magnetotactic bacteria. *Science* 1975;**190**:377–9.
17. Frankel RB, Blakemore RP, Wolfe RS. Magnetite in freshwater magnetotactic bacteria. *Science* 1979;**203**:1355–6.
18. Farina M, Esquivel DMS, de Barros HGP. Magnetic iron-sulphur crystals from a magnetotactic microorganism. *Nature* 1990;**343**:256.
19. Frankel RB, Blakemore RP, Wolfe RS. Biomineralization of ferrimagnetic greigite (Fe_3S_4) and iron pyrite (FeS_2) in a magnetotactic bacterium. *Nature* 1990;**343**:258–61.
20. Balkwill DL, Maratea D, Blakemore RP. Ultrastructure of a magnetotactic spirillum. *J Bacteriol* 1980;**141**:1399–408.
21. Bazylinski DA, Frankel RB. Magnetosome formation in prokaryotes. *Nat Rev Microbiol* 2004;**2**:217–30.
22. Gorby YA, Beveridge TJ, Blakemore RP. Characterization of the bacterial magnetosome membrane. *J Bacteriol* 1988;**170**:834–41.
23. Grünberg K, Wawer C, Tebo BM, Schüler D. Biochemical and proteomic analysis of the magnetosome membrane in Magnetospirillum gryphiswaldense. *Appl Environ Microbiol* 2004;**70**:1040–50.
24. Tanaka M, Okamura Y, Arakaki A, Tanaka T, Takeyama H, Matsunaga T. Origin of magnetosome membrane: proteomic analysis of magnetosome membrane and comparison with cytoplasmic membrane. *Proteomics* 2006;**6**:5234–47.
25. Bazylinski D, Garratt-Reed J, Frankel RB. Electron microscopic studies of magnetosomes in magnetotactic bacteria. *Micros Res Tech* 1994;**27**:389–401.

26. Butler RF, Banerjee SK. Theoretical single-domain grain size range in magnetite and titanomagnetite. *J Geophys Res* 1975;**80**:4049–58.
27. Dunin-Borkowski RE, McCartney MR, Frankel RB, Bazylinski DA, Posfai M, Buseck PR. Magnetic microstructure of magnetotactic bacteria by electron holography. *Science* 1998;**282**:1868–70.
28. Frankel RB, Bazylinski DA, Johnson MS, Taylor BL. Magneto-aerotaxis in marine coccoid bacteria. *Biophys J* 1997;**73**:994–1000.
29. Frankel RB, Williams T, Bazylinski D. Magneto-aerotaxis. In: Schüler D, editor. *Magnetoreception and magnetosomes in bacteria*. Springer; 2007. p. 1–24.
30. Maratea D, Blakemore RP. Aquaspirillum-magnetotacticum Sp-Nov, a magnetic spirillum. *Int J Syst Bacteriol* 1981;**31**:452–5.
31. Kawaguchi R, Burgess JG, Matsunaga T. Phylogeny and 16s ribosomal-RNA sequence of magnetospirillum Sp Amb-1, an aerobic magnetic bacterium. *Nucleic Acids Res* 1992;**20**:1140.
32. Schleifer KH, Schuler D, Spring S, Wiezenegger M, Amann R, Ludwig W, et al. The genus magnetospirillum Gen-Nov—description of magnetospirillum-gryphiswaldense Sp-Nov and transfer of aquaspirillum-magnetotacticum to magnetospirillum-magnetotacticum comb-Nov. *Syst Appl Microbiol* 1991;**14**:379–85.
33. Bazylinski DA, Williams TJ, Lefèvre CT, Trubitsyn D, Fang J, Beveridge TJ, et al. Magnetovibrio blakemorei gen. nov., sp nov., a magnetotactic bacterium (Alphaproteobacteria: Rhodospirillaceae) isolated from a salt marsh. *Int J Syst Evol Microbiol* 2013;**63**:1824–33.
34. Bazylinski DA, Williams TJ, Lefèvre CT, Berg RJ, Zhang CL, Bowser SS, et al. Magnetococcus marinus gen. nov., sp nov., a marine, magnetotactic bacterium that represents a novel lineage (Magnetococcaceae fam. nov., Magnetococcales ord. nov.) at the base of the Alphaproteobacteria. *Int J Syst Evol Microbiol* 2013;**63**:801–8.
35. Lefevre CT, Bazylinski DA. *Ecology*, diversity, and evolution of magnetotactic bacteria. *Microbiol Mol Biol Rev* 2013;**77**:497–526.
36. Sakaguchi T, Arakaki A, Matsunaga T. Desulfovibrio magneticus sp nov., a novel sulfate-reducing bacterium that produces intracellular single-domain-sized magnetite particles. *Int J Syst Evol Microbiol* 2002;**52**:215–21.
37. Nakazawa H, Arakaki A, Narita-Yamada S, Yashiro I, Jinno K, Aoki N, et al. Whole genome sequence of Desulfovibrio magneticus strain RS-1 revealed common gene clusters in magnetotactic bacteria. *Genome Res* 2009;**19**:1801–8.
38. Lefèvre CT, Viloria N, Schmidt ML, Pósfai M, Frankel RB, Bazylinski DA. Novel magnetite-producing magnetotactic bacteria belonging to the Gammaproteobacteria. *ISME J* 2012;**6**:440–50.
39. Abreu F, Martins JL, Silveira TS, Keim CN, de Barros HG, Filho FJ, et al. 'Candidatus Magnetoglobus multicellularis', a multicellular, magnetotactic prokaryote from a hypersaline environment. *Int J Syst Evol Microbiol* 2007;**57**:1318–22.
40. Komeili A, Vali H, Beveridge TJ, Newman DK. Magnetosome vesicles are present before magnetite formation, and MamA is required for their activation. *Proc Natl Acad Sci* 2004;**101**:3839–44.
41. Komeili A, Li Z, Newman DK, Jensen GJ. Magnetosomes are cell membrane invaginations organized by the actin-like protein MamK. *Science* 2006;**311**:242–5.
42. MacLean D, Jones JDG, Studholme DJ. Application of 'next-generation' sequencing technologies to microbial genetics. *Nat Rev Microbiol* 2009;**7**:287–96.
43. Richter M, Kube M, Bazylinski DA, Lombardot T, Glöckner FO, Reinhardt R, et al. Comparative genome analysis of four magnetotactic bacteria reveals a complex set of group-specific genes implicated in magnetosome biomineralization and function. *J Bacteriol* 2007;**189**:4899–910.

44. Schübbe S, Kube M, Scheffel A, Wawer C, Heyen U, Meyerdierks A, et al. Characterization of a spontaneous nonmagnetic mutant of Magnetospirillum gryphiswaldense reveals a large deletion comprising a putative magnetosome island. *J Bacteriol* 2003;**185**:5779–90.
45. Fukuda Y, Okamura Y, Takeyama H, Matsunaga T. Dynamic analysis of a genomic island in Magnetospirillum sp strain AMB-1 reveals how magnetosome synthesis developed. *FEBS Lett* 2006;**580**:801–12.
46. Lefèvre CT, Trubitsyn D, Abreu F, Kolinko S, de Almeida LG, de Vasconcelos AT, et al. Monophyletic origin of magnetotaxis and the first magnetosomes. *Environ Microbiol* 2013;**15**:2267–74.
47. Lohsse A, Ullrich S, Katzmann E, Borg S, Wanner G, Richter M, et al. Functional analysis of the magnetosome island in Magnetospirillum gryphiswaldense: the mamAB operon is sufficient for magnetite biomineralization. *PLoS One* 2011;**6**.
48. Grünberg K, Müller EC, Otto A, Reszka R, Linder D, Kube M, et al. Biochemical and proteomic analysis of the magnetosome membrane in Magnetospitillum gryphiswaldense. *Appl Environ Microbiol* 2004;**70**:1040–50.
49. Arakaki A, Webb J, Matsunaga T. A novel protein tightly bound to bacterial magnetic particles in Magnetospirillum magneticum strain AMB-1. *J Biol Chem* 2003;**278**:8745–50.
50. Uebe R, Junge K, Henn V, Poxleitner G, Katzmann E, Plitzko JM, et al. The cation diffusion facilitator proteins MamB and MamM of Magnetospirillum gryphiswaldense have distinct and complex functions, and are involved in magnetite biomineralization and magnetosome membrane assembly. *Mol Microbiol* 2011;**82**:818–35.
51. Grass G, Otto M, Fricke B, Haney CJ, Rensing C, Nies DH, et al. FieF (YiiP) from Escherichia coli mediates decreased cellular accumulation of iron and relieves iron stress. *Arch Microbiol* 2005;**183**(1):9–18.
52. Rong C, Huang Y, Zhang W, Jiang W, Li Y, Li J. Ferrous iron transport protein B gene (feoB1) plays an accessory role in magnetosome formation in Magnetospirillum gryphiswaldense strain MSR-1. *Res Microbiol* 2008;**159**:530–6.
53. Rong C, Zhang C, Zhang Y, Qi L, Yang J, Guan G, et al. FeoB2 Functions in magnetosome formation and oxidative stress protection in Magnetospirillum gryphiswaldense strain MSR-1. *J Bacteriol* 2012;**194**:3972–6.
54. Carballido-Lopez R. The bacterial actin-like cytoskeleton. *Microbiol Mol Biol Rev* 2006;**70**:888–909.
55. Scheffel A, Gruska M, Faivre D, Linaroudis A, Plitzko JM, Schüler D. An acidic protein aligns magnetosomes along a filamentous structure in magnetotactic bacteria. *Nature* 2006;**440**:110–4.
56. Feng S, Wang L, Palo P, Liu X, Mallapragada SK, Nilsen-Hamilton M. Integrated self-assembly of the Mms6 magnetosome protein to form an iron-responsive structure. *Int J Mol Sci* 2013;**14**:14594–606.
57. Amemiya Y, Arakaki A, Staniland SS, Tanaka T, Matsunaga T. Controlled formation of magnetite crystal by partial oxidation of ferrous hydroxide in the presence of recombinant magnetotactic bacterial protein Mms6. *Biomaterials* 2007;**28**:5381–9.
58. Wang L, Prozorov T, Palo PE, Liu X, Vaknin D, Prozorov R, et al. Self-assembly and biphasic iron-binding characteristics of Mms6, a bacterial protein that promotes the formation of superparamagnetic magnetite nanoparticles of uniform size and shape. *Biomacromolecules* 2012;**13**:98–105.
59. Tanaka M, Mazuyama E, Arakaki A, Matsunaga T. Mms6 protein regulates crystal morphology during nano-sized magnetite biomineralization in vivo. *J Biol Chem* 2011;**286**:6386–92.
60. Murat D, Falahati V, Bertinetti L, Csencsits R, Körnig A, Downing K, et al. The magnetosome membrane protein, MmsF, is a major regulator of magnetite biomineralization in Magnetospirillum magneticum AMB-1. *Mol Microbiol* 2012;**85**:684–99.

61. Tanaka M, Arakaki A, Matsunaga T. Identification and functional characterization of liposome tubulation protein from magnetotactic bacteria. *Mol Microbiol* 2010;**76**:480–8.
62. Okamura Y, Takeyama H, Matsunaga T. A magnetosome-specific GTPase from the magnetic bacterium magnetospirillum magneticum AMB-1. *J Biol Chem* 2001;**276**:48183–8.
63. Siponen MI, Adryanczyk G, Ginet N, Arnoux P, Pignol D. Magnetochrome: a c-type cytochrome domain specific to magnetotatic bacteria. *Biochem Soc Trans* 2012;**40**:1319–23.
64. Yang J, Li S, Huang X, Li J, Li L, Pan Y, et al. MamX encoded by the mamXY operon is involved in control of magnetosome maturation in Magnetospirillum gryphiswaldense MSR-1. *BMC Microbiol* 2013;**13**:203.
65. Xie J, Chen K, Chen X. Production, modification and bio-applications of magnetic nanoparticles gestated by Magnetotactic bacteria. *Nano Res* 2009;**2**:261–78.
66. Ceyhan B, Alhorn P, Lang C, Schüler D, Niemeyer CM. Semisynthetic biogenic magnetosome nanoparticles for the detection of proteins and nucleic acids. *Small* 2006;**2**:1251–5.
67. Maeda Y, Yoshino T, Takahashi M, Ginya H, Asahina J, Tajima H, et al. Noncovalent immobilization of streptavidin on in vitro- and in vivo-biotinylated bacterial magnetic particles. *Appl Environ Microbiol* 2008;**74**:5139–45.
68. Sun JB, Duan JH, Dai SL, Ren J, Zhang YD, Tian JS, et al. In vitro and in vivo antitumor effects of doxorubicin loaded with bacterial magnetosomes (DBMs) on H22 cells: the magnetic bio-nanoparticles as drug carriers. *Cancer Lett* 2007;**258**:109–17.
69. Hu LL, Zhang F, Wang Z, et al. Comparison of the ˆ1H NMR relaxation enhancement produced by bacterial magnetosomes and synthetic iron oxide nanoparticles for potential use as MR molecular probes. *IEEE Trans Appl Supercond* 2010;**20**:822–5.
70. Alphandéry E, Faure S, Raison L, Duguet E, Howse PA, Bazylinski DA. Heat production by bacterial magnetosomes exposed to an oscillating magnetic field. *J Phys Chem C* 2011;**115**:18–22.
71. Hergt R, Hiergeist R, Zeisberger M, et al. Magnetic properties of bacterial magnetosomes as potential diagnostic and therapeutic tools. *J Magn Magn Mater* 2005;**293**:80–6.
72. Heyen U, Schuler D. Growth and magnetosome formation by microaerophilic Magnetospirillum strains in an oxygen-controlled fermentor. *Appl Microbiol Biotechnol* 2003;**61**:536–44.
73. Sun JB, Zhao F, Tang T, Jiang W, Tian JS, Li Y, et al. High-yield growth and magnetosome formation by Magnetospirillum gryphiswaldense MSR-1 in an oxygen-controlled fermentor supplied solely with air. *Appl Microbiol Biotechnol* 2008;**79**:389–97.
74. Staniland S, Williams W, Telling N, Van Der Laan G, Harrison A, Ward B. Controlled cobalt doping of magnetosomes in vivo. *Nat Nanotechnol* 2008;**3**:158–62.
75. Tanaka M, Brown R, Hondow N, Arakaki A, Matsunaga T, Staniland S. Highest levels of Cu, Mn and Co doped into nanomagnetic magnetosomes through optimized biomineralisation. *J Mater Chem* 2012;**22**(24):11919–21.
76. Rawlings AE, Bramble JP, Staniland SS. Innovation through imitation: biomimetic, bioinspired and biokleptic research. *Soft Matter* 2012;**8**:6675–9.
77. Galloway JM, Arakaki A, Masuda F, Tanaka T, Matsunaga T, et al. Magnetic bacterial protein Mms6 controls morphology, crystallinity and magnetism of cobalt-doped magnetite nanoparticles in vitro. *J Mater Chem* 2011;**21**:15244–54.
78. Galloway JM, Bird SM, Bramble JP, Critchley K, Staniland SS. Biotemplating magnetic nanoparticles on patterned surfaces for potential use in data storage. *MRS Proc* 2013;**1569**.
79. Galloway JM, Bramble JP, Rawlings AE, Burnell G, Evans SD, Staniland SS. Nanomagnetic arrays formed with the biomineralization protein Mms6. *J Nano Res* 2012;**17**:127–46.
80. Wang L, Schultz PG. Expanding the genetic code. *Chem Comm* 2002;1–11.

80a. Kolinko I, Lohße A, Borg S, Raschdorf O, Jogler C, Tu Q, et al. Biosynthesis of magnetic nanostructures in a foreign organism by transfer of bacterial magnetosome gene clusters. *Nat Nanotechnol* 2014;**9**:193–7. http://dx.doi.org/10.1038/nnano.2014.13.
81. Scott RWJ, Wilson OW, Crooks RM. *J Phys Chem B* 2005;**109**:692–704.
82. Myers VS, Weir MG, Carino EV, Yancey DF, Pande S, Crooks RM. Dendrimer-encapsulated nanoparticles: new synthetic and characterization methods and catalytic applications. *Chem Sci* 2011;**2**:1632–46.
83. Vögtle F, Richard G, Werner N. *Dendrimer chemistry: concepts, syntheses, properties, applications*. Weinheim: Wiley-VCH; 2009.
84. Campagna S, Ceroni P, Puntoriero F, editors. *Designing dendrimers*. New York, NY: Wiley; 2011.
85. Caminade A-M, Turrin C-D, Laurent R, Ouali A, Delavaux-Nicot B, editors. *Dendrimers: towards catalytic, material and biomedical uses*. Chichester: Wiley; 2011.
86. Baars MWPL, Meijer EW, editors. *Dendrimers II: host-guest chemistry of dendritic molecules*. Berlin: Springer; 2000.
87. Lee H, Ooya T. *Chem Eur J* 2012;**18**:10624–9.
88. Majoros IJ, Baker Jr JR, editors. *Dendrimer-based nanomedicine*. Singapore: Pan Stanford Publishing; 2008.
89. Boas U, Christensen JB, Heegaard PMH. *Dendrimers in medicine and biotechnology. New molecular tools*. Cambridge: RSC Publishing; 2006.
90. Mignani S, El Kazzouli S, Bousmina M, Majoral J-P. *Prog Polym Sci* 2013;**38**:993–1008.
91. Dendritech Inc. *via* Sigma–Aldrich supplies worldwide PAMAM Dendrimers from generations 0 to 10 with 5 different core types and 10 functional surface groups.
92. Garcia-Martinez JC, Crooks RM. *J Am Chem Soc* 2004;**126**:16170–8.
93. Bronstein LM, Shifrina ZB. *Chem Rev* 2011;**111**:5301–44.
94. Sathish Kumar K, Amutha R, Arumugam P, Berchmans S. *ACS Appl Mater Interfaces* 2011;**3**:1418–25.
95. Scott RWJ, Datye AK, Crooks RM. *J Am Chem Soc* 2003;**125**:3708–9.
96. Scott RWJ, Sivadinarayana C, Wilson OM, Yan Z, Goodman DW, Crooks RM. *J Am Chem Soc* 2005;**127**:1380–1.
97. Wilson OMS, Scott RWJ, Garcia-Martinez JC, Crooks RM. *J Am Chem Soc* 2005;**127**:1015–24.
98. Hoover NN, Auten BJ, Chandler BD. *J Phys Chem B* 2006;**110**:8606–12.
99. Chung Y-M, Rhee H-K. *J Mol Catal A Chem* 2003;**206**:291–8.
100. Knecht MR, Garcia-Martinez JC, Crooks RM. *Langmuir* 2005;**21**:11981–6.
101. Knecht MR, Garcia-Martinez JC, Crooks RM. Synthesis, characterization, and magnetic properties of dendrimer-encapsulated nickel nanoparticles containing <150 atoms. *Chem Mater* 2006;**18**:5039–44.
102. Knecht MR, Crooks RM. Magnetic properties of dendrimer-encapsulated iron nanoparticles containing an average of 55 and 147 atoms. *New J Chem* 2007;**31**:1349–53.
103. The size calculation for monometallic nanoparticles is based on the equation $n=4\pi r^3/3V_g$, where n is the number of metal atoms, r is the radius of the nanoparticle, and V_g is the volume of one atom derived from the molar volume.
104. Bozorth RM. *Ferromagnetism*. New York, NY: D. Van Nostrand Company, Inc.; 1951
105. Lesli-Pelecky DL, Rieke RD. *Chem Mater* 1996;**8**:1770–83.
106. Ma Z, Wu R, Han Q, Chen R, Gu Z, Korean J. *Chem Eng* 2011;**28**:717–22.
107. Wang SH, Si X, Antwerp MV, Cao Z, Swanson SD, Bi X, et al. *Adv Funct Mater* 2007;**17**:3043–50.

108. Choi HC, Kim W, Wang D, Dai H. *J Phys Chem B* 2002;**106**:12361–5.
109. Goubert-Renaudin SNS, Wieckowski A. *J Electroanal Chem* 2011;**652**:44–51.
110. Atwater JE, Akse JR, Holtsnider JT. *Mater Lett* 2008;**62**:3131–4.
111. Mitran E, Dellinger B, McCarley RL. *Chem Mater* 2010;**22**:6555–63.
112. Wu L, Ling J, Wu Z-Q. *Adv Synth Catal* 2011;**353**:1452–6.
113. Wu L, Zhang X, Tao Z. *Catal Sci Technol* 2012;**2**:707–10.
114. Peng X, Pan Q, Rempel GL. *Chem Soc Rev* 2008;**37**:1619–28.
115. Hultgren R, Orr RL, Anderson PD, Kelly KK. *Selected values of thermodynamic properties of metal and alloys*. New York, NY: John Wiley & Sons Inc; 1963.
116. Auten BJ, Hahn BP, Vijayaraghavan G, Stevenson KJ, Chandler BD. *J Phys Chem C* 2008;**112**:5365–72.
117. Chandler BD, Long CG, Gilbertson JD, Pursell CJ, Vijayaraghavan G, Stevenson KJ. *J Phys Chem C* 2010;**114**:11498–508.
118. Luo J, Maye MM, Petkov V, Karuiki NN, Wang L, Kjoki P, et al. *Chem Mater* 2005;**17**:3086–91.
119. Hills CW, Mack NH, Nuzzo RG. *J Phys Chem B* 2003;**107**:2626–36.
120. Garcia-Martinez JC, Scott RWJ, Crooks RM. *J Am Chem Soc* 2003;**125**:11190–1.
121. Wilson OM, Scott RM, Garcia-Martinez JC, Crooks RM. *Chem Mater* 2004;**16**:4202–4.
122. Bai L, Wan H, Street SC. *Colloid Surfaces A* 2009;**349**:23–8.
123. Wan H, Shi S, Bai L, Shamsuzzoha M, Harrell JW, Street SC. *J Nanosci Nanotechnol* 2010;**10**:5089–92.
124. Gates AT, Nettleton EG, Myers VS, Crooks RM. *Langmuir* 2010;**26**:12994–9.
125. Castillo VA, Kuhn JN. *J Phys Chem C* 2012;**116**:8627–33.
126. Marvin KA, Thadani NN, Atkinson CA, Keller EL, Stevenson KJ. *Chem Commun* 2012;**48**:6289–91.
127. Sattler K, Mühlback J, Recknagel E. Generation of metal clusters containing from 2 to 500 atoms. *Phys Rev Lett* 1980;**45**:821–4.
128. Dietz TG, Duncan MA, Powers DE, Smalley RE. Laser production of supersonic metal cluster beams. *J Chem Phys* 1981;**74**:6511–2.
129. Baker SH, Thornton SC, Edmonds KW, Maher MJ, Norris C, Binns C. Characterisation of a gas aggregation source for the preparation of size-selected nanoscale transition metal clusters. *Rev Sci Instr* 2000;**71**:3178.
130. Barborini E, Siviero F, Vinati S, Lenardi C, Piseri P, Milani P. A portable ultrahigh vacuum apparatus for the production and *in situ* characterization of clusters and cluster-assembled materials. *Rev Sci Instr* 2002;**73**:2060–6.
131. Bouwen W, Thoen P, Vanhoutte F, Bouckaert S, Despa F, Weidele H, et al. Production of bimetallic clusters by a dual-target dual-laser vaporization source. *Rev Sci Instrum* 2000;**71**:54–8.
132. Haberland H, Karrais M, Mall M, Thurner Y. Thin films from energetic cluster impact: a feasibility study. *J Vac Sci Technol A* 1992;**10**:3266–71.
133. Siekmann HR, Luder Ch, Faehrmann J, Lutz HO, Meiwes-Broer K-H. The pulsed arc cluster ion source (PACIS). *Z Phys D* 1991;**20**:417–20.
134. Methling RP, Senz V, Klinkenberg D, Diederich T, Tiggesbaumker J, Holzhuter G, et al. Magnetic studies on mass-selected iron particles. *Eur Phys J D* 2001;**16**:173–6.
135. Briehl B, Urbassek HM. Monte Carlo simulation of growth and decay processes in a cluster aggregation source. *J Vac Sci Technol A* 1999;**17**:256–65.
136. Granqvist CG, Buhrman RA. Ultrafine metal particles. *J Appl Phys* 1976;**47**:2200–19. Milani P, Iannotta S. *Cluster beam synthesis of nanostructured materials*. Berlin: Springer; 1999.
137. Binns C. Nanoclusters deposited on surfaces. *Surf Sci Rep* 2001;**44**:1–50.

138. Wegner K, Piseri P, Tafreshi HV, Milani P. Cluster beam deposition: a tool for nanoscale science and technology. *J Phys D* 2006;**39**:439–59.
139. Binns C. Magnetism in nanoclusters. In: Nalwa Hari S, editor. *Encyclopaedia of nanoscience and nanotechnology*, vol. 4. Valencia, California: American Scientific Publishers; 2004. p. 899–923.
140. Toennies JP, Vilesov AF. Superfluid helium droplets: a uniquely cold nanomatrix for molecules and molecular complexes. *Angew Chem Int Ed* 2004;**43**:2622–48.
141. Close JD, Federmann F, Hoffmann K, Quaas N. *J Low Temp Phys* 1998;**111**:661.
142. An der Lan L, Bartl P, Leidlmair C, Schöbel H, Jochum R, Denifl S, et al. *J Chem Phys* 2011;**135**:044309.
143. An der Lan L, Bartl P, Leidlmair C, Schöbel H, Jochum R, Denifl S, et al. *Phys Rev B* 2012;**85**:115414.
144. Tiggesbäumker J, Stienkemeier F. *Phys Chem Chem Phys* 2007;**9**:4748.
145. Diederich T, Döppner T, Fennel T, Tiggesbäumker J, Meiwes-Broer K-H. *Phys Rev A* 2005;**72**:23203.
146. Loginov E, Gomez LF, Vilesov AF. *J Phys Chem A* 2011;**115**:7199.
147. Loginov E, Gomez LF, Chiang N, Halder A, Guggemos N, Kresin VV, et al. *Phys Rev Lett* 2011;**106**:233401.
148. Boatwright A, Feng C, Spence D, Latimer E, Binns C, Ellis AM, et al. Helium droplets: a new route to nanoparticles. *Faraday Discuss* 2013;**162**:113–24.
149. Krishnan G, Verheijen MA, ten Brink GH, Palasantzas G, Kooi BJ. Tuning structural motifs and alloying of bulk immiscible Mo-Cu bimetallic nanoparticles by gas-phase synthesis. *Nanoscale* 2013;**5**:5375–83.
150. Bansmann J, Baker SH, Binns C, Blackman JA, Bucher J-P, Dorantes-Dávila J, et al. Magnetic and structural properties of isolated and assembled clusters. *Surf Sci Rep* 2005;**56**: 189–275.
151. Dobrynin AN, Ievlev DN, Verschoren G, Swerts J, Van Bael MJ, Temst K, et al. Atomic-scale modification of hybrid FePt cluster-assembled films. *Phys Rev B* 2006;**73**:104421.
152. Bhattacharyya S, Thanh NT, De Haeck J, Hansen K, Lievens P, Janssens E. Mass-selected photodissociation studies of AlPbn+ clusters (n=7-16): evidence for the extraordinary stability of AlPb10+ and AlPb12+. *Phys Rev B* 2013;**87**:054103.
153. Binns C, Qureshi MT, Peddis D, Baker SH, Howes PB, Boatwright A, et al. Exchange bias in Fe@Cr core–shell nanoparticles. *Nanoletters* 2013;**13**:3334–9.
154. Jianmin B, Wang J-P. High-magnetic-moment core-shell-type FeCo-Au/Ag nanoparticles. *Appl Phys Lett* 2005;**87**:152502.
155. Binns C, Maher MJ, Pankhurst QA, Kechrakos D, Trohidou KN. Magnetic behaviour of nanostructured films assembled from pre-formed Fe clusters embedded in Ag. *Phys Rev B* 2002;**66**:184413.
156. Binns C, Maher MJ. Magnetic behaviour of thin films produced by depositing pre-formed Fe and Co nanoclusters. *New J Phys* 2002;**4**:85.1.
157. Allia P, Coisson M, Tiberto P, Vinai F, Knobel M, Novak MA, et al. *Phys Rev B* 2001;**64**:144420.
158. Chudnovsky EM. Random anisotropy in amorphous alloys. In: Fernandez-Baca JA, Ching Wai-Yim, editors. *The magnetism of amorphous metals and alloys*. Singapore: World Scientific; 1995, Chapter 3.
159. Baker SH, Roy M, Gurman SJ, Louch S, Bleloch A, Binns C. Atomic structure of Fe clusters embedded in matrices. *J Phys Condens Matt* 2004;**16**:7813.

160. Baker SH, Asaduzzaman AM, Roy M, Gurman SJ, Binns C, Blackman JA, et al. Atomic structure and magnetic moments in cluster-assembled nanocomposite Fe/Cu films. *Phys Rev B* 2008;**78**:014422.
161. Baker SH, Roy M, Louch S, Binns C. Atomic structure in magnetic cluster assembled Fe/Co films as determined from extended absorption fine structure. *J Phys Condens Matt* 2006;**18**:2385.
162. Baker SH, Roy M, Thornton SC Binns C. Realizing high magnetic moments in fcc Fe nanoparticles through atomic structure stretch. *J Phys Condens Matter* 2012;**24**:176001.
163. Parent F, Tuaillon J, Stern LB, Dupuis V, Prevel B, Perez A, et al. Giant magnetoresistance in Co-Ag granular films prepared by low-energy cluster beam deposition. *Phys Rev B* 1997;**55**:3683–7.
164. Binns C, Edmonds KW, Baker SH, Maher MJ, Thornton SC, Upward MD. Magnetic behaviour of exposed and embedded nanoclusters. *Scr Mater* 2001;**44**:1303–7.
165. Laureti S, Peddis D, Del Bianco L, Testa AM, Varvaro G, Agostinelli E, et al. Exchange bias and magnetothermal properties in Fe@Mn nanocomposites. *J Magn Magn Mater* 2012;**324**:3503–7.
166. Binns C, Domingo N, Testa AM, Fiorani D, Trohidou KN, Vasilakaki M, et al. Interface exchange coupling in Co nanoparticles dispersed in a Mn matrix. *J Phys Condens Matt* 2010;**22**:436005.
167. Domingo N, Fiorani D, Testa AM, Binns S, Baker SH, Tejada J. Exchange bias in a superspin glass system of Co particles in Mn matrix. *J Phys D* 2008;**41**:134009.
168. Binns C, Prieto P, Baker SH, Howes PB, Dondi R, Burley G, et al. Preparation of hydrosol suspensions of elemental and core-shell nanoparticles by Co-deposition with water vapour from the gas-phase in ultra-high vacuum conditions. *J Nanoparticle Res* 2012;**14**:1136.

Chapter 4

Magnetic Interactions: A Tool to Modify the Magnetic Properties of Materials Based on Nanoparticles

Davide Peddis*, Petra E. Jönsson[†], Sara Laureti* and Gaspare Varvaro*
**Institute of Structure of Matter, National Research Council (CNR) Via Salaria, Monterotondo Scalo, Rome, Italy*
[†]Department of Physics and Astronomy, Uppsala University, Box 516, Uppsala, Sweden

1 INTRODUCTION

In condensed matter, many physical properties depend on the size of an object when that size is comparable to a dimension relevant to the properties in question. In magnetism, typical sizes, such as the dimensions of magnetic domains or the lengths of exchange coupling interactions, are in the nanometer range. For this reason, on entering the nanometer scale regime, the magnetic properties of condensed matter show substantial differences with respect to the bulk state, leading to new physics[1–3] and applications.[4–8] Among nanostructured magnetic materials, nanoparticles are unique complex physical objects. In fact, at the nanoscale, a multidomain organization is energetically unfavourable and single magnetic domain particles are formed (see Chapter 1, Section 4). A magnetic monodomain particle can be considered as a "superspin" with a magnetic moment, depending on the particle volume, in the range of 10^3–10^5 μ_B.[9,10] Depending on the type (e.g. dipolar interaction, exchange interaction) and the strength of the interactions among the magnetic entities, the magnetic behaviour of an assembly of superspins evolves from ferromagnetic (FM)-like behaviour to paramagnetic-like behaviour, through spin glass like behaviour. Due to the enhanced time and magnetization scale with respect to atomic systems, the suffix super has been used to define these magnetic states, for example, superferromagnetism (SFM),[11] superparamagnetism (SPM)[2] and superspinglass (SSG).[11,12] In addition, the magnetism of nanoparticle assemblies has been often called *supermagnetism*.[9] In nanoparticle-based materials, the exchange interaction between two

Frontiers of Nanoscience, Vol. 6. http://dx.doi.org/10.1016/B978-0-08-098353-0.00004-X

different phases can also strongly influence the physics of the system. As an example, in ferromagnetic (FM) or ferrimagnetic (FMi) magnetic nanoparticles surrounded by an antiferromagnetic (AFM) shell or embedded in an AFM matrix, the exchange interaction across the FM (FMi)/AFM interface gives rise to an additional source of anisotropy, which affects the magnetization reversal process of the whole system, giving rise, after cooling under a magnetic field, to the typical horizontal shift of the hysteresis loop which is usually accompanied by an enhancement of the coercivity (exchange bias, EB).[13–15]

The significant advances in synthetic strategies for the preparation of nanoparticles based materials open up the possibility to grow suitable magnetic nano-architectures alloying to tune interparticle interaction. In this view, both the preparation of nanocomposite materials[16] (e.g. nanoparticles embedded in host matrix or shell) and suitable surface functionalization of the particles are efficient approaches to modulate interparticle interactions.[17]

The nature and strength of magnetic interactions depend both on the volume fraction of the magnetic nano-entities, and on the structural, textural and magnetic properties of the matrix/shell.[18–22] In addition, the possibility to modify the magnetic nature of the matrix/shell enable the introduction of a new source of anisotropy, allowing a further degree of freedom in the design of new materials. In this framework, the aim of this chapter is to describe, by discussing some examples reported in literature, how the ability to synthesize suitable nano-architectures, makes interparticle interactions a fundamental tool to modify magnetic properties of nanoparticles based materials. After a short introduction about general concepts in nanoparticle magnetism (Section 2.1), several kinds of magnetic interactions will be discussed, together with a phenomenological introduction to EB in nanoscaled systems. The first part of the chapter will be concluded presenting some experimental protocols to study interparticle interactions in nanoparticles based materials. In Section 3, the supermagnetic behaviour of the nanoparticles will be described, showing how, by synthesis of a suitable nano-architecture, it is possible to evolve the magnetic behaviour from superparamagnetic to superferromagnetic, through SSG states. The interplay between interparticle interactions and interface exchange coupling (EB) in nanoparticle-based systems will be also discussed. The last section of this chapter will be dedicated to some examples of applications, where the interparticle interaction plays a fundamental role in improving the performances of nanostructured materials. In this part, some perspective about magnetic interacting nanostructures will be given, showing some idealized nano-architectures for specific applications.

2 MAGNETIC INTERACTION IN NANOSCALED SYSTEMS

In condensed matter, the atomic magnetic moments can mutually act together (*cooperative magnetism*) leading to a different behaviour from what would be observed if all the magnetic moments were isolated (*non-cooperative magnetism*). The two basic types of non-cooperative magnetism are diamagnetism and paramagnetism. Diamagnetism is due to the effect of an external magnetic field on the motion of the atomic inner electrons and the magnetic moment induced by the field is very weak and opposite to the field direction. If unpaired electrons are present in the atomic outer shells, the material is paramagnetic. Paramagnetism arises from identical, uncoupled atomic magnetic moments located in isotropic surroundings: in a paramagnetic material, there is no long-range order and, under an external magnetic field, the magnetic moments partially align along the field direction. The paramagnetic susceptibility is much larger than the diamagnetic one and opposite in sign.[23,24] In the cooperative magnetism, the interactions between adjacent magnetic moments determine the magnetic order in the material. Two classes of interactions can be distinguished: *direct exchange* interaction and *indirect exchange* interaction. *Direct exchange* is due to the overlap of the wavefunctions of nearest-neighbour magnetic atoms, whereas *indirect exchange* arises when the atomic magnetic moments are coupled over relatively large distances; it can be also mediated by a non-magnetic ion which is placed in between the magnetic ions (e.g. oxygen atoms in oxides) and in this case it is called *super exchange*. The exchange interaction between magnetic centres in a three-dimensional solid can be quantitatively described by the Heisenberg spin Hamiltonian[23,24]:

$$H_{\text{exch}} = 2\sum_{ij} J_{ij}\vec{S}_i \cdot \vec{S}_j \tag{4.1}$$

where J_{ij} is the exchange integral describing the magnitude of coupling between the spins $\vec{S}_i$ and $\vec{S}_j$. If the exchange integral has a positive value, below a critical temperature T_C (*Curie temperature*), the magnetic moments are aligned parallel to each other and the substance is called FM. Above T_C, the material is no longer FM but becomes paramagnetic because the thermal energy becomes higher than the exchange energy, destroying the magnetic order. If J_{ij} is negative, the spins are aligned antiparallel to each other and below a critical temperature the material is said to be AFM (if all the magnetic moments are equal) or FMi (if the opposed moments have a different magnitude leading to a net moment). This temperature is called Néel temperature (T_N) and T_C for antiferromagnetic and ferromagnetic materials respectively.

Then, magnetic properties in condensed matter can be understood and controlled through magnetic coupling, which is closely related to the chemical composition and crystalline structure of the materials. Minimization of energy provides a basis for predicting the direction of events in physical systems. This is why a ferromagnet, in order to minimize its magnetostatic energy, organizes itself in small regions, with different size and shape, called domains (i.e. uniformly magnetized regions having all atomic magnetic moments oriented in the same direction—see Chapter 1, Section 2). The magnetic domains have such directions so that the sum of overall domain moments is essentially zero.[24] Two adjacent domains are separated by a transition region, called a domain wall (see Chapter 1, Section 3), in which the spin direction gradually rotates, from one domain to the other.

2.1 Magnetism in Nanoparticles: An Introduction

In nanostructures, where the FM material size is reduced below a critical value, a magnetic single-domain state is energetically more favourable with respect to multidomain one. As first calculated by Frenkel and Dorfman, there is a minimum domain size below which a FM material is expected to consist of a single magnetic domain.[25] A more rigorous estimate of the critical radius r_c below which a magnetic nanoparticle with uniaxial anisotropy will be a single domain is given by[26]:

$$r_c \approx 9\frac{(AK_a)^{1/2}}{\mu_0 M_s^2} \tag{4.2}$$

where K_a is the anisotropy constant, M_s is the saturation magnetization, A is the exchange stiffness[10] and μ_0 is the vacuum permeability. r_c depends on the material and is normally in the range of 5–1000 nm. For example, the critical radius for Fe is around 6 nm and for Fe_3O_4 it is around 60 nm, while for $SmCo_5$ it is as large as 750 nm.[9,10] A magnetic monodomain particles (in this chapter defined as a particle that is in a state of uniform magnetization at any field[2]) can be considered as a large magnetic unit (superspin for short) with a magnetic moment (μ_n) is proportional to the volume ($\vec{\mu}_n = M_s \times V$) expected in the range 10^3–10^5 Bohr magnetons (μ_B). However, such nanoparticle is also characterized by the instability of magnetization due to thermal agitation. For an isolated spherical particle with a uniaxial anisotropy, this results in the flip of magnetization between two antiparallel easy directions, separated by an energy barrier ΔE_a, that is proportional to the anisotropy constant K and to the volume V. Then, according to the Néel–Brown theory,[1,3] above a certain temperature (blocking temperature, T_B) and on a certain time scale, the particle moment can make a thermally activated transition (see Chapter 1, Section 6). This phenomenon is analogous to paramagnetism but, involving superspins, it is characterized by different time and magnetization scales, and for this reason it is called SPM. The energy of a magnetic particle is dependent on the magnetization direction, and for a uniaxial anisotropy nanoparticle with the external magnetic field (H) aligned along the easy direction, can be written as[3,27,28]:

$$E = K_a V \sin^2\theta - M_s V H \sin\theta \tag{4.3}$$

where θ is the angle between the magnetization vector and the field direction and the other parameters have the same meanings as previously defined.

Below T_B, the particle's moment is blocked and unable to overcome the barrier in the time of a measurement; the system preserves the FM behaviour and the hysteresis is observed.

From Eq. (4.3) the height of the energy barrier can be derived:

$$\Delta E_a = K_a V \left(1 \pm \frac{H}{H_K}\right)^2 \quad \text{where } H_K = \frac{2K_a}{M_s}. \tag{4.4}$$

From Eqs. (4.3)–(4.4) it emerges that the energy barrier ΔE_a, strictly dependent on the volume and on the anisotropy, is a key parameter in order to understand and to tune magnetic properties of nanoparticle-based magnetic materials.

2.2 Magnetic Interactions at the Nanoscale

The magnetic behaviour of an assembly of superspins is strongly affected by different kinds of magnetic interparticle interactions existing among them. Their physics is governed by similar laws to those governing spin interactions in bulk samples, but on a different scale. The magnetic moment of the individual spin is replaced by the giant moment of the susperspin and depending on the distance between the particles (d_i) different interactions mechanism come into play. Here, a brief qualitative discussion of some types of magnetic interaction will be given, with particular attention being paid to the most important interactions at the nanoscale. A complete description of magnetic interactions in bulk and nanostructured systems can be found elsewhere.[9,10,29]

If the particles are in close contact, exchange interactions between the surface atoms of neighbouring particles (direct exchange) may be significant. Interparticle exchange interactions, that can be described by using the Heisenberg spin Hamiltonian (Eq. 4.1), are responsible for range magnetic order.[9] Superferromagnetic behaviour is expected for an assembly of exchange coupled superspins.

In order to avoid aggregation phenomena and to control morphostructural properties, the nanoparticles are often dispersed in a host matrix. If both the matrix and the particles are metallic, RKKY (Ruderman–Kittel–Kasuya–Yosida) interactions[30–32] between nanoparticles, mediated by the spin polarization of the conduction electron in the host matrix, occur and depend on separation as $1/d_i^3$. If the host matrix is insulating, super-exchange interactions can occur via intermediate atoms or ions (e.g. oxygen) depending on the structure and the nature of the matrix and the bonding at the particle–matrix interface.[9]

In nanoparticle assemblies, dipolar interactions are also very important since they affect the magnetic behaviour of the whole system.[3,33,34] The dipolar interaction energy (E_d) between two magnetic dipoles μ_i and μ_j is given by

$$E_d = \frac{\mu_0}{4\pi}\left[\frac{\left(\vec{\mu}_i \cdot \vec{\mu}_j\right)}{r_{ij}^3} - \frac{3\left(\vec{\mu}_i \cdot \vec{r}_{ij}\right)\left(\vec{\mu}_j \cdot \vec{r}_{ij}\right)}{r_{ij}^5}\right] \quad (4.5)$$

where r_{ij} is the vector connecting the centres of the two dipoles and μ_0 is the vacuum permeability. In an assembly of nanoparticles with a random distribution of easy axes with an average magnetic $\langle\mu_n\rangle$ and an average separation distance $\langle d_i\rangle$, the average value of E_d is approximately

$$\langle E_d\rangle \cong \frac{\mu_0\langle\mu^2\rangle}{4\pi d_i^3} \quad (4.6)$$

The order of magnitude of E_d for two magnetic atoms, with magnetic moments of few μ_B, in a crystal lattice, separated by a distance between 0.1 and 1 nm, is around 10^{-23} J ($\cong$1 K). It is clear that the dipolar interaction is too weak to account for the FM or AFM ordering in condensed matter, since most of the magnetic materials have ordering temperature that are much higher. On the other hand, in an assembly of superspins where each nanoparticle has a moment of about 10^3–10^5 μ_B, and the interparticle distance is in the nm range, E_d is of the order of hundreds of Kelvin, thus playing a key role in the magnetic ordering of nanoparticle-based materials.

2.2.1 Interface Interactions

The magnetic interaction between two different magnetic phases at the nanoscale represents an additional source of anisotropy and can be exploited to modify and improve the magnetic properties of a material.[35] An important example is given by the interfacial exchange anisotropy (EB) between an FM and an AFM material, which is manifested by the shift of the hysteresis loop along the magnetic field axis directions.[14,36,37] EB was first observed by Meiklejohn and Bean in oxidized Co nanoparticles exhibiting[13] hysteresis loop displaced along the field axis after cooling the sample from room temperature under unapplied magnetic field. They attributed this phenomenon to the exchange interaction at the interface between the FM Co core and the AFM CoO shell. This interpretation was generally accepted as a phenomenological explanation of EB but many models have been proposed to explain it from a microscopic point of view and many controversial results have been presented in the literatures.[38–53]

In a simple and intuitive picture, the EB phenomenon can be described in terms of an alignment of the AFM spins at the FM–AFM interface parallel to the FM spins occurring during the field cooling procedure. The coupling between the AFM and FM spins at the interface exerts an additional torque on the FM spins, which the external field has to overcome. If a magnetic field is applied at a temperature T so that $T_N < T < T_C$ and the field is large enough, all the spins in the FM will align parallel to H, that is, the FM will be saturated. Meanwhile, the spins in the AFM will remain random (paramagnetic state), since $T > T_N$. When the FM–AFM couple is cooled through T_N, the magnetic order in the AFM is set up. During the cooling the spins at the FM–AFM interface interact with each other by exchange coupling; the first layer of spins in the AFM will tend to align parallel to the spins in the FM (assuming a FM interaction at the interface), while the adjacent layers in the AFM will orient antiparallel to each other, so as to give a zero net magnetization in the AFM. The intuitive spin configuration, for a FM–AFM couple, is shown schematically in Figure 4.1 for different stages of a hysteresis loop. After the field cooling process, the spins in both the FM and the AFM lie parallel to each other at the interface (Figure 4.1, step 2). When the magnetic field is reversed, the spins in the FM start to rotate. However, if the AFM anisotropy, K_{AFM}, is large enough, as is often the case, the spins in the AFM will remain fixed. Consequently, due to the interface coupling, they will exert a microscopic torque on the spins in the FM, trying to keep them in their original position (Figure 4.1, step 3). Thus, the magnetic field required to completely reverse the magnetization in the FM will be higher than the uncoupled case, that is, a higher magnetic field will be required to overcome the microscopic torque exerted by the spins in the AFM. As a result, the coercive field in the negative field branch increases (Figure 4.1, step 4). Conversely, when the magnetic field is reversed back to positive values, the rotation of spins in the FM will be easier than in the a uncoupled FM, since the interaction with the spins in the AFM will now favour the magnetization reversal, that is, the microscopic torque exerted by the AFM will be in the same direction as the applied magnetic field (Figure 4.1, step 5). Therefore, the coercive field in the positive field branch

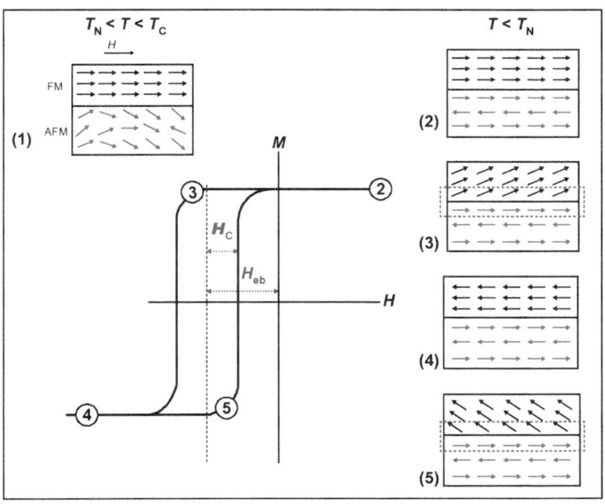

FIGURE 4.1 An intuitive picture of the spin configuration of an FM–AFM couple (AFM having a large anisotropy) at different stages of an exchange biased hysteresis loop.

will be reduced. The net effect will be a shift of the hysteresis loop along the magnetic field axis, usually measured by the so-called EB field $H_{ex} = -(H_{c\text{-right}} + H_{c\text{-left}})$. In other words, the spins in the coupled FM have only one stable configuration (i.e. unidirectional anisotropy).

To fully understand and most importantly to tune the EB represents a challenging topic in the development of nanostructured material. For this purpose, many experimental studies have been carried out on different FM/AFM nanostructured systems[36,43–57] in order to assess the effect of the size reduction on the interface exchange coupling. Indeed, EB models predict that the strength of the EB field depends on the ratio between some characteristic length over which the uncompensated AF moment and the sample dimensions are defined, but an unequivocal interpretation of results is still missing. Experimentally, the properties of Co/CoO nanostructures were examined in many papers such as Refs. 49,58. For square nanostructures, above 200-nm size, it was found that the bias field and the coercive field tend to increase with decreasing size. In other systems, no clear dependence of H_{eb} and H_c on dot size could be established and it was suggested that this was related to a competition between the increased intrinsic coupling energy and the reduced blocking temperature in nanostructures. Recently, Laureti and coworkers[59] presented a systematic study of Co/CoO nanostructures of dimension $l \times 3l$ at small ($\cong$6, 10 nm) and large Co thicknesses ($\cong$23 nm). Samples of three categories were prepared by e-beam lithography, each being characterized by the nanostructure's lateral size: (i) small nanostructures (samples S) (with dimension $30 \times 90\ \text{nm}^2$), (ii) medium nanostructures (samples M) ($60 \times 180\ \text{nm}^2$), and (iii) large nanostructures (samples L) ($120 \times 360\ \text{nm}^2$), respectively (Figure 4.2). As confirmed by magnetic force microscopy analysis,[60] the nanostructures are essentially in the single-domain state, and their rectangular shape guarantees that a unique magnetization direction is defined. After field cooling from room temperature under $H_{appl} = 1$ T, a strong increase in the bias field and the associated coercive field were found in the 5 K hysteresis loop, when going from $l \cong 120$ to $l \cong 30$ nm.

Assuming a continuous AFM single crystalline layer, Malozemoff[38] has shown that the exchange field due to the FM layer acts as a random field in the AFM one. Consequently, correlated regions form in the antiferromagnet with characteristic dimension D_{AF}. The experimental results obtained by Laureti and coworkers indicate that D_{AF}, which governs EB effects, is the nanostructure size l. At larger Co thickness (~23 nm), the EB field does not depend on the nanostructure size, implying that D_{AF} is smaller than the dimensions of the dot. On the other hand, in 25-nm-thick Co nanostructures, the bias field is not dependent on the nanostructure size and it is equal to the bias field found in continuous films.

At the nanoscale, the thickness of the AFM layer (t_{AF}) appears to be fundamental in order to tune the EB, and thus controlling t_{AF} accurately is essential. Recently, Binns and coworkers[47] presented an interesting study on magnetic properties of Fe@Cr core–shell nanoparticles varying Cr shell

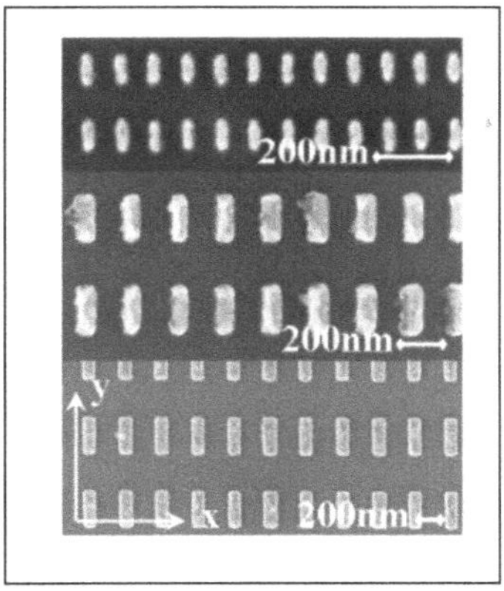

FIGURE 4.2 Scanning electron microscope images of S, M, and L samples. *Reproduced with permission from Ref. 59.*

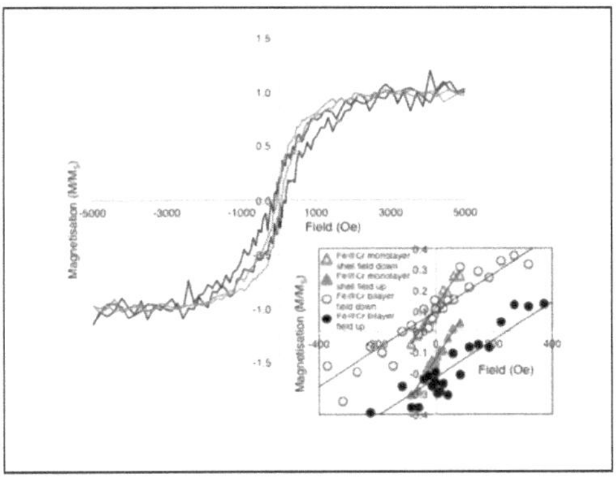

FIGURE 4.3 Magnetization loops at 5 K of Fe@Cr monolayer shell (red line) and Fe@Cr bilayer shell (bold line) nanoparticles in Ag measured by a SQUID after field cooling at +1000 Oe. The inset shows details around zero field and demonstrates clearly the onset of EB and increase in coercivity on increasing the Cr shell thickness from one to monolayer shell (full and empty triangles) and Fe@Cr bilayer shell (full and empty circles) nanoparticles in Ag measured by a SQUID after field cooling at +1000 Oe. The inset shows details around zero field and demonstrates clearly the onset of EB and increase in coercivity on increasing the Cr shell thickness from one to two layers. *Reproduced with permission from Ref. 47.*

thickness between 1 (Fe@Cr_1) and 2 monolayers (Fe@Cr_2). After field cooling, the Fe@Cr_2 sample shows EB, while no EB is observed in Fe@Cr_1 (Figure 4.3). Also an increase of coercivity is observed going from Fe@Cr_1 to Fe@Cr_2. The authors ascribed both these effects to F/AF interface interaction, concluding that at least two Cr monolayers are necessary to create AFM order and then to produce EB.

2.3 Experimental Protocols to Study Magnetic Interactions in Nanostructured Materials

The study of the nature and strength of interparticle interactions is a key point in understanding the physics of nanostructured magnetic materials and for this purpose several experimental approaches have been proposed. This paragraph will focus on two experimental approaches based on the measurement of the remanent magnetization and time dependence of magnetization that, to a first approximation, allow giving an indication about the nature of interparticle interactions and provide semi-quantitative information about their strength.

2.3.1 *Remanence Plot Technique*

Interaction effects in assemblies of nano-objects can be estimated by the field dependence of remanent magnetization measurements. The technique is based on the comparison of the isothermal remanent magnetization curve, IRM, and the Dirent Current Demagnetization remanence curve, DCD. The IRM curve is obtained starting from a totally demagnetized state by applying a positive magnetic field H, which is then removed, and the remanence $M_r^{IRM}(H)$ is measured at zero field; the process is repeated by increasing the field H until the sample reaches the saturation and the remanence takes the saturation value $M_r^{IRM}(\infty)$ (Figure 4.4A, lower part). A DCD experiment is basically similar to that of an IRM one. The samples is saturated, then a small external field in the opposite direction of magnetization is applied and, after a short time is switched off and the $M_r^{DCD}(H)$ is measured. This is repeated, increasing the field until saturation in the opposite direction is reached (Figure 4.4B, lower part); the differentiated remanence curve consists of the derivative of M_r with respect to H and then it is a susceptibility of the irreversible magnetization ($\chi_{irr} = dm_r/dH$). This quantity can be considered as a measure of the energy barrier distribution. In a nanoparticles system, the energy barrier distribution is associated with a distribution of particles' coercivities and is generally called the switching field distribution.[61] In IRM and DCD remanence curves, it is possible to define a remanence coercivity as the field where DCD is equal to zero (H_{cr}^{DCD}) and the field where IRM (H_{cr}^{IRM}) achieves its half maximum value.[61] These quantities are often assumed to be the maximum of the respective switching field distribution.[62]

For an ideal assembly of non-interacting single-domain particles with uniaxial anisotropy that reverse their magnetization by coherent rotation, both $M_r^{IRM}(H)$ and $M_r^{DCD}(H)$ account for the same energy barriers, as clearly expressed by the so-called Wohlfarth relation[63]:

$$m_r^{DCD}(H) = 1 - 2m_r^{IRM}(H) \tag{4.7}$$

where $m_r^{DCD}(H)$ and $m_r^{IRM}(H)$ denote the reduced terms $M_r^{DCD}(H)/M_r^{DCD}(\infty)$ and $M_r^{IRM}(H)/M_r^{IRM}(\infty)$, respectively. Following Eq (4.7), the two irreversible susceptibilities are related by

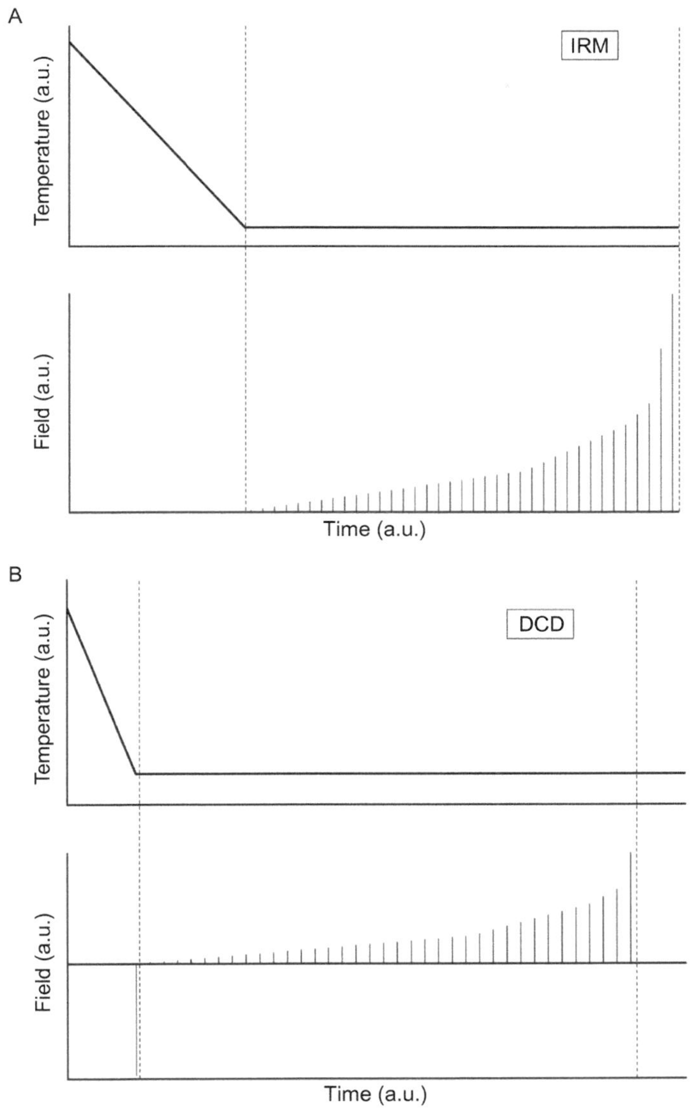

FIGURE 4.4 Schematic representation of IRM (A) and DCD (B) protocols for nanoparticles systems. In Nanoparticles systems both IRM and DCD measurement should be performed after cooling the sample in zero field to temperature lower than blocking temperature. *Reproduced with permission from G. Muscas, Supermagnetismo in Ossidi Ferritici Nanostrutturati, Master Degree thesis in Materials Science at the University of Cagliari.*

$$\left|\frac{\mathrm{d}m_{\mathrm{r}}^{\mathrm{DCD}}}{\mathrm{d}H}\right| = 2\frac{\mathrm{d}m_{\mathrm{r}}^{\mathrm{IRM}}}{\mathrm{d}H} \tag{4.8}$$

with a maximum at the same reverse field. Any deviation from such behaviour can be ascribed to the effect of interactions, which can be quantified in terms of the so-called interaction field[62]:

$$H_{\mathrm{int}} = \frac{H_{\mathrm{cr}}^{\mathrm{DCD}} - H_{\mathrm{cr}}^{\mathrm{IRM}}}{2} \tag{4.9}$$

where $H_{\mathrm{cr}}^{\mathrm{DCD}}$ and $H_{\mathrm{cr}}^{\mathrm{DCD}}$ are previously defined.

Kelly et al.[64] rewrote the Wohlfarth's relationship in order to show more clearly the deviation of a real system from the non-interacting case:

$$\delta m(H) = m_{\mathrm{r}}^{\mathrm{DCD}} - \left[1 - 2m_{\mathrm{r}}^{\mathrm{IRM}}\right] \tag{4.10}$$

where the δm parameter is zero in the absence of interactions. Positive values of δm are due to interactions promoting the magnetized state (exchange and other positive interactions), whereas negative values of δm are due to demagnetizing interactions (e.g. dipole–dipole interactions). A different method to provide information on the nature of the interactions consists in plotting $m_{\mathrm{r}}^{\mathrm{DCD}}(H)$ against $m_{\mathrm{r}}^{\mathrm{IRM}}(H)$ (*Henkel plot*).[65] A negative curvature indicates demagnetizing interactions, while a positive curvature can be attributed to interactions promoting the magnetized state.

It is important to underline that by δm/Henkel plot analysis is just possible to determine what kind of interactions are predominant and affect the magnetic behaviour of the system. In fact, magnetizing and demagnetizing interactions, simultaneously present, contribute in an opposite way to the shape of the curve, thus resulting in a balance between the different contributions.

An interesting examples of the applications of Henkel plots is reported by Zeng and coworkers in the study of chemically ordered fct ($L1_0$) FePt nanoparticles deposited on a silicon substrate.[66] The $L1_0$ $Fe_{50}Pt_{50}$ alloy is characterized by a high saturation magnetization ($M_s = 1100$ emu/cm^3) and a large magnetocrystalline anisotropy ($K_u > 5 \times 10^7$ erg/cm^3), which arise from the peculiar structure of alternating planes of pure Pt and Fe and is not present in the chemically disordered fcc phase.[8,67,68] These characteristics make nanostructured FePt systems attractive for several application as permanent magnet materials and magnetic recording media.[8,69,70] In their paper, Zeng and co-authors[66] present a system of monodispersed FePt nanoparticles produced by high-temperature solution phase decomposition of $Fe(CO)_5$ and reduction of $Pt(acac)_2$, controlling the composition by the molar ratio of the precursors.[70] Depositing the nanoparticle dispersion on a solid substrate and controlling the solvent evaporation lead to FePt nanoparticle assemblies. Thermal annealing results in a phase transformation from the as prepared chemically disordered fcc to the chemically ordered fct structure with high magnetic anisotropy. For this purpose, all the deposited samples were annealed in a flow of an Ar+H_2 (5%) gas mixture at different annealing temperatures (T_a) between 500 and 800 °C for 1 h. It is interesting to observe that increasing T_a leads to a decrease of interparticle distance as evidenced in Figure 4.5A and B, showing FePt nanoparticles deposited in a silicon substrate treated at 530 and 600 °C. In the samples treated at 600 °C, aggregation occurs resulting in a dramatic change of interparticle interactions, as show by the δm plots, reported in Figure 4.5. For an annealing temperature of 550 °C, δm shows a negative peak, indicating that the predominant interparticle interactions are dipolar. This is consistent with the TEM observation that reveals the particles are well separated at this annealing temperature. At T_a=600 °C, a small positive peak appears, and at 700 °C the δm, plot indicates the prevalence of strong exchange interactions between particles. It is interesting to underline that the high-resolution transmission electron microscopy images confirm this picture, showing that at T_a=600 °C the particles not only touch each other but also they share common grain boundaries. The appearance of

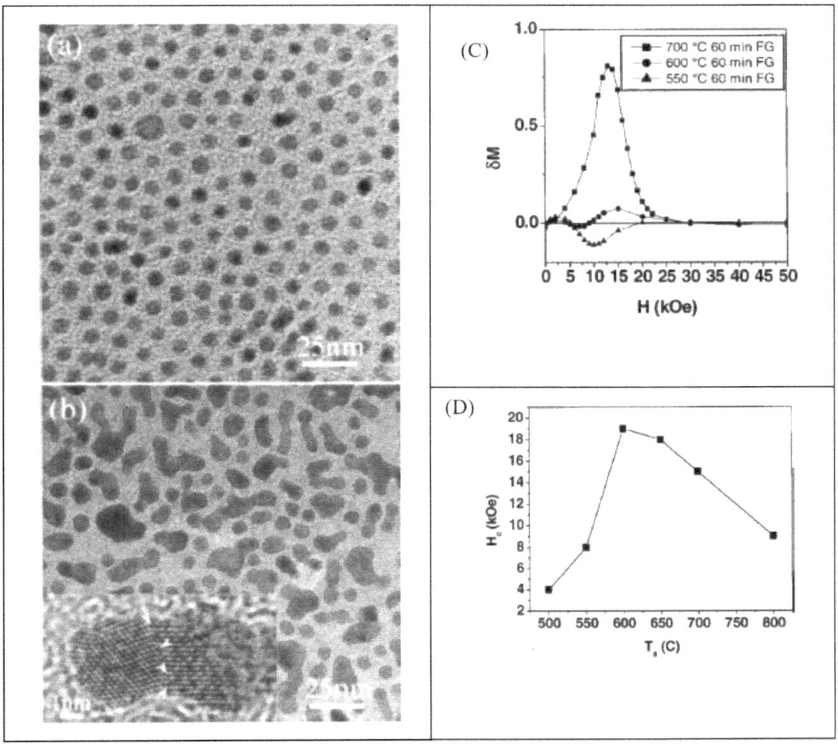

FIGURE 4.5 TEM images of 4 nm FePt nanoparticles deposited on a silicon substrate after annealing to 530 °C (A) and 600 °C (B). Inset (B) high-resolution TEM of two FePt particles; (C) δm plots for samples annealed at 550, 600, and 700 °C, respectively; (D) coercivity as a function of the annealing temperature T_a. *Reproduced with permission from Ref. 66.*

exchange interactions leads to an increase of the coercive field (Figure 4.5D), although the further increase of interactions induces a reduction of H_c. This results show that optimal magnetic hardening is achieved in the highly chemically ordered FePt nanoparticles for thermal treatment at 600 °C. In this system, the magnetization rotation of individual particles yields high coercivity. Moderate exchange coupling caused by moderate aggregation of the particles gives a high remanence ratio and coercive squareness, which is required for some applications (e.g. permanent magnets).[66]

Remanence magnetization curves are also an ideal approach to study ordered nanostructured superlattices (SPls), which represent a great opportunity for making a new generation of tailored materials with completely new physical properties.[71–73] Indeed, the major role in the collective physical behaviour of SPls is played by magnetic interactions among nanoparticles. An interesting examples is given by Cannas *et al.*[74] who prepared spherical aggregates of hexagonal $CoFe_2O_4$ nanoparticles (inset of Figure 4.6(right)) of about 50–60 nm (Figure 4.6(left)) in diameter covered by a 7–8 nm silica shell. TEM images show crystallites with an average size of 7.3 nm (with a polydispersity of 14%) forming twin crystals with shared edges. The structure of the aggregates shows worm-like nanopores with a pore size of about 2–3 nm.[74] The high-resolution images show that parallel lattice fringes are uniformly extended over each aggregate, indicating that primary nanoparticles are organized into an iso-oriented structure. This is further confirmed by the FFT image showing symmetrically aligned spots. The hysteresis curve at 5 K shows a coercivity around 6.6 kOe and a value of 0.75 for the reduced remanent magnetization (M_r/M_s). The M_r/M_s value suggests that particles have a cubic anisotropy,[75] this value being very close to the theoretical value of 0.83.[76] In Figure 4.7A and B are reported the IRM/DCD curves and the irreversible susceptibility obtained from the two curves, respectively. The interaction field calculated according to Eq. (4.8) suggests the presence of weak

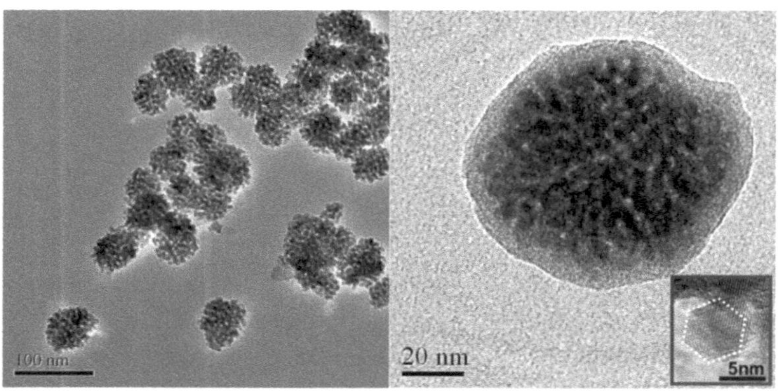

FIGURE 4.6 HRTEM image of the sample at different magnifications. Inset on right side: cubic $CoFe_2O_4$. *Reproduced with permission from Ref. 33.*

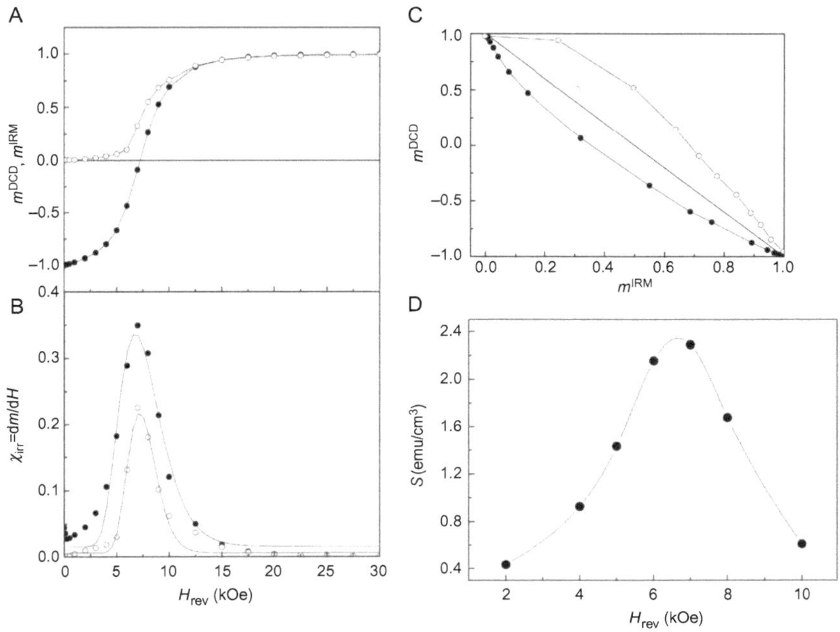

FIGURE 4.7 (A) Remanent magnetization curves (●) measured by DCD (full symbols) and IRM (empty symbols); (B) irreversible susceptibility. (C) Henkel plots experimental data (full circles); calculated data for non-interacting uniaxial Stoner–Wohlfarth particles and (continuous lines) non-interacting particles with cubic anisotropy (empty circles). (D) Viscosity S as a function of the reverse field (H_{rev}). *Reproduced with permission from Ref. 33.*

demagnetizing interactions. It is worth noting that H_{int} is of the same order of magnitude as the maximum dipolar field H_{dip} acting on a mean particle due to nearest-neighbours calculated starting from Eq. (4.9) ($\cong$0.7 kOe). The presence of negative interactions acting among nanoparticles is also confirmed by the Henkel plot (Figure 4.7C), showing that experimental data (full circles) lie below the Wohlfarth linear relationship. This example allows us to discuss an important point in the application of Henkel plots. In fact in the case of particle systems with cubic anisotropy, the Wohlfarth relation could no longer be applicable. In single-domain particle systems with cubic anisotropy, the behaviour of the Henkel plot has been studied by using extensive Monte Carlo (MC) simulations.[77,78] According to the authors, in non-interacting systems, a positive deviation (downwards concave curve) is expected, which is not associated with magnetizing interactions. The effect is counterbalanced by negative dipolar interactions, resulting in a negative curvature (upwards concave curve). At intermediate levels of interparticle interactions, the Henkel plot may resemble a straight line, which can lead to the incorrect definition of a non-interacting system. Therefore, the conclusions drawn from the analysis of the Henkel plots are not exhaustive and further investigation using other methods is required for a proper evaluation of the nature of interparticle interactions.

2.3.2 Relaxation Measurements

In order to have a deeper insight into the nature of interparticle interactions, the determination of the so-called "*activation volume*" (V_{act}) represents a complementary study of the IRM/DCD analysis because it allows one to discriminate between dipolar and exchange interactions. The activation volume is defined as the volume of a material involved in the process of overcoming the energy barrier; in other words, it is the smallest unit reversing its magnetization during a switching process. V_{act} can be determined by magnetization time-dependence measurements.

The time relaxation of magnetization is a well-known phenomenon observed in many systems and arises from the thermally activated decay of metastable states over energy barriers. For a collection of particles, characterized by a distribution of energy barriers (or activation energies), the decay of $M(t)$ is often found to be logarithmic,

$$M(t) = \text{const.} \pm S \ln\left(\frac{t}{t_0}\right) \tag{4.11}$$

where S is the time-dependence coefficient (or viscosity coefficient) and the plus or minus sign describes whether M is increasing or decreasing with time. In general, the viscosity of a real system of particles depends on a number of parameters, such as the reversal mechanism, interparticle interactions, particle volume, anisotropy field and easy axis distributions.[79,80] Moreover, it is a function of the applied field H and the temperature T. At a constant temperature, S increases with H up to a maximum close to the coercive field followed by a monotonic decrease as H is further increased.

In a nanoparticles assembly, the field dependence of magnetic viscosity is recorded at low temperature, in order to avoid superparamagnetic relaxation and provides important information about the switching process of nanoparticles magnetization. In a typical experiment, the sample is brought to a negative saturation field; then a reverse field is applied and the time dependence of magnetization is measured. This experiment is repeated for several reverse field, around the coercivity value, in order to determine the maximum value of viscosity (S_{max}).

S_{max} is related to the fluctuation field H_f defined as

$$H_f = \frac{S_{max}}{\chi_{irr}} \tag{4.12}$$

where χ_{irr} is the irreversible susceptibility of the material.[33] The fluctuation field, first suggested by Néel,[1] is used to describe the effects of thermal activation on the magnetization reversal processes, that is, a field description of k_BT. The fluctuation field H_f allows the calculation of the activation volume,

V_{act}, that, for a single-domain particle with uniaxial anisotropy, is given by the following expression[63]:

$$V_{act} = \frac{k_B T}{M_s H_f} \quad (4.13)$$

where M_s is the saturation magnetization. In the case of nanoparticles with cubic anisotropy, the energy barrier is reduced with respect to the uniaxial anisotropy case by a factor that depends on the sign of the anisotropy constant.[81]

The comparison between the activation volume and the physical volume V provides information about the interactions. For small single domain weakly interacting nanoparticles, the activation volume corresponds to the entire particle physical volume V ($V_{act}(u) = V$), whereas larger values ($V_{act}(u) > V$) are expected if positive exchange interactions are present. Therefore, the determination of activation volume by magnetization time-dependence measurements provide information about the interactions in assemblies of nano-objects.

Relaxation measurements and determination of activation volume in spherical aggregates of hexagonal $CoFe_2O_4$ nanoparticles,[33] discussed in the previous paragraph, complete the description of interparticles interactions in that complex system.

After saturation at 5 T, the time dependence of magnetization was measured for different reverse fields in the range between 2 and 10 kOe. The viscosity coefficient S, whose field dependence is reported in Figure 4.7D, has been calculated by fitting the experimental data using Eq. (4.10). The maximum value of S (2.3 emu cm^{-3}) was found to be at $H_{rev} = -6.5$ kOe, that is, very close to the system coercivity ($h_c = 6.6$ kOe). For this system, the value of V_{act}, calculated considering the correction for a cubic anisotropy is 275 nm^3, corresponding to a cubic particle with an edge of 6.5 nm. This value is very close to the physical volume of the primary nanoparticles inside the aggregates measured using HRTEM images, indicating that exchange interactions are absent despite the intimate contact among particles and the iso-oriented structure. This would be associated with the high degree of porosity of the aggregates, as also evidenced by the high BET specific surface area (160 m^2/g). These spherical porous aggregates of iso-oriented nanoparticles present advantages over 3D SPl such as high magnetic moments, and the improved magnetic properties of small particles such as high anisotropy (H_c (5 K) = 0.66 T).

3 MAGNETIC INTERACTION: A TOOL TO MODIFY MAGNETIC PROPERTIES OF NANOPARTICLES-BASED MATERIALS

In the physics of nanostructured magnetic materials, interactions between nano-objects play an important role. The strong advances in synthetic strategies for growing suitable magnetic nano-architectures able to tune interparticle interactions, introduce the possibility of designing new materials for specific applications. Both preparation of nanocomposite materials (e.g. nanoparticles embedded in host material) and suitable surface functionalization of the particles are efficient approaches to modulate interparticle interactions. The term "nanocomposite", which was first coined by Komarneni and colleagues during the period 1982–1983,[16] refers to materials with more than one solid phase where at least one dimension is in the nanometer range. Indeed, embedding magnetic nano-objects in a magnetic or non-magnetic matrix not only allows one to control the morphological and the structural properties of the material but also represents an effective method to tune interparticle interactions. The nature and strength of magnetic interactions depend both on the volume fraction of the magnetic nano-entities, and on the structural, textural and magnetic properties of the matrix.[18–22] Encapsulation of nanoparticles in an nanometric shell (i.e. a core–shell structure) represents another important class of nanocomposite materials. In this particular kind of nano-architecture, the capability to properly modulate the material and thickness of the shell represents a powerful tool to tune magnetic properties of the materials. Finally, the functionalization of nanoparticle surfaces represents a fundamental step for the application of these materials in the biomedical field (drug delivery, tumour targeting), and as sensors. Indeed, the properties of the magnetic nanoparticles may be tailored by adapting their surface molecular coating in order to meet a variety of requirements (e.g. colloidal stability, bioconjugation).[17,82,83] Nanoparticle surface functionalization can induce important modifications on surface magnetic anisotropy and on interparticle interactions.

In summary, the ability to create suitable nano-architecture allows one to modulate magnetic properties of nanostructured materials by tuning interparticle interactions.[84] In the following section, the dependence of static and dynamic magnetic properties on interparticle interactions will be discussed, showing that these properties can be tuned by the design of a suitable magnetic nano-architecture.

3.1 Supermagnetism

In nanoparticle-based materials, several magnetic regimes can be distinguished with increasing interparticle interactions. For non-interacting particles, a pure superparamagnetic regime is observed that has been already shortly discussed at the beginning of this chapter and in chapter 1, section 6. For weakly interacting particles, a SPM regime in which it is still possible to define the energy barrier for a given particle, is modified by the presence of interactions. This is often called "*modified superparamagnetism*" or "*interacting superparamagnetism*" (ISPM) as discussed in Chapter 1, Section 10. For an intermediate strength of magnetic interaction and in presence of randomness and frustration, spinglass-like features can be observed,[85–90] due to the random distribution of particle moments correlated at long range. This was called a SSG state by analogy with the low temperature collective disordered spin arrangement in spinglasses.[11,12]

A further increase of interparticle interactions can lead to a kind of FM domain state. FM-like correlation will arise between the "supermoments" of the nanoparticles in addition to those between the atomic moments within the particles. The behaviour of nanoparticle in this FM-like state has been called SFM.[11]

Hence, the effects of interparticle interactions of different nature and strengths give rise to a variety of magnetic states such as the SPM, ISPM, SSG and SFM that are collectively referred to as *supermagnetism*.[9]

3.1.1 Interacting Superparamagnetism

Until the early 1990s, the study of the effect of interparticle interactions was restricted to the weak interactions regime. In an assembly of magnetic nanoparticles, this is the state where the interparticle dipolar interaction energy, contributing to the total effective anisotropy energy, is much lower than the individual particle anisotropy energy. The superparamagnetic model basically still holds, provided that the actual contribution of interparticle interactions to the total anisotropy energy is taken into account. In such conditions, the effect of interparticle interactions was considered just a perturbation to the SPM state, leading to the so-called "*interacting SPM*" regime, where the static and dynamical properties of individual particles are partially modified by the interaction with the neighbouring particles.[91] In the pure *SPM* regime, where interparticle interactions are absent, the static properties of the particles are described by the Langevin function. Allia *et al.*[92] proposed a modified Langevin function to account for the presence of weak interparticle interactions (see also Chapter 1, Section 10). The dynamical properties for an assembly of non-interacting nano-entities with uniaxial symmetry were described by the Néel model,[1] which predicts a temperature and anisotropy energy dependence of the relaxation time (τ) according to an Arrhenius law. This law implies a $T \ln t/\tau_0$ scaling in the whole SP regime.[93] The effect of interparticle interactions on the dynamical properties was described by modifying the Arrhenius law for SP systems and replacing it by the phenomenological Fulcher law:

$$\tau = \tau_0 \exp\left(\frac{K_a V}{k_B (T - T_0)}\right) \tag{4.14}$$

where the effect of weak interparticle interactions is accounted for by the temperature T_0, giving an estimation of their strength.[94] A more general description of the dynamical properties was given by Dormanni *et al.*[95] with a model describing the effect of magnetic dipolar interactions on the relaxation time of the particle moments.[34,88]

The possibility to modulate the strength of interparticle interactions in a range in which SPM model basically holds, introduces the interesting possibility to modify dynamical and quasi-static properties of superparamagnetic materials. In this view, embedded nanoparticles in diamagnetic matrix or shell (e.g. SiO_2, Ag) allow to easily tune interparticle interaction. In particular, silica nanocomposites represent an interesting class of nanostructured materials for application in several field (e.g. biomedicine, catalysis). In fact, silica is a good noncytotoxic and biocompatible material; it is optically transparent, chemically inert and protects the surface of the nanoparticles from oxidation.[96,97] In addition, silica surfaces are easy to functionalize[98]; decoration of Silica with functional groups including thiol, amine and carboxylate, facilitates the solubility of the spheres in different solvents. Very dilute magnetic particles/silica nanocomposite can show superparamagnetic blocking with negligible interparticle interactions. As an example, the magnetic behaviour of small $CoFe_2O_4$ nanoparticles ($\langle D_{TEM} \rangle \cong 3$ nm) dispersed in a silica matrix with a weight fraction of 5% will be discussed (Figure 4.8A). Details about synthesis, structural and morphological characterization are reported elsewhere.[99–101] General information about the dynamics of nanoparticle-based materials can be obtained by DC magnetization measurements. Usually, the thermal dependence of the magnetization is studied by zero-field-cooled (ZFC), field-cooled protocols (FC). In the ZFC protocol, the sample is first cooled down from high temperature to 5 K in zero magnetic field, then a static magnetic field is applied and M_{ZFC} is measured during warming up. In a FC measurement, the sample is cooled down to 5 K under a constant magnetic field and M_{FC} is measured during a subsequent warming up. ZFC/FC magnetization measurements, recorded with an applied field of 5 mT, of 3 nm $CoFe_2O_4$ nanoparticles (Figure 4.8B) exhibit a blocking process typical of an assembly of very weakly interacting single-domain magnetic particles with a distribution of blocking temperatures.[101] The ZFC magnetization curves present a maximum at a temperature (T_{max}), which is related to the average blocking temperature ($T_b \propto \beta T_{max}$) where β is a proportionality constant, depending on the type of size distribution. For a log-normal distribution, the

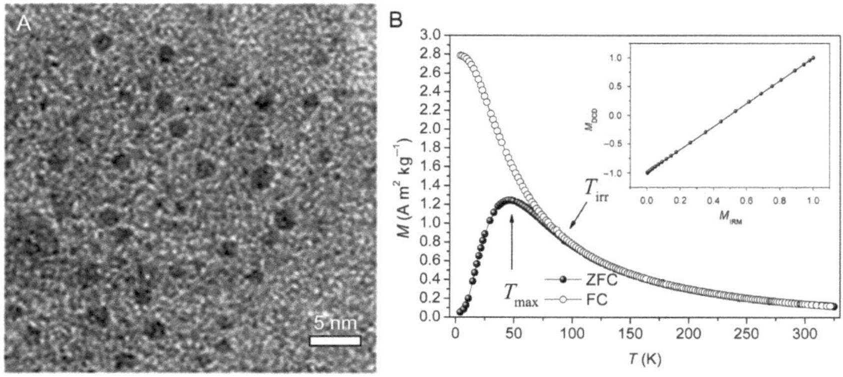

FIGURE 4.8 (A) TEM images of $CoFe_2O_4$ nanoparticles embedded in silica matrix. (B) Zero field cooled (full circles) and field cooled (empty circles) magnetization ($H_{appl} = 5$ mT). Inset: Henkel plot recorded at 5 K.

value of β is typically in the range of 1.5–2.5.[102] The temperature below which the ZFC and FC curves show an irreversible behaviour (T_{irr}) is associated with the blocking of the biggest particles,[103] assuming that the anisotropy energy barrier is determined by the magnetocrystalline anisotropy. Above this temperature a fully superparamagnetic regime is entered. The continuous increase of M_{FC} with decreasing temperature indicates that interparticle interactions, if present, are weak.[104] Henkel plot (inset Figure 4.8B) indicates that in this sample interparticle interaction are not present and the activation volume, determined by relaxation measurements, is equal to the physical particle volume determined by TEM measurements.[101] AC susceptibility measurements indicate that the dynamics of the nanoparticle assembly is well described by the Néel–Brown model,[100] confirming that, for a low density of the magnetic phase, the preparation of silica nanocomposite enables a reduction of interparticle interactions. Embedding nanoparticles in a diamagnetic silica shell thus allow one to regulate interparticle interactions by tuning the thickness of the shell (i.e. interparticle distance). In a recent paper, Cannas and coworkers[105] showed that it is possible to modulate dipolar interaction and the blocking temperature in cobalt ferrite nanoparticles, using a shell of diamagnetic amorphous silica to decrease interactions. $CoFe_2O_4$ oleic acid-capped nanoparticles ($\langle D_{TEM}$ core$\rangle$=8 nm) prepared by high thermal decomposition of a metal organic precursor,[106] were uniformly covered by a silica shell. The presence of oleic acid at the nanoparticle surface keeps them isolated from each other by a coating layer of ~2 nm. The presence of the silica shell increases the interparticle distance up to 30 nm, leading to a decrease of dipolar interactions (Figure 4.9 left side). This picture is well confirmed by DCD/IRM analysis. δm plots (Figure 4.9 right side) indicate the prevalence of dipolar interactions, confirming the morphological features evidenced by TEM: oleic and silica coatings are homogeneous in core and core/shell samples, respectively, preventing any kind of aggregations among nanoparticles. Beyond observing the tendency to lower interparticle interactions in core–shell sample, it is difficult to extract quantitative information form this analysis due to the tendency to cubic anisotropy showed by these particles (M_r/M_s 5 K$\cong$0.9). On the other hand, the decrease of interparticle interactions was corroborated by evaluating the magnetic dipolar energy assuming a point-dipole model (Eq. 4.6) which gives E_{dip}/k_B equal to about 60 and 4 K for core and core/shell samples, respectively. The modulation of interparticle interactions enables the modification of dynamic and quasi-static properties of the materials. The blocking temperature is reduced from 262 to 235 K in core and core–shell nanoparticles, respectively. It is worth noting that for both samples, E_{dip}/k_B is lower than T_{max}, indicating that the transition to superparamagnetic blocking is mainly governed by the magnetic anisotropy energy of the nanoparticles. In addition, the decrease of interactions influences the reversal of magnetization leading to an increase of coercivity in the core–shell sample (H_c core = 1.7 T; H_c core/shell = 1.98 T).[107,108] In summary, for

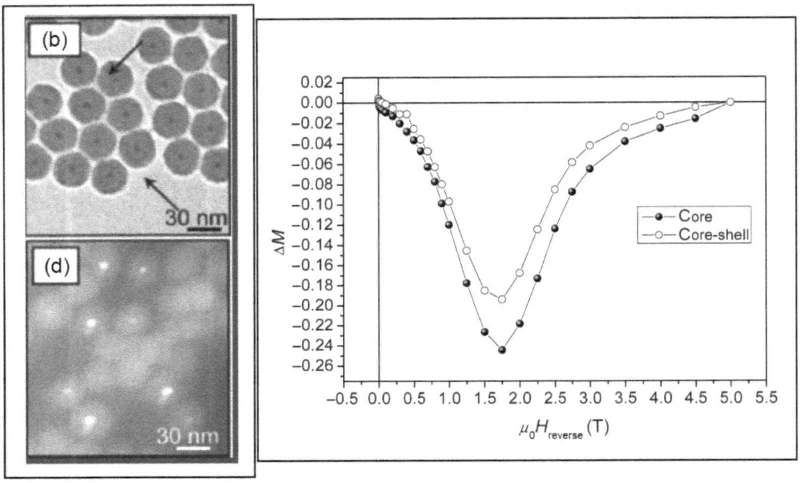

FIGURE 4.9 Left side: bright (b) and dark (d) field image at low magnification of a $CoFe_2O_4$/-SiO_2 core–shell sample. Right side: ΔM plots for $CoFe_2O_4$ core (full circles) and $CoFe_2O_4/SiO_2$ core–shell (empty circles). *Reproduced with permission from Ref. 105.*

relative weakly interacting system, the suitable modulation of interparticle interactions allows one to modulate the blocking temperature and quasi-static properties such as the coercive field.[19,108]

3.1.2 Superspinglass

In a particles system with a sufficiently narrow size distribution, sufficiently strong interparticle interactions, randomness in the distribution of particle positions and orientation of anisotropy axes, a SSG state can be observed. This is characterized by a collective freezing of particle moments below a characteristic glass temperature (T_g). Both the required ingredients of the SG state are also present in the SSG one, that is, randomness and frustration of interactions between particle moments. Being the collective freezing between particle moments rather than between atomic spins, the dynamics in a SSG is much slower, as the microscopic flip time of one superspin (of the order of 10^{-9} s at room temperature and up to 10^{-6} in the frozen state at low temperature) is much longer than an atomic spin flip time (of the order of 10^{-12} s). The growth of dynamical correlation length is slower in a SSG and thus the slower dynamics of these systems is of particular interest because shorter time scale are experimentally accessible.[104]

As a good example, of the several papers presented in literature[34,88–90,95,109–112] on the SSG state and on the analogy between the SG and the SSG state, recently De Toro and coworkers present a study on "*a nanoparticle replica of the spinglass state*".[113] Highly crystalline bare maghemite (γ-Fe_2O_3) nanoparticles with a very narrow size distribution

($\langle D_{TEM}\rangle \cong 8$ nm) were synthesized by thermal decomposition of iron pentacarbonyl followed by oxidation with trimethylamine N-oxide at high temperature.[114,115] In order to obtain an effective random-close packed (RCP) configuration, after a mild homogenization using agate mortar and pebble, the dry powder was uniaxially pressed under approximately 0.7 GPa to form a disc, with a volume filling factor (VFF) of 67%.

Figure 4.10A displays M_{ZFC}/H and M_{FC}/H curves measured under different applied magnetic fields for the RCP sample. The temperature dependence of the low field magnetization (black points in Figure 4.10A) is characterized by a maximum of M_{ZFC}/H and a broad maximum of the M_{FC}/H, followed by a plateau with decreasing temperature. The M_{FC} decreases with magnetic field and becomes weakly temperature dependent at low temperature; ZFC and FC curves merge at lower temperature with increasing field. For comparison, Figure 4.10B shows M_{ZFC}/H and M_{FC}/H curves recorded at different applied fields for $Fe_{0.5}Mn_{0.5}TiO_3$, which can be considered as a canonical spin glass.[116] Nanoparticles mimic the SG, although the RCP system shows a stronger field dependence than the SG (as expected from the larger Zeeman energy term in a NP system). In addition its ZFC curves appear to merge and flatten out at the lowest temperatures to approach $T=0$ K with zero slope (due to the divergence of the individual particle relaxation time), while in the SG the low temperature ZFC M/H curves increase approximately linearly with temperature.[113]

The SG system show a critical behaviour in the temperature dependence of the relaxation time with decreasing temperature, diverging at T_g, according to a power law:

$$\tau = \tau_0 \left(\frac{T_g}{T - T_g} \right)^{-z\nu} \tag{4.15}$$

Where z is the dynamical critical exponent and ν is the critical exponent driving the correlation length, ξ^{ν}. The $z\nu$ value was found to be dependent on the anisotropy of the SG system, as shown for Ising ($Fe_{0.5}Mn_{0.5}TiO_3$; $z\nu = 10.5$) and Heisenberg-like SG ($CdCr_{1.7}In_{0.3}$; $z\nu = 7$).[117] Also, the dynamics in SSG system is well described by a power law with critical exponents having values quite similar to those reported for a SG ($z\nu = 8$–10).[9,44,85,118,119] A critical slowing down analysis of ac data from the RCP sample yields a good fit to Eq. (4.15), with $T_g \cong 140$ and $z\nu \cong 9.5$.

Other ways to provide evidence for a SG/SSG transition are the critical behaviour in the temperature dependence of the non-linear susceptibility (χ_2), diverging at T_g (e.g. Fe_3N,[120] γ-Fe_2O_3,[121] Fe–C[118] nanoparticles) and the so-called dynamic scaling, that is, the collapse onto a master curve of the values of the imaginary component of the dynamic susceptibility (χ'') measured at different frequencies and temperatures.[85,118,121]

Finally, in SG/SSG system the dynamical properties below T_g are characterized by ageing and memory effects, associated with randomness and frustration of the interparticle interactions. These effects can be satisfactorily be

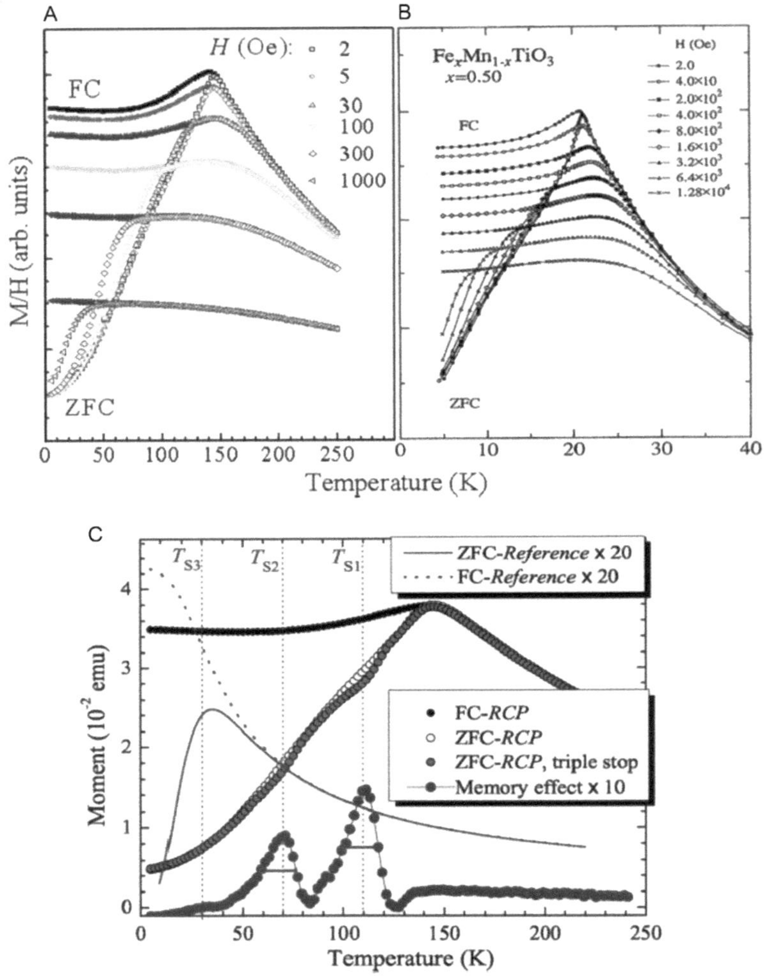

FIGURE 4.10 Comparison of dc field effects between the RCP system (A) and a model spin-glass (B), and (C) ZFC and FC curves for RCP (67% VFF) and reference sample (0.4% VFF) recorded under an applied field of 5 Oe. For the RCP, the sample memory effect is reported as described in the text. *(A) and (B) Reprinted with permission from Ref. 116 and (C) reproduced with permission from Ref. 113.*

explained both by the model of a hierarchical system of energy minima in phase space, by Parisi[122] and by the droplet model of Fisher and Huse.[123] Anyway, in this context, just a phenomenological description will be given, highlighting that many books, papers and reviews have comprehensively described this phenomenon.[124,125]

At high temperature, the spin dynamics in a SSG system is dominated by thermal fluctuation and when the system is cooled down to lower temperatures, the independent spins slow down and form locally correlated units, denoted as clusters, droplets or domains. If the system is further cooled the glassy correlations between the spins freeze at T_f ($>T_g$). The freezing temperature depends on the cooling rate and at this temperature the system is trapped in metastable configurations from which the ground state can only be reached asymptotically slowly. This is the basis for the ageing of a SSG in the glassy phase[121] and for the observation of the so-called memory effect upon ageing, an important signature—sometimes even regarded as a "proof"—of true spin glass behaviour.[9] Figure 4.10C reports ZFC/FC data recorded at low fields for the RCP sample. Memory effects, probed by the so-called "stop-and-wait" experimental protocol during the ZFC process, are observed. In this protocol, the ZFC cooling down is interrupted at a given temperature below T_g, the system is maintained at this temperature for a certain time and then the cooling continues down to the lowest temperature. In this case, three stops for 4 h were performed at $T=110$, 70 and 30 K during the ZFC cooling and a dip at the stop temperature when the magnetization is subsequently recorded upon heating is produced.[113] This is a clear indication of the non-equilibrium dynamics typical of SSG systems.

In order to study the effect of interparticle interactions De Toro and coworkers studied the evolution of magnetic properties, starting from an RCP assembly, of γ-Fe_2O_3/silica core–shell nanoparticles as a function of the shell thickness. All the different dry powders were uniaxially pressed under approximately 0.7 GPa to form five compacted samples, with different maghemite Volume Filling Fraction (VFF) in the range of 67–0.4% (VVF 67, bare particles, 56%, 27%, 16%, 0.4%). The decrease of VVF corresponds to an increase of shell thickness and thus of the interparticle distance (i.e. a decrease of interparticle interactions). As an example, Figure 4.11 left side shows a TEM

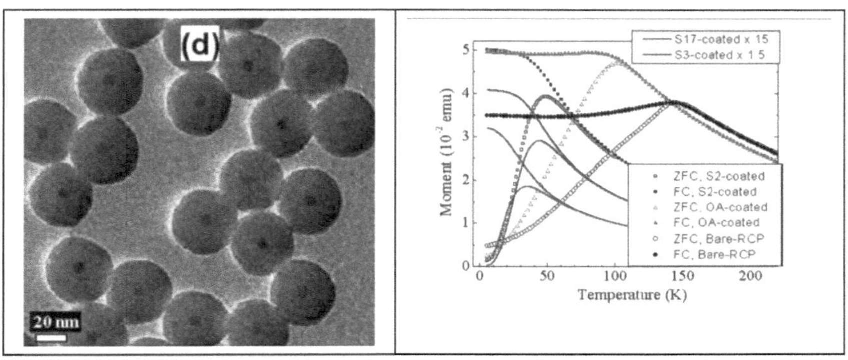

FIGURE 4.11 Left side: Transmission electron micrograph of γ-Fe_2O_3/silica core–shell nanoparticles (VVF 0.4%, thickness of silica shell ≅ 17 nm). Right side: ZFC/FC magnetization curves measured in an applied field of $H=5$ Oe in each sample. *Both figures are reproduced with permission from Ref. 114.*

image of the most dilute sample (VVF 0.4%). The core–shell structure is clearly visible, with a mean shell thickness of ~17 nm. ZFC FC curves of this sample are reported in Figure 4.10C (continuous line) showing a typical blocking process of weakly interacting nanoparticles, in strong contrast with the RCP sample where the data indicates a SSG transition. Figure 4.11 right side shows the ZFC and FC magnetization of the samples with different VVF. Decreasing the shell thickness causes the magnetic properties to evolve form a typical superparamagnetic behaviour (VVF 0.4%) to a SSG in the RCP sample. In the RCP sample, the FC curve flattens out below T_{max} and the FC/ZFC irreversibility begins very close to the ZFC maximum. These qualitative features, together with the increase in T_{max}, reflect the increasingly strong interactions between the maghemite cores due to the decrease of the shell thickness (i.e. interparticle distance). Only the three most concentrated samples (VFF 67%, 53%, 26%) exhibit memory behaviour, indicating a non-equilibrium dynamics typical of SSG systems. Figure 4.12A shows the concentration dependence of the temperature corresponding to the maximum in ZFC. For low values of VFF, a moderate variation of T_{max}, which can still be related to a modified blocking temperature (ISPM), is observed. For a VFF higher than 27%, a steeper sloped linear dependence is found showing a cooperative freezing essentially due to the increase of interparticle interactions. Within this picture, the VFF 27% sample shows a week memory effect and it can be considered as a crossover between the two regimes. It should be underlined that this crossover point will depend on the relative strength of the dipolar interactions.[126] It is interesting to observe that for VFF > 27% T_{max} is closely proportional to the percentage of magnetic phase and thus to the

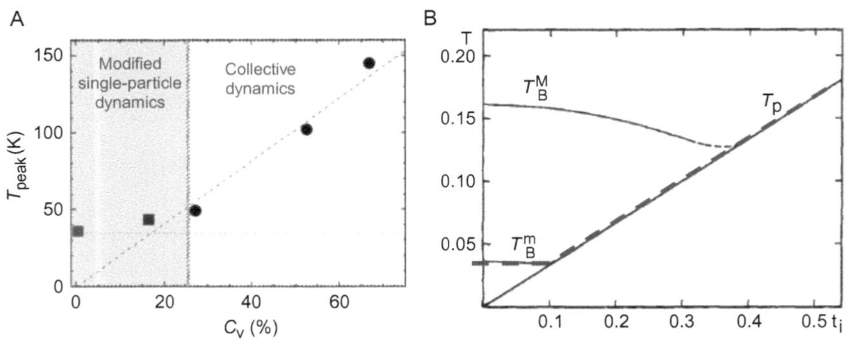

FIGURE 4.12 (A) ZFC peak temperatures of all the samples plotted as a function of maghemite core concentration. (B) The schematic phase diagram proposed illustrating the appearance of an assembly of random nanoparticles as a function of the interaction strength where t ($=k_BT/KV$) is the reduced temperature, $t_i(=\langle E_i^2\rangle^{1/2}/KV)$; T_b^m and T_b^M are the reduced blocking temperatures extracted from magnetization measurements and Mössbauer spectroscopy, respectively. T_p is the transition temperature. *(A) Reproduced with permission from Ref. 114 and (B) reproduced with permission from Ref. 12.*

strength of interparticle interactions. The results obtained by the very complete study of De Tore and coworkers can be partially, qualitatively discussed in the theoretical framework defined by Mørup.[12] Figure 4.12B shows a schematic phase diagram proposed by the Danish scientist to describe the magnetic state, observed on the magnetization time scale and Mössbauer time scale, of an assembly of random nanoparticles as a function of the interactions strength. De Toro and coworkers studied the dynamic of the particles just by magnetization measurements and their results reproduce qualitatively the trend predicted by Mørup for this time scale (dashed line in Figure 4.12B). The study of the magnetic dynamics of these samples in the Mössbauer time scale could be extremely interesting. In summary, the ability to fine tune interparticle distance by preparing core–shell nano-architecture enables the evolution of the sample magnetic behaviour from the SPM to the SSG state.

As already discussed, the magnetic nature of the matrix can strongly influence the magnetic behaviour of the hosted magnetic nano-entities. In this view, a detailed investigation were performed by Binns and coworkers on FM nanoparticles (Fe and Co) embedded in diamagnetic (Ag) and AFM matrix (Cr and Mn) in different proportion.[22,44,47,51,127] In this framework, comparison between magnetic properties of dilute nanocomposites of FM Co nanoparticles dispersed in diamagnetic Ag and AFM Mn matrices, clarifies the effect of the matrix on interparticle interactions. Co particles embedded in Mn (Co@Mn) and Ag matrix (Co@Ag) with ~5% volume filling fraction (VFF) were prepared in thin-film form by co-deposition using a gas aggregation cluster source and a molecular beam epitaxy (MBE) source.[22,44,51] The cluster source produces a log-normal distribution of particle sizes, with a mean size of 1.8 nm, as measured *in situ* by an axially mounted quadrupole filter. The comparison between ZFC and FC curves (Figure 4.13A) provides clear evidence for different behaviour in the two samples. Co@Ag shows an independent blocking of particle moments, with ZFC showing a maximum around 17 K and FC increasing, for decreasing temperature (i.e. Curie-like behaviour). On the other hand, Co nanoparticles embedded in a Mn matrix show correlated freezing of particle moments of the SSG type. Both ZFC and FC show a maximum at ~65 K, with FC that decreases above T_{max} and becomes temperature independent below ~30 K. This landscape is confirmed by AC susceptibility measurements: in Co@Ag, the frequency dependence of T_{max} is well described by an Arrhenius law, whereas in Co@Mn by a power law ($z\nu \cong 8.2$).[44] The Co@Mn sample exhibits a non-equilibrium glassy dynamics, as shown the by the stop-and-wait experiment using the ZFC protocol (Figure 4.13B). At low temperature (5 K), Co@Mn shows EB ($H_{ex} \cong 750$ Oe), together with an increase in coercive field and remanent magnetization. In ZFC, $H_C \cong 1600$ Oe and $M_r/M_s \cong 0.25$, whereas after field cooling $H_C \cong 2250$ Oe and $M_r/M_s \cong 0.5$. A much lower H_C (~275 Oe) was found for Co@Ag at 5 K, for which, as expected, no shift of the hysteresis cycle is observed after FC.[44] The strong

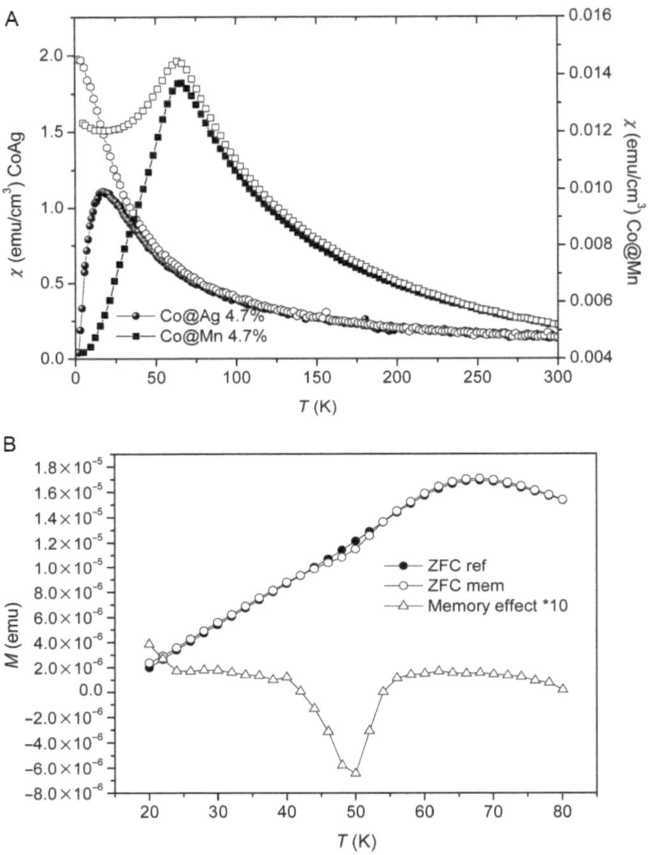

FIGURE 4.13 (A) ZFC (full circles/squares) and FC (empty circles/squares) susceptibility for Co@Ag (circles) and Co@Mn (squares) samples with a Co VFF of 5%. (B) Upper curve: ZFC-reference curve (full circles) and ZFC-memory curve (empty circles squares): for a stop-and-wait of 10.000 s at $T/T_f \approx 0.8$ ($T = 50$ K); lower curve: difference, multiplied by a factor 10, between ZFC-reference and ZFC-memory curve (empty triangles).

difference in magnetic behaviour between the Co@Mn and the Co@Ag sample can be ascribed to the role of the matrix. Interparticle interactions actually originate from the interface exchange interaction between the Co particles and the Mn matrix and propagate through its AFM structure. RKKY and dipole–dipole interactions should play a minor role, as they are also present in the Co@Ag system, which behaves as an assembly of very weakly interacting particles.[44,51] In order to clarify the relative weight of EB and RKKY/dipole interactions role non-equilibrium dynamics in more concentrated Co@Mn (VFF 9.8%) sample was careful investigated.[127,128]

Figure 4.14A shows ZFC magnetization data after a stop-and-wait measurement during cooling for waiting times from 3×10^2 to 3×10^4 s on a

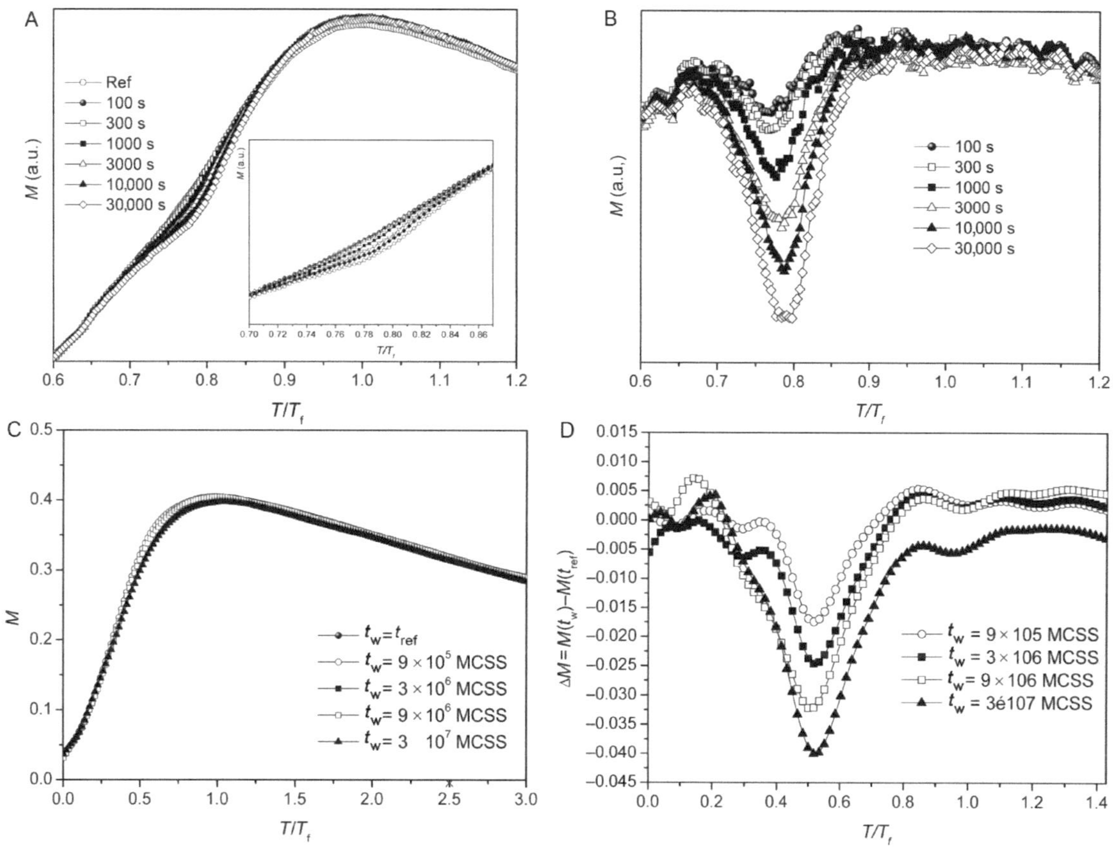

FIGURE 4.14 (A) Experimental data of the ZFC-reference curve (open circles) and ZFC-memory curves for a stop-and-wait between100 and 30,000 s at $T = 55$ K ($\sim 0.8T_f$) for 9.8% VFF of a system of Co nanoparticles in a Mn matrix. (B) Difference between ZFC-reference and ZFC-memory curve with waiting times between 100 and 30,000 s. (C) Monte Carlo simulation of ZFC-reference curve ($t_{ref} = 3 \times 10^3$ MCSS) and ZFC-memory curves after stop-and-wait for different waiting times at $T/T_f = 0.05$ for 9.8% VFF and $H = 0.20$ JC/g μ_B. (D) difference between ZFC-reference and ZFC-memory curve as a function of temperature for different waiting times t_w. *(B) Reproduced with permission from Ref. 127 and (D) reproduced with permission from Ref. 128.*

normalized temperature scale. The difference plot (Figure 4.14B) $\Delta(M/H)=(M/H)_{Mem}-(M/H)_{Ref}$ illustrates the logarithmic character of the slow dynamics typical of SSG system. For a logarithmic spacing of waiting times as those used in the experiments, equally spaced depths in the measured memory dips are observed. In Figure 4.14C, Monte Carlo (MC) simulations of ZFC curves for different waiting times $t_w=300$, 1000, 3000 and 10,000 s are reported on a normalized temperature scale. The differences between the reference and memory curves for different t_w are reported in Figure 4.14D. Simulated ΔT curves confirm the logarithmic character of the slow dynamics. In addition, for a logarithmic spacing of waiting times in our simulations, we obtain equally spaced depths in the calculated memory dips. In summary, in this system the interface exchange coupling between Co FM particles and the AFM Mn matrix plays a very important role in transmitting long-range interparticle interactions. Both dipolar interparticle interactions and the interface exchange coupling with the Mn matrix contribute to stabilize a SSG phase and its dynamical behaviour. Simulations including both contributions reproduce well the memory effect for different t_w. On the other hand, simulations without including dipolar interactions still show the memory effect, although weaker, confirming the role played by interface exchange coupling between FM particles and AFM matrix.[128]

3.1.3 Superferromagnetism

As described in the previous paragraph for an intermediate strength of magnetic interactions, randomness of particle positions and sufficiently narrow size distribution, SSG magnetic behaviour is observed. A further increase of interparticle interactions, prior to the percolation threshold of the system, can lead to a FM-like correlations between "supermoments" (Figure 4.15A). This magnetic state is called superferromagnetism. "*superferromagnetism or superspinglass behaviour?*" was the rhetoric question asked by Mørup and coworker in 1983, regarding the interpretation of the temperature dependence of the magnetic hyperfine field of microcrystalline goethite.[11] The Danish scientist introduced for the first time the term Superferromagnetism, referring to a model based on a modified Weiss mean field theory for these interacting particles. After that, this term has been used with reference to different magnetic systems, although experimental evidence of SFM domains in nanoparticle-based materials is not so common.[9,130,131] On the other hand, theoretical studies of the electron transport properties of SFM nanoparticle-based materials open interesting perspectives. Glatz and coworkers[129,132] did an interesting study on electronic transport properties of FM nanoparticles in the limit of weak coupling. Figure 4.15B upper part shows the paramagnetic/SPM and SPM/SFM transition in a 3D nanoparticle assembly. Figure 4.15B bottom part shows calculated temperature dependence resistivity ρ, in the absence and presence of a magnetic field B aligned with the magnetization of the SFM. For zero magnetic field, the resistivity shows a peak at the SPM/SFM transition.

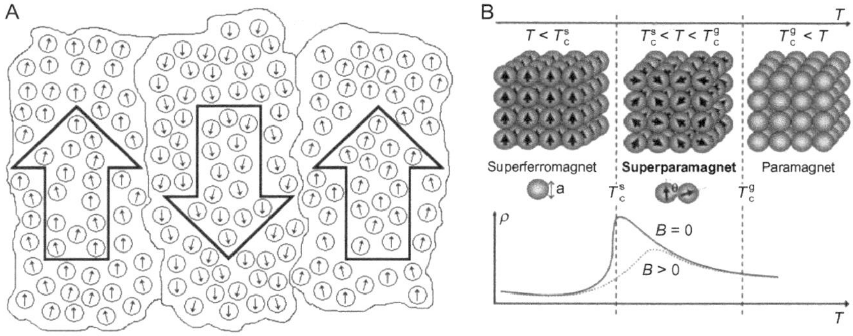

FIGURE 4.15 (A) Sketch of the FM-like correlations, where the magnetic particles are FM coupled in correlated regions, while neighbouring FM regions are antiferromagnetically realigned. (B) Upper part: sketch of a 3*d* nanoparticles assembly showing the different states (PM, SPM and SFM), bottom part: schematic behaviour of the temperature dependence of resistivity (ρ), in the different states, in the absence ($B=0$; continuous line line) and presence ($B>0$; dotted line) of a magnetic field B aligned with the magnetization. *(A) Reproduced with permission from G. Muscas, Supermagnetismo in ossidi ferritic nanostruturati, Master thesis and (B) reproduced with permission from Ref. 129.*

In addition, magnetoresistance shows a double peak in the SFM state, which coalesces into a single peak in the SPM state. These results seem to be applicable to many emerging materials, including artificially self-assembled nanoparticle arrays, where localization effects within the clusters can be neglected.[9]

Among the relative few papers on SFM behaviour in nanoparticle-based materials, an interesting example is given by Sankar and coworker,[130] showing FM-like correlations between nanoparticles in a non-percolated Co–SiO_2 nanocomposite. Two compositions of $Co_x(SiO_2)_{1-x}$ ($x=0.41$, $x=0.32$) granular films (5000 Å thick) were co-sputtered from separate Co (DC) and SiO (RF) targets onto Si(100) substrates mounted on a rotating table. TEM cross-sectional electron micrographs (Figure 4.16A and B) show non-percolated particles for both samples. For $Co_{32}(SiO_2)_{0.68}$, the average particle diameter is 2.5–3.0 nm with 1 nm separation. For $Co_{41}(SiO_2)_{0.59}$, the mean particle size is around 3 nm, while particle separation remains constant (~1 nm). DC magnetization measurements confirm this landscape, indicating the presence of isolated particles. Small-angle neutron scattering (SANS), obtained by saturating the sample in a large magnetic field, shows a peak corresponding to centre-to-centre particle separation of 10 nm. The structural (high-field) data were subtracted from the ZFC data to obtain the magnetic contribution to the scattering and, consequently, the magnetic correlation length.[130] FM-like correlation were observed only for most concentrated nanocomposite, $Co_{41}(SiO_2)_{0.59}$ (Figure 4.16C) in the ZFC condition. The nanoparticles are coupled ferromagnetically in regions of ~55 nm, and neighbouring ordered regions are aligned antiferromagnetically over a correlation length of 160 nm (Figure 4.16D). Any SFM correlation has been observed in the diluted nanocomposite, suggesting that the SFM state originates from

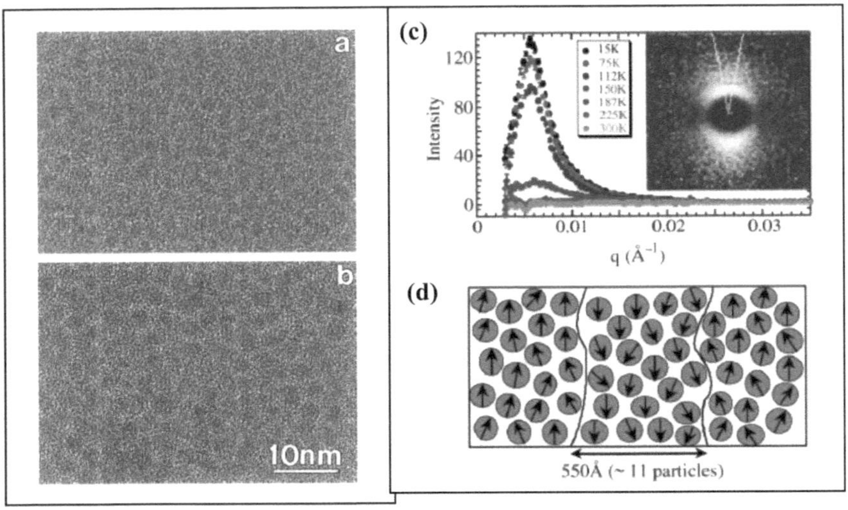

FIGURE 4.16 Cross-sectional electron micrographs of a $Co_x(SiO_2)_{1-x}$ granular film with $x=32$ (A) and $x=41$ (B); (C) sector-averaged intensity versus wave vector q for ZFC magnetic scattering for $Co_{41}(SiO_2)_{59}$ at varying temperatures. (Inset) 2D-projected magnetic SANS scattering for $Co_{41}(SiO_2)_{59}$ at 15 K. (D) Illustration of the ZFC magnetic superferromagnetic state. *(D) Reproduced with permission from Ref. 130.*

dipolar interactions among the Co particles rather than from exchange interactions due to the tunneling electrons. The role of dipolar interactions in SFM order is one the most controversial points in the theoretical framework describing the SFM state and should be discussed. Indeed, in a single-domain nanoparticle assembly, dipolar interactions are preponderant, but these interactions should not be able to generate a SFM ground state because their non-uniform directional dependence. For this reason, a model system to describe a SFM should be a system in which only exchange interactions are present. In a perfect AFM material, the net magnetization is equal to zero because the sublattice magnetizations have identical size but opposite directions, and for this reason, dipolar interaction is equal to zero. However, in nanoparticles, a small net magnetic moment is always present due to the uncompensated spins in the surface and/or the interior of the particles.[56,133–138] Anyway, this uncompensated magnetic moment is usually so small that dipole interactions can be considered negligible with respect to exchange. In this view, the AFM nanoparticles appear to be an ideal system to discuss the theoretical framework of SFM.[11,139,140]

Considering a system of strongly interacting AFM nanoparticles, the magnetic energy of a particle is given by:

$$E = KV\sin^2\theta - \sum_{ij} J_{ij}\vec{S}_i\vec{S}_J \qquad (4.16)$$

In Eq. (4.16), the first term represents the magnetic anisotropy energy, while the second term is due to the interaction energy, where all the factors have the meaning as previously defined. The influence of interparticle interactions may be described using a mean field model, in which the summation in the last term in Eq. (4.16) is replaced by an effective interaction field:

$$E = KV\sin^2\theta - J_{\text{eff}}\vec{M}(T)\cdot\left\langle\vec{M}(T)\right\rangle \tag{4.17}$$

where $M(T)$ represents the sublattice magnetization vector of the particle at temperature T and J_{eff} is an effective exchange coupling constant, such that $J_{\text{eff}}\,M(T)$ is the effective interaction field acting on $M(T)$.

In a recent study, Frandsen and coworkers[139] showed that AFM hematite nanoparticles (α-Fe_2O_3) can, in some cases, be attached with a common orientation such that both the crystallographic and the magnetic order continue across the interface. HRTEM images (Figure 4.17A and B) show individual particles, 5–10 nm in diameter, having rounded shapes without pronounced facets. In addition, HRTEM analysis shows that particles are organized in chains, typically of two to five entities; inside the chains, particles tend to be aligned along their common [001] axis. In most cases, the lattice planes are continued from one particle into its neighbour suggesting an epitaxial growth. X-ray diffraction and neutron diffraction pattern (Figure 4.17C) show broad reflections, due to the crystallographic and the magnetic correlation lengths as described by the Scherrer formula. It is worth noting that the purely magnetic [003] reflection in the NPD pattern is considerably narrower than the other reflections. This shows that the magnetic correlation length in this direction is larger than the particle size: the magnetic (and the crystallographic) correlation extends over several particles. A sketch of the magnetic order is shown in Figure 4.17D. It should be underlined that after gentle grinding, neutron diffraction studies showed that the width of the (003) peak becomes similar to those of the other peaks, indicating that the oriented attachment is destroyed.[139] If particles are attached with a common

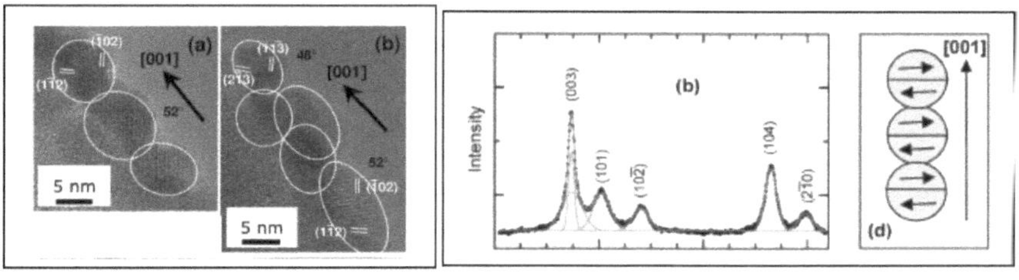

FIGURE 4.17 *Left side*: HRTEM images showing that the particles are attached to one another along their common [001] axis. The arrows and angles indicate the direction of the [001] axis of the particles relative to the plane of the paper. The arrows point into the paper. Figure 4.15: *right side* Neutron diffraction data for α-Fe_2O_3 particles obtained at 20 K. Schematic illustration of the antiferromagnetic correlation across the particle interfaces. *Adapted from Ref. 139.*

orientation, it may be a good first order approximation to assume that the interaction field and the anisotropy field are parallel. Starting from this viewpoint and considering magnetic anisotropy energy negligible compared to the interaction energy, it is possible, from Eq. (4.17), to derive an expression for the ordering temperature of nanoparticles in the SFM state:

$$T_{\mathrm{ord}} \cong \frac{J_{\mathrm{eff}} M^2 (T_{\mathrm{ord}})}{3k_{\mathrm{B}}} \tag{4.18}$$

Detail on the derivation of Eq. (4.18) can be found in Refs. 29,140. This superferromagnetic model has been successfully used by Mørup to describe the magnetic behaviour of interacting AFM nanoparticles.

3.2 Interface Exchange Coupling

In most of the applications in which a key role is played by magnetic nanoparticle-based materials (e.g. magnetic recording, energy transformation), the most important requirement is that the nanoparticles maintain their magnetization stable with time and temperature.

However, with decreasing particle size, the magnetic anisotropy energy becomes comparable to the thermal energy, and superparamagnetic relaxation occurs. In this regard, the question of "*beating the superparamagnetic limit*"[35] represents a key point in the design of new materials for specific applications. As described in Section 2.2.1, the exchange interaction across the ferro (ferri) magnetic and AFM interface (EB) gives rise to an additional anisotropy term, which affects the magnetization reversal process of the whole system, producing significant changes in coercivity and thermal stability of the materials. In this framework, if the comprehension of the size confinement on the FM/AFM exchange coupling mechanism needs further fundamental studies, a much stronger effort should be applied to understand the effect of magnetic interactions between ferro(ferri)magnetic entities on the EB phenomenon.

A fundamental example in this field has been given by Skumryev and coworkers,[35] who demonstrated how the EB phenomenon can be exploited to beat the superparamagnetic limit in a nanoparticle system.

They studied Co/CoO core–shell particles (Figure 4.18) embedded either in a paramagnetic (Al_2O_3) or in an AFM matrix (CoO). The system, obtained by sequential deposition of the matrix and nanoparticles, is made of fcc Co nanoparticles with a roughly spherical core ($\langle D \rangle \cong$ 3–4 nm) and an fcc CoO (Figure 4.18C) shell with a thickness around 1 nm, as shown by HRTEM images (Figure 4.18A and C). ZFC/FC measurements of Co/CoO core–shell nanoparticles (Figure 4.19A, diamonds) embedded in a paramagnetic Al_2O_3 matrix show a blocking process typical of an assembly of very weakly interacting single-domain magnetic particles with a sharp distribution of blocking temperatures whose maximum is around 10 K. The hysteresis loop measured below the superparamagnetic blocking temperature is centred with respect to

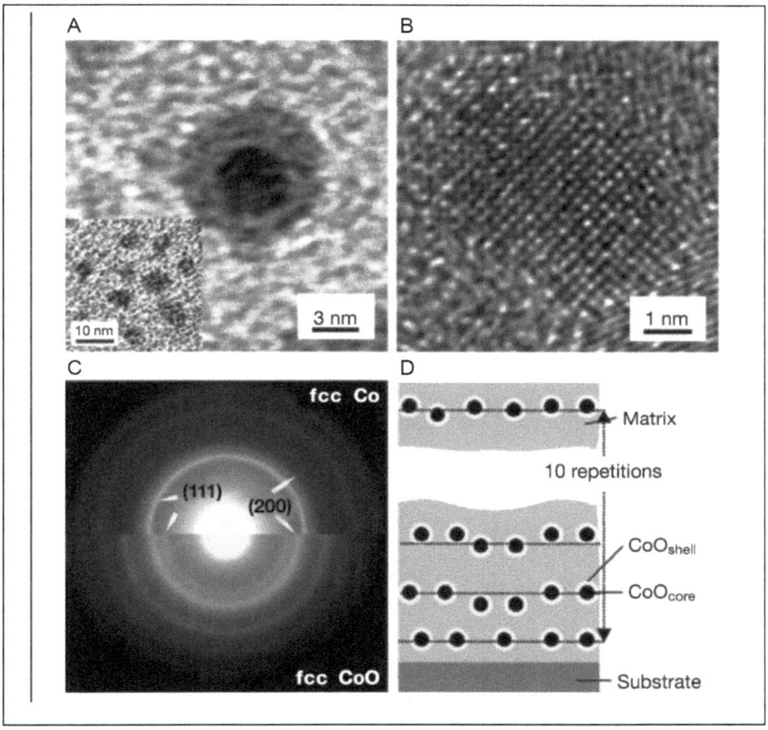

FIGURE 4.18 (A) High-magnification bright field image Co/CoO core–shell nanoparticles, inset: plan view distribution of nanoparticles; (B) high-resolution lattice image of a nanoparticle with a [001] fcc crystallographic orientation; (C) electron diffraction patterns, showing fcc Co and fcc CoO reflections; (D) schematic drawing of the samples cross section, showing Co cores (black), CoO shell (white) and matrix (grey). *Reproduced with permission from Ref. 35.*

the field axis (i.e. no EB is present) with a low value of coercive field. On the other hand, the magnetic behaviour of Co/CoO nanoparticles drastically changes when they are embedded in an AFM CoO matrix. ZFC/FC curves (Figure 4.19A, circles) clearly show that the Co core retains a FM behaviour up to the Néel temperature of CoO (T_N CoO$\cong$290 K). Below T_N, the Co cores are prevented from flipping over the energy barrier due to an extra source of anisotropy. The hysteresis loop after field cooling below T_N in Co/CoO nanoparticles embedded in a CoO matrix highlight the presence of the EB anisotropy, evidenced by the horizontal shift of the loop that supports the magnetization thermal behaviour shown by the ZFC/FC measurements. The improved magnetic stability of Co/CoO nanoparticles when embedded in CoO with respect to Al_2O_3 is further highlighted by the temperature dependence of coercivity and remanence (Figure 4.19B). It is important to underline that in this system the mean interparticle distance is around 12 nm and the calculated maximum dipolar field ($\mu_0 H \cong 0.07$) cannot produce the observed

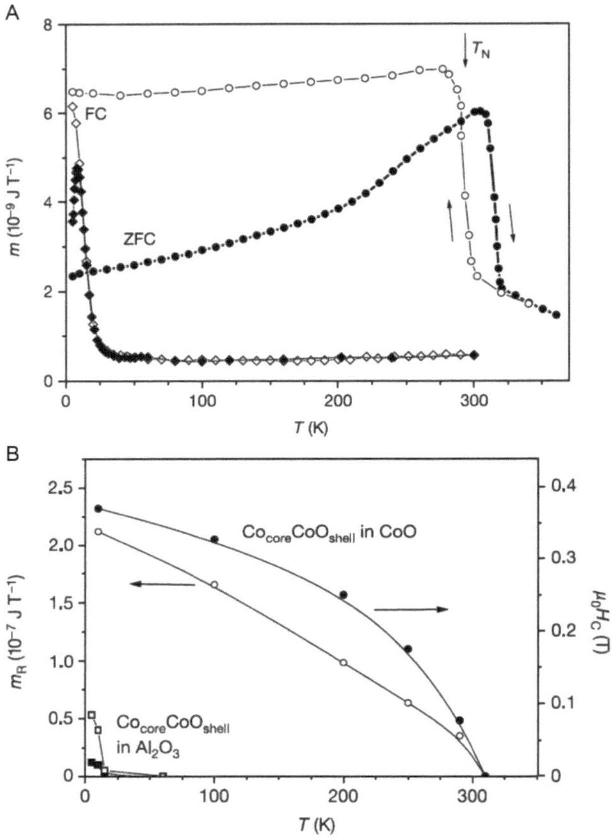

FIGURE 4.19 (A) ZFC (filled symbols) and FC (open symbols) of Co/CoO core–shell nanoparticles embedded in a paramagnetic (Al_2O_3) matrix (diamonds) or in an AFM (CoO) matrix (circles). The measuring field is $\mu_0 H = 0.01$ T; (B) temperature dependence of the coercivity, H_c (filled symbols) and remanence m_r (open symbols), after ZFC, of Co/CoO core–shell nanoparticles embedded in an Al_2O_3 matrix (squares) and CoO matrix (circles). *(B) Reproduced with permission from Ref. 35.*

increase of anisotropy. An interesting approach to the study of the effect of interparticle interactions on the EB phenomenon has been presented by Nogues and coworkers.[141] In this paper, the magnetic properties of Co–CoO core–shell nanoparticles were investigated as a function of the in-plane coverage density of Co/CoO core–shell nanoparticles ($\langle D_{core} \rangle \cong 4$ nm; $\langle t_{shell} \rangle \cong 1$ nm) embedded in an Al_2O_3 matrix. Controlling the nanoparticle deposition conditions produced three samples with different coverage densities of 3.5% (S1), 5.5% (S2), and 15% (S3) A "fully dense" sample, which contained no Al_2O_3 matrix layer, labelled SD, was also grown as a reference. Figure 4.20A and B is shown ZFC/FC curves for samples S1 and S3,

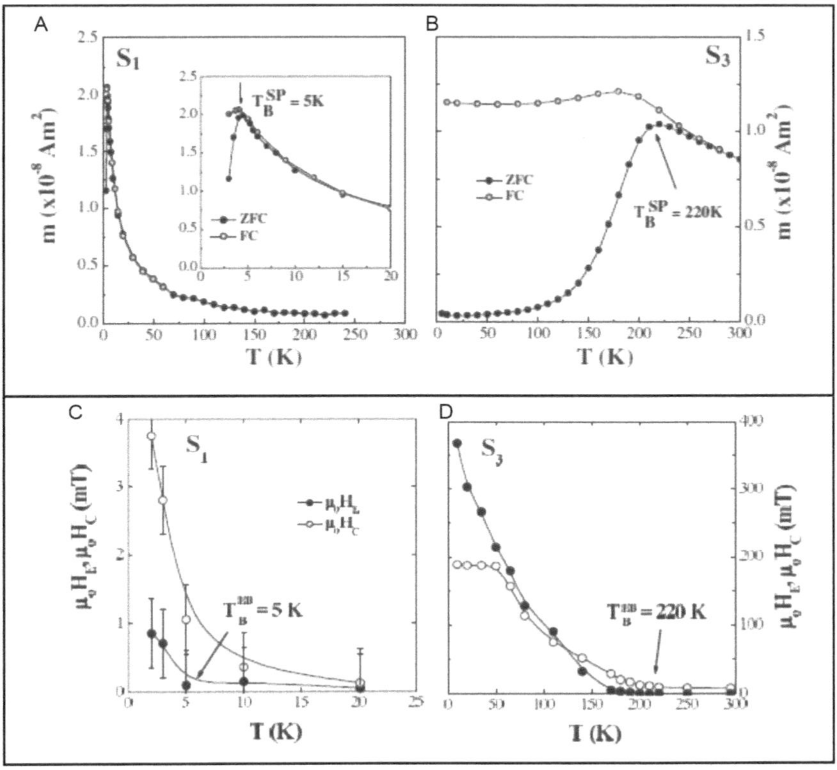

FIGURE 4.20 ZFC and FC magnetization curves ($\mu_0 H_{appl} = 5$ mT) for samples S1 and S3. Shown in the inset of (A) is a low-temperature magnification of the magnetization. Temperature dependence of the coercivity and exchange bias field for S1 (C) and S3 (D) sample. *(C and D) Reproduced with permission from Ref. 141.*

respectively. A strong increase of T_{max} was observed in the sample with the highest nanoparticles coverage ($T_{max} = 220$ K) as compared to sample S1 (3.5% coverage) whose ZFC maximum is at 5 K. The investigation of EB properties shows that both the coercive field and the EB field at low temperature (2 K) are much larger for the sample with the highest Co/CoO core–shell coverage (S3: $\mu_0 H_c \cong 187$ mT; $\mu_0 H_{ex} \cong 367$ mT; S1: $\mu_0 H_c \cong 3.8$ mT; $\mu_0 H_{ex} \cong 0.8$ mT). The same trend is observed in the temperature dependence of EB field. In sample S3, EB appears at around 200 K (Figure 4.20D), while in S1 EB is clearly visible only below 5 K (Figure 4.20C). At first sight, the drastic variation in the T_{max} and in the EB properties can be simply interpreted as due to the interparticle dipolar interactions, which increase with increasing nanoparticle density. A careful comparison with the existing experimental[142,143] and theoretical[144] data, clearly indicates that the dipolar interactions cannot justify the high values of T_{max} and H_{ex}; for this reason, the authors propose that in most diluted sample a strong degradation of CoO shell properties is present while, in the high coverage sample, the exchange interaction between the shells can restore the AFM order, thus affecting the magnetic behaviour (i.e. when the particles are in contact, the CoO thickness is expected to be locally increased up to 2 nm). This picture is confirmed by the magnetic behaviour of the fully dense sample in which the absence of the Al_2O_3 matrix and the close contact between particles, completely restore the AFM order in CoO inducing a strong increase in T_{max} and EB properties (Figure 4.21).

Domingo and coworkers gained some insight into the effect of interparticle interactions on the EB phenomenon by investigating the magnetic properties of Co nanoparticles embedded in a manganese matrix at different concentrations. Samples with a volume filling fraction of 9.8%, 4.6% and 1.3% were prepared by co-deposition using a gas aggregation cluster source and a MBE source. The cluster source produces a log-normal distribution of cluster sizes, with a mean particle size of ~1.8 nm for all the samples. Basic magnetic characterization (ZFC/FC, Henkel plot) clearly indicates that interparticle interactions increase with increasing VFF.[51] The EB field also increases with increasing volume fraction from $H_{ex} \cong 475$ Oe (1.3% VFF) to $H_{ex} \cong 750$ Oe (4.7% VFF) and $H_{eb} \cong 1000$ Oe (9.2% VFF) at 5 K, and it decreases with increasing temperature, vanishing around 30 K. On the other hand, the presence of the EB effect also implies an increase of the coercivity when the samples are field cooled. This increase is of the order of 175% for the sample with the lowest concentration, while it is of the order of 145% for the samples with higher concentrations. In summary, the efficiency of the EB effect on the stabilization of the magnetization is higher for the lower concentration sample, and for higher concentration samples, the magnetic

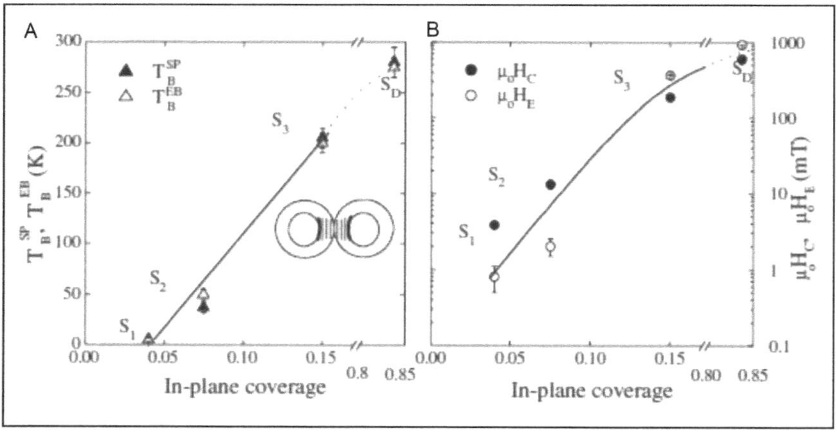

FIGURE 4.21 Dependence of the superparamagnetic and exchange bias blocking temperature (A) and the coercivity and exchange bias on the in-plane nanoparticle density (B). *(B) Reproduced with permission from Ref. 141.*

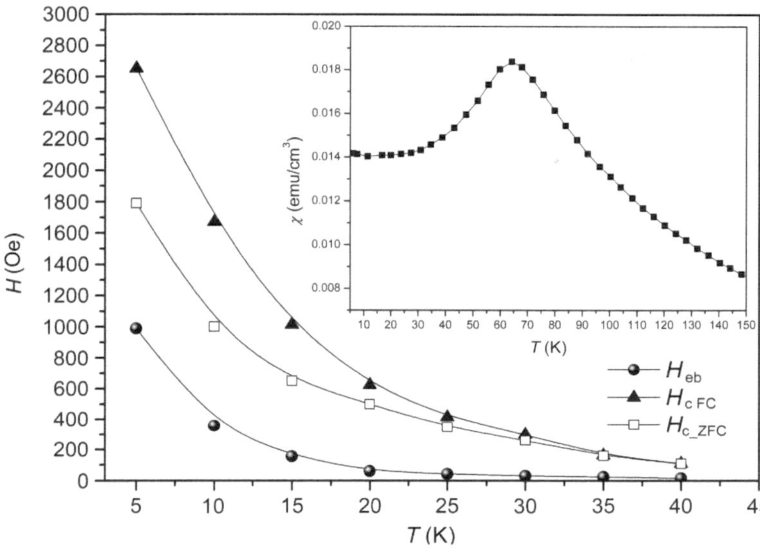

FIGURE 4.22 Temperature dependence of H_{ex}, H_{c_FC} and H_{C_ZFC} for Co@Mn_9.8%; inset: FC susceptibility in the temperature range 150–5 K.

properties become harder due to the stronger effect of long-range interaction of the FM nanoparticles through the AFM matrix. The temperature dependence of H_{ex}, H_{c_FC} and H_{C_ZFC} for the nanocomposites with a volume filling fraction of 9.8% is reported in Figure 4.22. A similar trend or blocking temperature and EB is observed for the most diluted nanocomposite. Equilibrium and non-equilibrium magnetization dynamics of these systems has been discussed in Section 3.1.2, showing a SSG behaviour. EB appears at around ~30 K, the temperature below which the FC magnetization shows a plateau, indicating that interface exchange coupling becomes effective in inducing unidirectional anisotropy when the whole-particle system becomes frozen.[44] EB is accompanied by an enhancement of anisotropy, as indicated by a more rapid increase in H_{c_FC} with respect to H_{c_ZFC} at temperatures below ~30 K. Due to the intrinsic energy structure of a SSG, characterized, according to the Parisi's model,[122,145] by a hierarchical organization of many energy minima separated by energy barriers of different height, EB involving a SSG phase provides an even better possibility of tuning the coercivity, since it is dependent on the energy barrier distribution.[44,146]

4 ROLE OF INTERACTIONS IN NANOSTRUCTURED MATERIALS FOR TECHNOLOGICAL APPLICATIONS

Magnetic nanostructured materials are readily used in a wide range of technological applications including, catalysis sensors, information storage media, biomedical and spintronic devises and are expected to become a key component for the development of novel advanced materials and future devices.[4,5,7,82,147–149]

Existing magnetic nanostructures are characterized by a wide variety of geometries, which cover the range from granular bulk materials to low-dimensional magnetic systems, such as nanoparticles, nanowires nanotubes, granular/continuous thin films and multilayers, dots, antidots, rings, etc. The magnetic behaviour of nanostructured magnetic materials and their performance for specific applications depends on their intrinsic magnetic properties and many other factors, including short- and long-range magnetic interactions, which can occur for example in assemblies of nano-objects or at the interface between two different magnetic phases.

The next two paragraphs will report on the important role played by magnetic interactions in determining the performance of current magnetic recording media for hard disk drives (HDDs), as well as of next generation nanocomposite permanent magnets.

4.1 Magnetic Recording Media

The demand for digital data storage devices is growing continuously and it seems to be endless, coming from every sector of modern society. A number of different storage technologies are currently available, such as HDDs, optical drives, magnetic tapes, flash and solid-state memories and novel technologies are being developed, such as spintronic and race-track memories, among others.[7,150–152,189,190] Each of these technologies has advantages and disadvantages making it more suitable for specific end-market applications in terms of capacity, performance, reliability and cost.

For massive data storage, magnetic recording is still the dominant technology and the HDD (Figure 4.23), which is the key component of this technology, is the most prevalent device with 577 million units sold in 2012. The main advantage of the HDD with respect to its competitors is the combination of a very large capacity and a very low cost per information bit (over 700 Gbits/in^2 with a cost/Gbyte $\sim$ 0.05$).[7,150] In the HDDs, the recording medium, where digital information is stored, consists of a disc-shaped substrate coated with a thin-film structure (Figure 4.24 left part) containing a semi-hard FM layer (recording layer), which can be permanently magnetized, thus allowing the information to be retained over time. In currently available HDDs, the magnetic layer of the recording medium includes a phase-segregated CoCrPt:Oxide granular magnetic thin film with perpendicular anisotropy ($K_u \sim 0.5$ MJ/m^3), which consists of columnar single-domain Co-rich magnetic grains (about 7 nm in diameter) separated by non-magnetic Cr and oxide (e.g. SiO_2) grain boundaries (Figure 4.24 right part).[7,150,154,155] A cluster of multiple grains is used to record a single bit, which is approximately 20 nm long and 100 nm wide in the most recent devices. The recording performance of a HDD has been optimized by simultaneously addressing

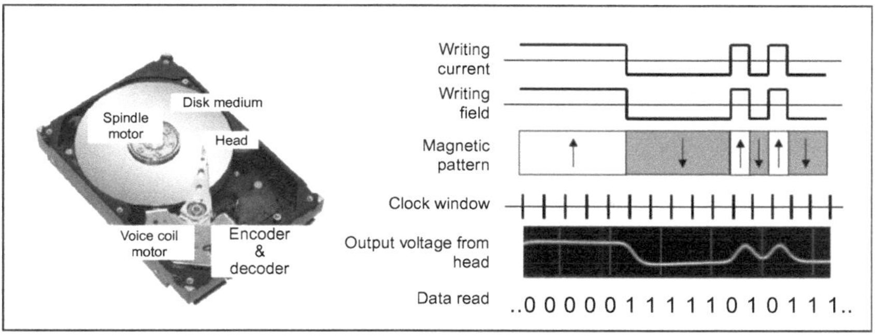

FIGURE 4.23 Left Part: Picture of a HDD showing the most important components. Right Part: Simplified schematic representation of the read/write process: an inductive recording head, which is fed by a pulsed current encoded to represent the data to be recorded, generates a pulsed magnetic field, which can magnetize the underlying region of the recording medium along two opposite directions, thus forming a pattern of opposite magnetized regions, each one representing the binary number "0" and "1"; a read sensor (based on the giant or tunnel magneto-resistance effect) is then used to convert the magnetic field from the recording medium to a voltage pulse, which can be synchronized to a clock pulse thus allowing the recovery of the stored information.

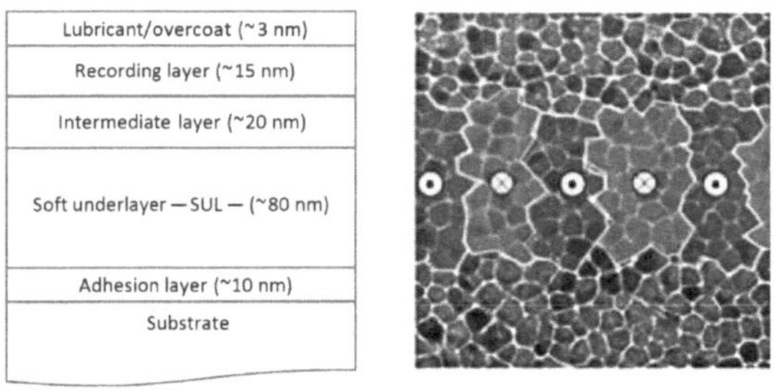

FIGURE 4.24 Left Part: Illustration of a typical multilayer structure of current perpendicular recording media. The adhesion layer (e.g. Al, Ti) improves the adhesion of the layers on the substrate (AlMg alloy or glass); the soft under layer—SUL—(e.g. CoTaZr, FeCoB amorphous alloys) helps to conduct the flux from the writing to the collector pole of the head; the intermediate layer (e.g. Ru) favours the epitaxial growth of the magnetic layer and exchange decouples it from the SUL; the recording layer (CGC CoCRPt-based structure) physically holds the information; the overcoat (DLC) and lubricant (PFPE) provide chemical/mechanical protection and reduces friction and wear during the head-disk contact, respectively. Right Part: TEM planar view of a CoCrPt:SiO_2 granular film, consisting of single-domain Co-rich grains separated by Cr and SiO_2 grain boundaries; red and blue area distinguish bits with opposite perpendicular magnetization corresponding to the binary number "0" and "1". *Reproduced with permission from Ref. 153.*

three different requirements[156,157]: the *signal-to-noise ratio* (SNR), which has to be larger than 25–30 dB to guarantee the retention of the information; the *thermal stability*, which requires $\Delta E_0 > 40\text{–}60\ k_B T_{300}$ for 10 years data recovery, where ΔE_0 is the energy barrier at zero external field between the two magnetic states (magnetization up and down) representing the binary number

"0" and "1" and $k_B T_{300}$ is the thermal energy at room temperature; the *writ ability*, which defines the maximum switching field (H_{sw}) that a storage material can possess in order that its magnetization can be reversed by the write field ($\mu_0 H_w$) generated by currently available write heads ($\mu_0 H_w < B_s$ with $B_{s,max} = 2.4$ T). The above-mentioned parameters depend on many factors, including the intergranular exchange interactions in the recording medium.[7] In particular, the exchange coupling should be decreased as much as possible to reduce the media noise and thus increase the SNR, even if an appropriate amount of intergranular exchange coupling is required for stabilizing the grains against thermal and demagnetizing effects, and then assuring an high thermal stability. To achieve a precise control of the exchange coupling, and thus of the recording performance, CoCrPt-based coupled granular continuous structures (CGCs) are used in the most recent recording media.[158] The CGC structure consists of two exchange coupled layers, that is, a granular layer, with ideally fully decoupled grains, which lead to high-SNR, and a continuous layer, which contributes to improved thermal stability. The CGC structure allows a finely tuned optimization of the overall exchange coupling by separately varying the microstructural and magnetic properties in the granular and continuous layers.

In the granular layer, grain isolation can be controlled by varying deposition parameters,[159,160] changing the amount and nature (crystalline or amorphous matrix) of the segregant materials[161–163] and adjusting the microstructure of interlayers and seed layers.[155] In addition, the structure can be controlled by ion irradiation with magnetic and non-magnetic ions,[164,165] inducing geometrical modification by means of pre-patterned substrates[166] and adjusting the magnetic layer thickness.[154,167,168] A first indication of the strength of the exchange interactions in granular films can be achieved by evaluating, from a hysteresis loop, the coercive squareness $S^* = 1 - M_r/\alpha H_c$, where M_r is the remanent magnetization, H_c is the coercivity and α is the hysteresis slope at the coercive field. It is generally assumed that large (small) values of S^* indicate strong (weak) intergranular exchange interactions. S^* values are typically affected by the demagnetizing field and temperature and also depend on the intrinsic distribution of switching fields, which is in turn a function of the angular distribution of easy axes as well as the dispersion of grain volume and anisotropy constant.[155,159,169–172] Hence, investigation of the S^* parameter alone could lead to some unreliable conclusions on the intergranular coupling.

To better investigate the degree of exchange coupling, the angular dependence of the switching field (i.e. H_{sw} vs. ϕ, where ϕ is the angle between the applied field H and the easy axis EA as defined in the inset of Figure 4.25), can be studied. The angular trend is expected to exhibit a behaviour between the two limits of the Stoner–Wohlfarth (S–W)[27] and Kondorsky[174] models which describe the magnetization reversal of a single and isolated uniaxial particle/grain (*coherent switching*) and of a continuous

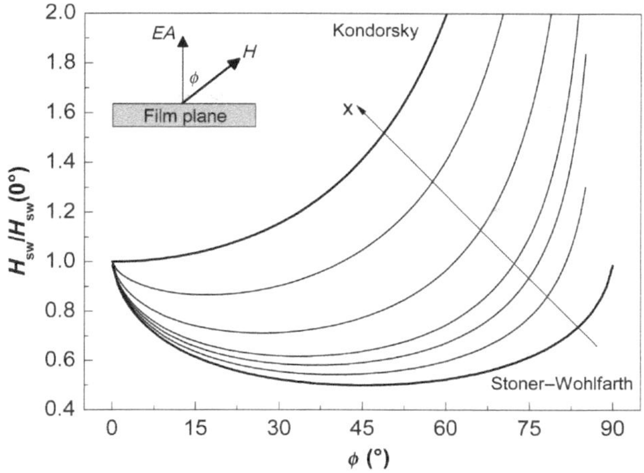

FIGURE 4.25 Angular trend of normalized switching field ($H_{SW}(\phi)/H_{SW}(0°)$) for Stoner–Wohlfarth (S–W, $1/[\cos(\phi)^{2/3} + \sin(\phi)^{2/3}]^{3/2}$ and Kondorsky ($1/\cos(\varphi)$) models of switching, where ϕ is the angle between the easy axis – EA – and the direction of the applied field (see inset). Reversal curves obtained from a weighted average of the two limit switching regimes are also reported ($x/\cos(\phi) + (1-x)H_{SW}(\phi)$), where x is the fraction of intergranularly coupled material). All the curves are calculated for $T = 0\,\text{K}$ and a 0° dispersion of easy axis. *Reproduced with permission from Ref. 173.*

or strongly exchange coupled film (*incoherent switching* by domain wall motion), respectively (Figure 4.25). Deviations of the S–W switching behaviour towards the Kondorsky behaviour can be an indication of the presence of intergranular exchange interactions. In particular, it was demonstrated that a shift of the angle at which the minimum switching field occurs (45° for a S–W system) is due to incoherent reversal induced by the intergranular exchange interactions, while a decrease of the depth of the S–W minimum ($H_{sw,45°} = 0.5H_{sw,0°}$) can arise from thermal activation. Other factors can also influence the angular behaviour such as the angular dispersion of the easy axis, which results in an increase of the minimum switching field.[173]

The best procedure to determine the angular dependence of the switching field consists in measuring at different ϕ values, a series of the so-called easy axis DCD curves by using a non-conventional procedure that allows a more intuitive correspondence between experiments and simple models, as described in Ref. 175. The easy axis DCD curves are collected by measuring at each ϕ angle, the component of the remanent magnetization along the easy-axis direction. From each easy axis DCD curve, the remanence coercivity H_{cr} is determined. According to Ref. 175, the so-deduced H_{cr} is more appropriate to investigate the switching process with respect to the remanence coercivity evaluated by using the conventional procedure where the DCD curves are collected along the field direction (see Section 3.1.1). Indeed, when the conventional procedure is used, H_{cr} is equal to the switching field H_{sw} only along the easy axis, whereas at other ϕ angles, H_{cr} and H_{sw} are not necessarily equivalent. On the other hand, when the non-conventional procedure is adopted, the equivalence between H_{cr} and H_{sw} is expected at each ϕ angle. Information on the intergranular exchange interactions can be also obtained by means of the investigation of the actual reversal volume (i.e. the activation volume V_{act} see Section 3.1.1), which is expected to be larger than the physical grain volume when intergranular exchange coupling interactions are present. Due to the perpendicular anisotropy of the recording layer, in order to compensate for the self-demagnetization fields, the activation volume can be better evaluated by using the waiting time method which is independent on the demagnetizing fields, thus allowing a more reliable and accurate analysis with respect to the more conventional magnetic viscosity method (described in Section 3.1.2) as explained in detail in Ref. 176. This method is based on the determination of the fluctuation field $H_f \cong \partial H/\partial \ln(t)|_{M_{irr}}$ from time-dependent measurements of the irreversible magnetization M_{irr} at different reversing field increments. Thus, at constant M_{irr} value, plots of H versus $\ln(t)$ will be linear and the slope of the variation gives H_f. Since the measurement of the fluctuation field is made at a constant value of the magnetization, the demagnetizing fields are constant and H_f can be determined independently. From the fluctuation field, the activation volume can be determined by using the equation $V_{act} \sim K_B T / \mu_0 H_f M_s$. Comparing the value of the activation and physical volumes provides information about the intergranular exchange coupling interactions.

Many experimental and theoretical studies have been devoted to understand how various relevant parameters affects the grains' interactions and how the latter contribute to the magnetic experimental observations and the recording performance. As an example, here we show how the film thickness can affect the intergranular coupling and the magnetic behaviour of granular thin films. For this purpose, we consider a set of CoCrPt:SiO_2 films deposited on Cr(2.5 nm)/Ru(8 nm)/Ru(12 nm)/HOYA-glass, which differ only in the film thickness (7.5, 10 and 17.5 nm), with the other parameters (e.g. SiO_2 content, grain size, easy axis spread, etc.) being kept constant as discussed in detail in Ref. 154. It has been demonstrated that the growth process leads to the occurrence of a non-uniform microstructure along the thickness, consisting of a very thin initial layer with poor granularity and large exchange coupling, on top of which, a layer with well-separated columnar grains grows; the portion of the region with well-separated grains raises with the increase of the film thickness, thus leading to a different magnetic behaviour for different film thicknesses.[154,167,168]

Figure 4.26A shows room temperature perpendicular hysteresis loops of the above-mentioned CoCrPt:SiO_2 films as a function of the layer thickness. The calculated S^* value reduces from 0.49 ± 0.01 to 0.17 ± 0.01 with the increase of layer thickness, suggesting that, on the whole, the intergranular exchange coupling progressively reduces as a consequence of the increase of the portion with well-separated grains. The occurrence of a non-uniform

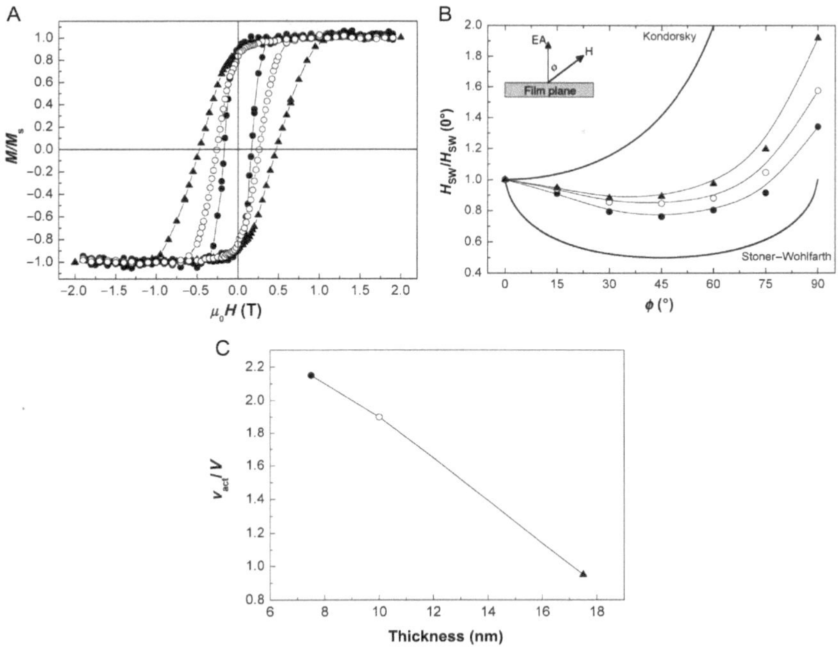

FIGURE 4.26 Magnetic behaviour of granular CoCrPt:SiO_2 films as a function of the layer thicknesses (●: 7.5 nm, ○: 10 nm, ▲: 17.5 nm). (A) Room temperature normalized (M/M_s) perpendicular magnetization loops. (B) Angular dependence of the normalized switching field ($H_{sw}(\varphi)/H_{sw}(0°)$); the theoretical Stoner–Wohlfarth (continuous line) and Kondorsky (dashed line) switching models are also reported. (C) V_{act}/V ratio as a function of the film thickness. *Reproduced with permission from Ref. 154.*

microstructure also affects the reversal mechanism as indicated by the evolution of the angular trend of the switching field with the film thickness (Figure 4.26B). Coherent and incoherent reversal processes coexist in all the samples and the degree of coherence increases with the increase of magnetic layer thickness. Incoherent switching behaviour is due to the initial layer consisting of grains in close contact; the increasing portion of the region with well-separated grains enhances the coherent behaviour with the increase of magnetic layer thickness. The raise of the incoherent switching behaviour with the decrease of film thickness also leads to an enlargement of the activation volume V_{act} with respect to the physical grain volume V as shown in Figure 4.26C. The δM plot technique (see Section 2.3.1 for details) can be also used to evaluate the intergranular interactions. However, for granular thin films with perpendicular anisotropy, neither DCD nor IRM curves can be directly measured by removing the external field (H_{ext}). This is because the actual field (H_{tot}) acting on the grains includes a demagnetizing contribution $H_d = -NM$, where M is the out-of-plane component of magnetization and N

is the demagnetizing factor ranging between 0 and 1, that is, $H_{tot} = H_{ext} - NM$. Hence, the effect of H_d should be take into account for a reliable analysis of the $\Delta M(H)$ curves. Mean field theory predicts a demagnetizing factor $N = 1$ for thin films with independent grains. However, when this value is used to correct the field axis of the remanence curves according to $H_{tot} = H_{ext} - NM$, misleading results can be obtained, as the effective demagnetizing factor can be smaller than 1 due to the presence of intergranular exchange coupling, which results in cooperative reversal. Different approaches have been proposed to take account of demagnetizing effects, such as the recoil loops method discussed by van de Veerdonk *et al.*,[177] which uses an iterative procedure to determine the effective demagnetizing factor, or the method described by Wu *et al.*,[179] which contemplates a direct correction of the demagnetizing field during the measurements. As an example, here we report the case of a series of CoCrPt samples with different thickness, whose intergranular interactions were investigated by measuring the $\Delta M(H)$ curve using the recoil loop method.[179] Figure 4.27 shows the $\Delta M(H)$ curves for different thicknesses.

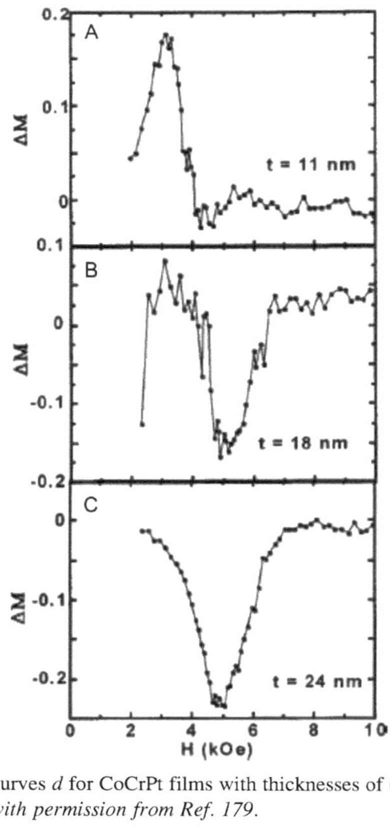

FIGURE 4.27 $\Delta M(H)$ curves *d* for CoCrPt films with thicknesses of (A) 11 nm, (B) 18 nm, and (C) 24 nm. *Reproduced with permission from Ref. 179.*

With increasing film thickness, the curve peak changes from positive to negative values, indicating a reduction of the interaction in the thicker films that was explained by the author as due to the presence an initial thin layer with a large intergranular exchange coupling.

4.2 Permanent Magnets

Permanent magnets are an essential part of our daily life being used in a number of applications including loudspeakers, magnetic field sources, actuators, levitation systems, medical devices and motors, among others. A permanent magnet is a material able to spontaneously create a magnetic field in the surrounding region. The performance of a permanent magnet is defined by the remanent magnetic flux density (B_r), which basically constitutes a measure of the intensity of the magnetic field; the coercivity, which determines the magnetic "hardness" against demagnetizing forces that arise during operation; and the Curie temperature, which defines the maximum operating temperature. The strength of a permanent magnet is quantified by the maximum energy product $(BH)_{max}$, a figure of merit calculated as the optimal product of the remanence flux density and the coercivity in the second quadrant of the (B vs. H) hysteresis loop (Figure 4.28). For an ideal permanent magnet, exhibiting a squared magnetization loop where the remanence magnetization M_r has the largest possible value (i.e. $\mu_0 M_r = \mu_0 M_s$), the maximum energy product is proportional to its squared saturation magnetization, that is, $(BH)_{max} = m_0 M_s^2/4$, if

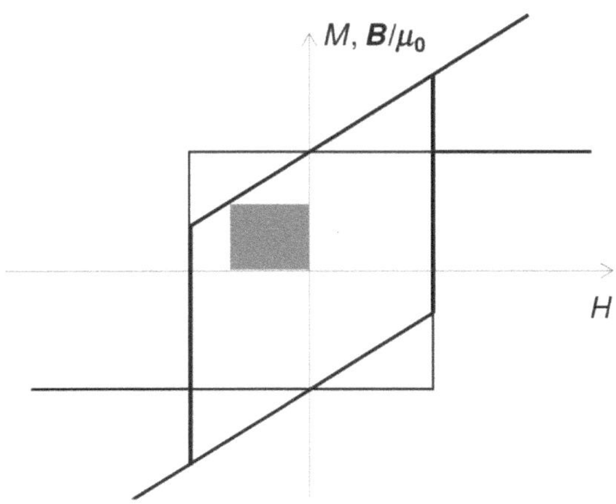

FIGURE 4.28 Hysteresis loop of an ideal permanent magnet: B vs. H; M vs. H. The magnetization (M) remains fully saturated until it reverses in the second quadrant. The coercivity should exceed half the magnetization if the maximum possible energy product is to be achieved. The corresponding $B(H)$ loop is indicated in bold; $[BH]_{max}$ is the shaded area. *Adapted from Ref. 70.*

$\mu_0 H_c$ is larger than $\mu_0 M_s/2$ i.e. the magnet maintains its saturation magnetization until the opposing field reaches the value $\mu_0 M_s/2$.[69,180]

To date, nanocrystalline permanent magnets based on the ternary $Nd_2Fe_{14}B$ compound exhibit the highest energy product at room temperature ($\sim$450 kJ/m^3). However, its relatively low T_C (600 K) limits the effectiveness of $Nd_2Fe_{14}B$ where high working temperatures are required, and Sm-Co alloys with T_C of approx. 1000 K are preferred at temperatures higher than 375 K, although they have a lower $(BH)_{max}$ (up to approx. 240 kJ/m^3).[69,180,181] These materials belong to the class of rare earth (RE)-based magnetic materials consisting of intermetallic compounds of a RE metal (e.g. Nd, Sm) and a 3*d* element (e.g. Co, Fe). Their exceptional properties originate from a favourable combination of the properties of the rare earth and the 3*d* sublattices, the former mainly providing the magnetic anisotropy and the latter giving large saturation magnetization and a high ordering temperature. Today, high-performance rare earth magnets cover 65% of permanent magnets applications, while ferrites, which are less expensive materials, provide another 34% of the total. The most pressing issue for the industry of permanent magnets is represented by the risks of supply of raw materials. Indeed, rare earth elements are experiencing increasing costs due to the extremely low numbers of raw materials extraction sites (97% of them being in China). Such global scenario has stimulated an intense activity for the search of new solutions to limit or avoid the use of RE elements, without compromising or even improving the performance.

A promising action for improving the properties, while reducing/avoiding the RE content, is based on the realization of textured nanocomposites consisting of intimately combined nanoscaled hard and soft FM phases, which are exchange coupled at the mutual interface.[182,183] The interphase magnetic exchange coupling yields a composite system that benefits from the best attributes of the constituent phases: the magnetically hard phase supplies high anisotropy and thus high coercivity, while the magnetically soft phase provides a large saturation magnetization and thus a high remanent flux density, leading to enhanced $(BH)_{max}$ values. The demagnetizing process of an exchange-spring magnet is shown schematically in Figure 4.29 for a one-dimensional system.[183] For a sufficiently small reverse field, most of the moments in the hard phase have not yet switched their direction due to the high anisotropy of the hard phase, while many more moments of the soft phase have already reversed; when the external filed is removed, the soft moments reversibly return to their original direction due to the exchange coupling with the neighbouring hard phase. This reversible behaviour is retained up to a critical external field, above which the non-uniform magnetization at the interface becomes energetically unfavourable and the whole system experiences an irreversible switching. The critical field for reversal of the hard phase is significantly higher than the coercivity of the uncoupled soft phase, but less than that of the hard phase; this enhanced coercivity together with the large saturation

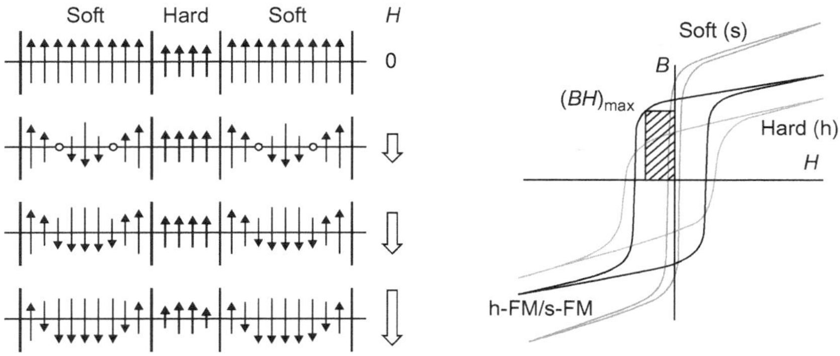

FIGURE 4.29 Upper part. Schematic one-dimensional model of the magnetic structure of an exchange-spring nanocomposite material. *Adapted from Ref. 183*. Lower part. Schematic representation of the hysteresis loop in exchange-spring nanocomposites. *Adapted from Ref. 180*.

magnetization of the soft phase, leads to an increase of the maximum energy product. The magnetic behaviour, previously described, resembles that of a mechanical spring, and such soft/hard nanocomposites are often refer as *exchange-spring* magnets.

According to the one-dimensional model of Kneller and Hawig,[183] for an optimum mutual exchange coupling and cooperative magnetization behaviour, the nanocomposite should be combined uniformly at the nanoscale such that the average size of the magnetically soft phase is twice the domain wall width of the magnetically hard phase $\delta = \pi(AK)^{0.5}$. Here, A and K are the exchange stiffness ($\sim 10^{-11}$ J/m for a FM material) and anisotropy constant of the hard phase, respectively. In a general way, a good magnetic coupling of the hard and soft components is achieved in nanostructured granular materials with grain sizes of about 10–15 nm, that is, twice the domain wall width of a typical hard magnetic material ($K \sim 1$–10 MJ/m^3).

Since the concept of exchange-spring magnet was introduced, many and more rigorous theoretical calculations have been performed by means of two- and three-dimensional models.[182,184,186] Numerical micromagnetic analysis and analytical calculations shown that, properly tuning the soft/hard volume ratio, the shape (layers, spheres, cylinders) and structure (soft inclusions in a hard matrix, hard inclusions in a soft matrix) of the two phases (Figure 4.30), a remanence enhancement and proper coercivity can be achieved. Thus, the maximum energy product would significantly surpass that provided by conventional uncoupled magnetic material. The highest values of $(BH)_{max}$ were predicted for multilayer systems consisting of alternating hard and soft thin layers. Although such multilayer systems are proved to be an effective model system for fundamental studies, they cannot be used for practical applications due to the impossibility of preparing macroscopic materials with such a structure. However, the most recent theoretical studies have shown

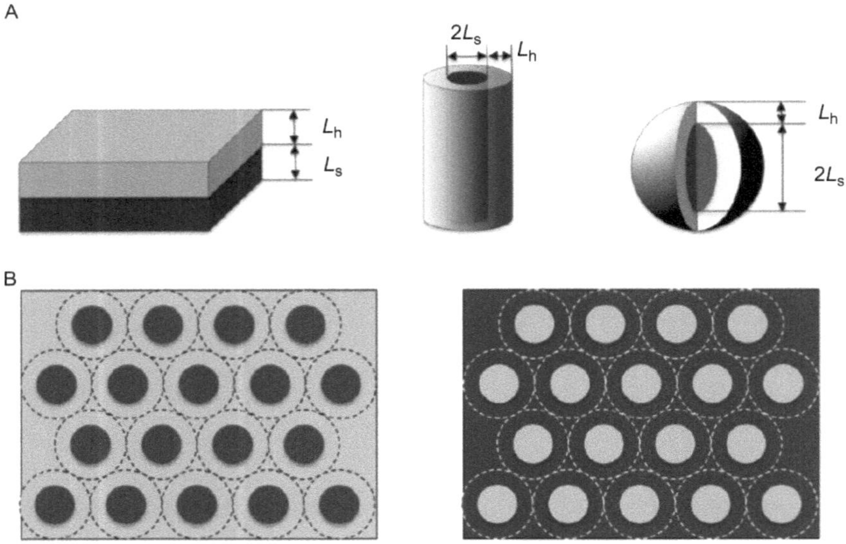

FIGURE 4.30 Schematic illustration of the bylayer, cylindrical and spherical core/shell nanocomposites and (b) of hexagonal close packing of exchange-spring elements. L_s and L_h indicate the size of the soft and hard phase, respectively. *Reproduced with permission from Ref. 182.*

that comparable values of maximum energy product can be achieved in exchange-spring permanent magnets consisting of hard cylindrical inclusions in a soft matrix, while lower values are predicted when using spherical nanoparticles, which however are more commonly obtained in the preparation processes.[185] Although theoretical predictions have shown the high potential of the exchange-spring nanocomposites for future permanent magnets, so far all the experimental results fall short of the theoretical predictions mainly because of the difficulty in obtaining a uniform microstructure of small grains (10–20 nm) by conventional processing techniques (e.g. melt spinning).[185] Bottom-up approaches based on synthesis by soft chemistry of magnetically soft and hard nanoparticles as building blocks have been proposed as an alternative route for the fabrication of exchange-spring permanent magnets, as they allow a fine control of the size of both soft and hard phases, in addition to being a valid method of fabricating a hierarchical material, also at the macroscale.[8,186] However, binary self-assembling of hard and soft magnetic nanoparticles leads to a random packing of soft and hard phases, which negatively affect the final properties of the nanocomposite. A more efficient alternative consists in using bi-magnetic core–shell particles, which are expected to improve dispersion and contact between the two phases following sintering. This approach is based on the chemical synthesis of soft (hard) magnetic nanoparticles onto pre-synthesized hard (soft) magnetic nanomaterials. Up to date, only few studies have been carried out, showing the feasibility of using such a method to prepare nanocomposite exchange-spring permanent magnets.[187,188] Although the results are promising, more investigation is necessary to produce exchange-spring magnets from core–shell nanoparticles with a optimal structure and enhanced $(BH)_{max}$ with respect to uncoupled systems.

ACKNOWLEDGEMENTS

This chapter partially reviews studies performed in the last years among several research institutions: Institute of Structure of the Matter (ISM), Italian National Research Council (CNR), Rome, Italy; University of Leicester (UL), UK; NCSR Demokritos (NCSRD)-Athens, Greece; University of Cagliari (UC), Italy; University of Barcelona (UB), Spain; University of Uppsala (UU). It is a sincere pleasure to thank Dr. K.N. Trohidou, Dr. M. Vasilakaki (NCSRD); Prof. C. Binns, Prof. S. Beaker (UK); Prof. P. Nordblad, Prof. R. Mathieu, Dr. M. Hudl (UU); Prof. J. Tejada, Dr. N. Domingo (UB); Prof. G. Piccaluga, Prof. A. Musinu, Prof. C. Cannas (University of Cagliari), Dr. D. Fiorani, Dr. E. Agostinelli, Dr. A.M. Testa (ISM-CNR), for their respective contributions in specific publications. D. Peddis, S. Laureti, G. Varvaro would like to acknowledge the financial support of MIUR under project FIRB2010 NANOREST. P.E. Jönsson would like to acknowledge financial support from the Swedish Research Council (VR) and the Göran Gustafson Foundation.

LIST OF ABBREVIATION

τ	relaxation time
μ_0	vacuum permeability
μ_B	Bohr magneton
μ_n	nanoparticle magnetic moments
A	exchange stiffness
AFM	antiferromagnetic
d_i	distance between particles
EB	exchange bias
E_d	dipolar energy
FM	ferromagnetic
FMi	ferrimagnetic
H_{cr}^{DCD}	reverse coercivity determined by IRM
H_{cr}^{IRM}	reverse coercivity determined by DCD
H_{ex}	exchange bias field defined as $H_{ex} = -(H_{c\text{-right}} + H_{c\text{-left}})$
H_f	fluctuation field
H_{int}	interaction field
J_{ij}	exchange integral
K_a	anisotropy constant
M_s	saturation magnetization
r_c	critical radius
S	magnetic viscosity
T_b	superparamagnetic blocking temperature

T_C	Curie temperature
T_{max}	temperature corresponding to the maximum in ZFC magnetization measurement
T_N	Néel temperature
V	particle
V_{act}	activation volume
χ_{irr}	irreversible susceptibility

REFERENCES

1. Néel L. Theorie Du Trainage Magnetique des feroomagnetiques en grain fins avec applications aux terres cuties. *Ann Geophys* 1949;**5**:99–136.
2. Bean CP, Livingston JD. Superparamagnetism. *J Appl Phys* 1959;**30**:S120–9.
3. Dormann JL, Fiorani D, Tronc E. *Advances Chemistry Physics*, XCVIII. Wiley; 1997.
4. Pankhurst QA, Thanh N, Jones S, Dobson J. Progress in applications of magnetic nanoparticles in biomedicine. *J Phys D Appl Phys* 2009;**42**:224001.
5. Pankhurst QA, Connelly J, Jones SK, Dobsonb J. Applications of magnetic nanoparticles in biomedicine. *J Phys D Appl Phys* 2003;**36**:R167–81.
6. Albrecht TR, Bedau D, Dobisz E, Gao H, Grobis M, Hellwig O, et al. Bit patterned media at 1 Tdot/in2 and beyond. *IEEE Trans Magn* 2013;**49**:773–8.
7. Piramanayagam SN, Chong TC. *Development in data storage.* Singapore: Wiley-IEEE Press; 2012.
8. Balamurugan B, Sellmyer DJ, Hadjipanayis GC, Skomski R. Prospects for nanoparticle-based permanent magnets. *Scr Mater* 2012;**67**:542–7.
9. Bedanta S, Kleemann W, Kleeman W. Supermagnetism. *J Phys D Appl Phys* 2009;**42**:013001.
10. Mørup S, Hansen MF, Frandsen C. Magnetic nanoparticles. In: Andrews D, Scholes G, Wiederrecht G, editors. *Comprehensive nanoscience and technology*, vol. 1. New York: Elsevier; 2011, p. 437–91.
11. Mørup S, Hansen MF, Frandsen C. Magnetic nanoparticles. In: Andrews D, Scholes G, Wiederrecht G, editors. *Comprehensive nanoscience and technology*, vol. 1. New York: Elsevier; 2011. p. 437–91.
12. Mørup S. Superparamagnetism and spin glass ordering in magnetic nanocomposites. *Europhys Lett* 1994;**28**:671–6.
13. Meiklejohn WH, Bean CP. New magnetic anisotropy. *Phys Rev* 1956;**102**:1413–4.
14. Grady KO, O'Grady K, Fernandez-Outon LEE, Vallejo-Fernandez G, O'Grady K. A new paradigm for exchange bias in polycrystalline thin films. *J Magn Magn Mater* 2010;**322**:883–99.
15. O'Grady K, Laidler H, Grady KO. The limits to magnetic recording—media considerations. *J Magn Magn Mater* 1999;**200**:616–33.
16. Komarneni S. Nanocomposites. *J Mater Chem* 1992;**2**:1219–30.
17. Caruso F, Caruso BF. Nanoengineering of particle surfaces. *Adv Mater* 2001;**13**:11–22.
18. Cannas C, Concas G, Gatteschi D, Musinu A, Piccaluga G, Sangregorio C. How to tailor maghemite particle size in γ-Fe_2O_3–SiO_2 nanocomposites. *J Mater Chem* 2002;**12**:3141–6.
19. Cannas C, Gatteschi D, Musinu A, Piccaluga G, Sangregorio C. Structural and magnetic properties of Fe_2O_3 nanoparticles dispersed over a silica matrix. *J Phys Chem B* 1998;**102**:7721–6.

20. Caizer C. Saturation magnetization of g-Fe_2O_3 nanoparticles dispersed in a silica matrix. *Phys B Condens Matter* 2003;**327**:27–33.
21. Laureti S, Peddis D, Del Bianco L, Testa AM, Varvaro G, Agostinelli E, et al. Exchange bias and magnetothermal properties in Fe @ Mn nanocomposites. *J Magn Magn Mater* 2012;**324**:3503–7.
22. Binns C, Domingo N, Testa AM, Fiorani D, Trohidou KN, Vasilakaki M, et al. Interface exchange coupling in Co nanoparticles dispersed in a Mn matrix. *J Phys Condens Matter* 2010;**22**:436005.
23. Coey JMD. *Magnetism and magnetic materials*. New York: Cambridge University Press; 2010.
24. Blundell S. *Magnetism in condensed matter*. New York: Cambridge University Press; 2003.
25. Frenkel J, Dorfman J. Spontaneous and induced magnetization in ferromagnetic bodies. *Nature* 1930;**126**:274.
26. Kittel C. Theory of the structure of ferromagnetic domains in films and small particles. *Phys Rev* 1946;**70**:965–71.
27. Stoner EC, Wohlfarth EP. A mechanism of magnetic hysteresis in heterogeneous alloys. *IEEE Trans Magn* 1991;**27**:3475–518.
28. O'Grady K, Chantrell RW, El-Hilo M. The ordering temperature in fine particles systems. *J Magn Magn Mater* 1992;**117**:21–8.
29. Mørup S, Hansen MF, Frandsen C, Morup S. Magnetic interactions between nanoparticles. *Beilstein J Nanotechnol* 2010;**1**:182–90.
30. Ruderman MA, Kittel C. Indirect exchange coupling of nuclear magnetic moments by conduction electrons. *Phys Rev* 1954;**96**:72–5.
31. Yosida K. Magnetic properties of Cu-Mn alloys. *Phys Rev* 1957;**106**:893–8.
32. Kasuya T. A theory of metallic ferro- and antiferromagnetism on Zener's model. *Prog Theor Phys* 1956;**16**:45–57.
33. Laureti S, Varvaro G, Testa AM, Fiorani D, Agostinelli E, Piccaluga G, et al. Magnetic interactions in silica coated nanoporous assemblies of $CoFe_2O_4$ nanoparticles with cubic magnetic anisotropy. *Nanotechnology* 2010;**21**:315701.
34. Fiorani D, Dormann JL, Cherkaoui R, Tronc E, Lucari F, D'Orazio F, et al. Collective magnetic state in nanoparticles systems. *J Magn Magn Mater* 1999;**196–197**:143–7.
35. Skumryev V, Stoyanov S, Zhang Y, Hadjipanayis G, Givord D, Nogues J. Beating the superparamagnetic limit with exchange bias. *Nature* 2003;**423**:850–3.
36. Nogués J, Schuller IK. Exchange bias. *J Magn Magn Mater* 1999;**192**:203–32.
37. Berkowitz AE, Takano K. Exchange anisotropy—a review. *J Magn Magn Mater* 1999;**200**:552–70.
38. Malozemoff AP. Random-field model of exchange anisotropy at rough ferromagnetic-antiferromagnetic interfaces. *Phys Rev B* 1987;**35**:3679–82.
39. Malozemoff AP. Mechanisms of exchange anisotropy (invited). *J Appl Phys* 1988;**63**:3874–9.
40. Mauri D, Siegmann HC, Bagus PS, Kay E. Simple model for thin ferromagnetic films exchange coupled to an antiferromagnetic substrate. *J Appl Phys* 1987;**62**:3047.
41. Geshev J. Analytical solutions for exchange bias and coercivity in ferromagnetic antiferromagnetic bilayers. *Phys Rev B* 2000;**62**:5627–33.
42. Koon NC. Calculations of exchange bias in thin films with ferromagnetic antiferromagnetic interfaces. *Phys Rev Lett* 1997;**78**:4865–8.
43. Cui WBB, Hu WJJ, Zhang YJJ, Liu XHH, Guo S, Yang F, et al. Cooling-field dependence of exchange bias in Mg-diluted $Ni_{1-x}Mg_xO$/Ni granular systems. *J Magn Magn Mater* 2009;**321**:1943–6.

44. Domingo N, Fiorani D, Testa AM, Binns C, Baker S, Tejada J. Exchange bias in a superspin glass system of Co particles in Mn matrix. *J Phys D Appl Phys* 2008;**41**:5, 134009.
45. Nguyen HM, Hsiao P-Y, Phan M-H. Monte Carlo study of a bilayer model for exchange bias using a magnetic glass exhibiting random magnetic anisotropy. *J Appl Phys* 2010;**107**, 09D706–09D713.
46. Ong QK, Wei A, Lin X-M. Exchange bias in Fe/Fe_3O_4 core-shell magnetic nanoparticles mediated by frozen interfacial spins. *Phys Rev B* 2009;**80**:134418.
47. Binns C, Qureshi MT, Peddis D, Baker SH, Howes PB, Boatwright A, et al. Exchange bias in Fe@Cr core-shell nanoparticles. *Nano Lett* 2013;**13**:3334–9.
48. Kovylina M, del Muro MG, Konstantinović Z, Varela M, Iglesias O, Labarta A, et al. Controlling exchange bias in Co-CoOx nanoparticles by oxygen content. *Nanotechnology* 2009;**20**:175702.
49. Dobrynin AN, Temst K, Lievens P, Margueritat J, Gonzalo J, Afonso CN, et al. Observation of Co/CoO nanoparticles below the critical size for exchange bias. *J Appl Phys* 2007;**101**:113913–7.
50. Suess D, Kirschner M, Schrefl T, Fidler J, Stamps RL, Kim JV. Exchange bias of polycrystalline antiferromagnets with perfectly compensated interfaces. *Phys Rev B* 2003;**67**:54419.
51. Domingo N, Testa AM, Fiorani D, Binns C, Baker S, Tejada J. Exchange bias in Co nanoparticles embedded in an Mn matrix. *J Magn Magn Mater* 2007;**316**:155–8.
52. Passamani EC, Larica C, Marques C, Proveti JR, Takeuchi AY, Sanchez FH. Exchange bias and anomalous vertical shift of the hysteresis loops in milled Fe/MnO_2 material. *J Magn Magn Mater* 2006;**299**:11–20.
53. Nogués J, Sort J, Langlais V, Skumryev V, Suriñach S, Muñoz JS, et al. Exchange bias in nanostructures. *Phys Rep* 2005;**422**:65–117.
54. Del Bianco L, Boscherini F, Fiorini AL, Tamisari M, Spizzo F, Antisari MV, et al. Exchange bias and structural disorder in the nanogranular Ni/NiO system produced by ball milling and hydrogen reduction. *Phys Rev B* 2008;**77**:94408–12.
55. Sun X, Frey Huls N, Sigdel A, Sun S. Tuning exchange bias in core/shell FeO/Fe_3O_4 nanoparticles. *Nano Lett* 2012;**12**:246–51.
56. Peddis D, Laureti S, Mansilla MVV, Agostinelli E, Varvaro G, Cannas C, et al. Exchange bias in $CoFe_2O_4$/NiO nanocomposites. *Superlattices Microstruct* 2009;**46**:125–9.
57. Ali SR, Janjua MB, Fecioru-Morariu M, Lott D, Smits CJP, Güntherodt G. Role of interface alloying in the exchange bias of Fe/Cr bilayers. *Phys Rev B* 2010;**82**:020402.
58. Girgis E, Portugal R, Loosvelt H, Van Bael M, Gordon I, Malfait M, et al. Enhanced asymmetric magnetization reversal in nanoscale Co/CoO arrays: competition between exchange bias and magnetostatic coupling. *Phys Rev Lett* 2003;**91**:187202.
59. Laureti S, Suck SY, Haas H, Prestat E, Bourgeois O, Givord D. Size dependence of exchange bias in Co/CoO nanostructures. *Phys Rev Lett* 2012;**108**:77205.
60. Suck SY, Neu V, Wolff U, Bahr S, Bourgeois O, Givord D. Magnetic force microscopy analysis of magnetization reversal in exchange-biased Co/CoO nanostructure arrays. *Appl Phys Lett* 2009;**95**:162503.
61. O'Grady K, Chantrelll RW. Remanence curves of fine particle system I: experimental studies. In: *International work. stud. magn. prop. fine part. their relev. to mater. sci*; 1991, p. 93–102.
62. Labarta A, Batlle X, Garcı M, García del Muro M. Interaction effects and energy barrier distribution on the magnetic relaxation of nanocrystalline hexagonal ferrites. *Phys Rev B* 1997;**55**:6440.
63. Wohlfarth EP. Relations between different modes of acquisition of the remanent magnetization of ferromagnetic particles. *J Appl Phys* 1958;**29**:595–6.

64. Kelly PE, Grady KO, Mayo PI, Chantrell RW. Switching mechanisms in cobalt-phosphorus thin films. *IEEE Trans Magn* 1989;**25**:3881–3.
65. Henkel O. Remanenzverhalten Und Wechselwirkungen in Hartmagnetischen Teilchenkollektiven. *Phys Stat Solidi* 1964;**7**:919–29.
66. Zeng H, Sun S, Vedantam TS, Liu JP, Dai ZR, Wang ZL. Exchange-coupled FePt nanoparticle assembly. *Appl Phys Lett* 2002;**80**:2583–5.
67. Fei L, Jing-han Z, Yang-long H, Song G. Chemical synthesis of magnetic nanocrystals: recent progress. *Chinese Physics B* 2013;**22**:1–19.
68. Poudyal N, Ping Liu J. Advances in nanostructured permanent magnets research. *J Phys D Appl Phys* 2013;**46**:043001.
69. Coey JMD. Advances in magnetics hard magnetic materials: a perspective. *IEEE Trans Magn* 2011;**47**:4671–81.
70. Sun S. Recent advances in chemical synthesis, self-assembly and applications of FePt nanoparticles. *Adv Mater* 2006;**18**:393–403.
71. Pileni MP. Self-assembly of inorganic magnetic nanocrystals: a new physics emerges. *J Phys D Appl Phys* 2008;**41**:134002.
72. Properties C, Pileni MP. Nanocrystal self-assemblies: fabrication and collective properties. *J Phys Chem B* 2001;**105**:3358–71.
73. Lalatonne Y, Richardi J, Pileni MP. Van der Waals versus dipolar forces controlling mesoscopic organizations of magnetic nanocrystals. *Nat Mater* 2004;**3**:121–5.
74. Cannas C, Ardu A, Musinu A, Peddis D, Piccaluga G. Spherical nanoporous assemblies of iso-oriented cobalt ferrite nanoparticles: synthesis, microstructure, and magnetic properties. *Chem Mater* 2008;**20**:6364–71.
75. Davies KJ, Wells S, Upadhyay RV, Charles SW, O'Grady K, El Hilo M, et al. The observation of multi-axial anisotropy in ultrafine cobalt ferrite particles used in magnetic fluids. *J Magn Magn Mater* 1995;**149**:14–8.
76. Walker M, Majo PI, O'Grady K, Charles SW, Chantrell RW. The magnetic properties of single-domain particles with cubic anisotropy. II. Remanence curves. *J Phys Condens Matter* 1993;**5**:2793.
77. Garcia-Otero J, Porto M, Rivas J, Introduction I. Henkel plots of single-domain ferromagnetic particles. *J Appl Phys* 2000;**87**:7376–81.
78. Geshev J, Mikhov M, Schmidt JE. Remanent magnetization plots of fine particles with competing cubic and uniaxial anisotropies. *J Appl Phys* 1999;**85**:7321–7.
79. Street R, Brown SD. Magnetic viscosity, fluctuation fields, and activation energies (invited). *J Appl Phys* 1994;**76**:6386–90.
80. Humfeld KD, Giri AK, Majestich AS, Venturini EL. Time dependent properties of iron nanoparticles. *IEEE Trans Magn* 2001;**37**:2194–6.
81. Walker M, Majo PI, O'Grady K, Charles SW, Chantrell RW. The magnetic properties of single-domain particles with cubic anisotropy. I. Hysteresis loops. *J Phys Condens Matter* 1993;**5**:2779.
82. Vékas L, Bica D, Avdeev MV, Ladislau V. Magnetic nanoparticles and concentrated magnetic nanofluids: synthesis, properties and some application. *Chin Particuol* 2007;**5**: 43–7.
83. Caruso F, Caruso R, Mohwald H. Nanoengineering of inorganic and hybrid hollow spheres by colloidal templating. *Science* 1998;**282**:1111–4.
84. Suber L, Peddis D. Approaches to synthesis and characterization of spherical and anisometric metal oxide magnetic nanomaterials, chapter 4. In: Kumar C, editor. *Nanomaterials for life science*. New York: Wiley.

85. Peddis D, Rinaldi D, Ennas G, Scano A, Agostinelli E, Fiorani D. Superparamagnetic blocking and superspin-glass freezing in ultra small δ-(Fe(0.67)Mn(0.33))OOH particles. *Phys Chem Chem Phys* 2012;**14**:3162–9.
86. Jonsson T, Mattsson J, Djurberg C, Khan FA, Nordblad P, Svedlindh P. Aging in a magnetic particle system. *Phys Rev Lett* 1995;**75**:4138–41.
87. Sahoo S, Petracic O, Kleemann W, Nordblad P, Cardoso S, Freitas PP. Non-equilibrium collective dynamics of a superspin glass. *J Magn Magn Mater* 2004;**272–276**:1316–8.
88. Dormann JL, Cherkaoui R, Spinu L, Noguès M, Lucari F, D'Orazio F, et al. From pure superparamagnetic regime to glass collective state of magnetic moments in [gamma]-Fe_2O_3 nanoparticle assemblies. *J Magn Magn Mater* 1998;**187**:L139–44.
89. Nakamae S, Tahri Y, Thibierge C, L'Hote D, Vincent E, Dupuis V, et al. Observation of superspin glass state in magnetically textured ferrofluid (γ-Fe_2O_3). *J Appl Phys* 2009;**105**:07E318.
90. Parker D, Dupuis V, Ladieu F, Bouchaud JP, Dubois E, Perzynski R, et al. Spin-glass behavior in an interacting g-Fe_2O_3 nanoparticle system. *Phys Rev B* 2008;**77**:104428–9.
91. Jönsson P, García-Palacios J. Thermodynamic perturbation theory for dipolar superparamagnets. *Phys Rev B* 2001;**64**:174416.
92. Allia P, Coisson M, Tiberto P, Vinai F, Knobel M, Novak MA, et al. Granular Cu-Co alloys as interacting superparamagnets. *Phys Rev B* 2001;**64**:144420.
93. Labarta A, Iglesias O, Balcells L, Badia F. Magnetic relaxation in small-particle systems: Ln(t/to) scaling. *Phys Rev B* 1993;**48**:10240.
94. Shtrikman S, Wohlfarth EP. The theory of the Vogel–Fulcher law of spin glasses. *Phys Lett* 1981;**85**:467–70.
95. Dormanni JL, Bessaist L, Fiorani D. A dynamic study of small interacting particles: superparamagnetic model and spin-glass laws. *J Phys C Solid State Phys* 1988;**21**:2015.
96. Wang Q, Iancu N, Seo D. Preparation of large transparent silica monoliths with embedded photoluminescent CdSe@ZnS core/shell quantum dots. *Chem Mater* 2005;**17**:4762–4.
97. Zhao W, Gu J, Zhang L, Chen H, Shi J, Mesoporous C, et al. Fabrication of uniform magnetic nanocomposite spheres with a magnetic core/mesoporous silica shell structure. *J Am Chem Soc* 2005;**127**:8916–7.
98. Selvan ST, Li C, Ando M, Murase N. Formation of luminescent CdTe-silica nanoparticles through an inverse microemulsion technique. *Chem Lett* 2004;**33**:434–5.
99. Cannas C, Musinu A, Peddis D, Piccaluga G. Synthesis and characterization of $CoFe_2O_4$ nanoparticles dispersed in a silica matrix by a sol-gel autocombustion method. *Chem Mater* 2006;**18**:3835–42.
100. Peddis D, Mansilla MV, Mørup S, Cannas C, Musinu A, Piccaluga G, et al. Spin-canting and magnetic anisotropy in ultrasmall $CoFe_2O_4$ nanoparticles. *J Phys Chem B* 2008;**112**:8507–13.
101. Peddis D, Cannas C, Piccaluga G, Agostinelli E, Fiorani D. Spin-glass-like freezing and enhanced magnetization in ultra-small $CoFe_2O_4$ nanoparticles. *Nanotechnology* 2010;**21**:125705.
102. Gittleman JI, Abeles B, Bozowski S. Superparamagnetism and relaxation effects in granular Ni-SiO_2 and Ni-Al_2O_3 films. *Phys Rev B* 1974;**9**:3891–7.
103. Hansen MF, Mørup S. Estimation of blocking temperatures from ZFC/FC curves. *J Magn Magn Mater* 1999;**203**:214–6.
104. Sasaki M, Takayama H, Mamiya H, Jönsson PE. Aging and memory effects in superparamagnets and superspin glasses. *Phys Rev B* 2005;**71**:104405.
105. Cannas C, Musinu A, Ardu A, Orrù F, Peddis D, Casu M, et al. $CoFe_2O_4$ and $CoFe_2O_4/SiO_2$ core/shell nanoparticles: magnetic and spectroscopic study. *Chem Mater* 2010;**22**:3353–61.

106. Sun S, Zeng H, Robinson DB, Raoux S, Rice PM, Wang SX, et al. Monodisperse MFe_2O_4 (M = Fe, Co, Mn) nanoparticles. *J Am Chem Soc* 2004;**126**:273–9.
107. Zysler RD, Ramos CA, De Biasi E, Romero H, Ortega A, Fiorani D. Effect of interparticle interactions in $(Fe_{0.26}Ni_{0.74})_{50}B_{50}$ magnetic nanoparticles. *J Magn Magn Mater* 2000;**221**:37–44.
108. Cannas C, Falqui A, Musinu A, Peddis D, Piccaluga G. $CoFe_2O_4$ nanocrystalline powders prepared by citrate-gel methods: synthesis, structure and magnetic properties. *J Nanopart Res* 2006;**8**:255–67.
109. Nakamae S, Crauste-Thibierge C, L'HÔte D, Vincent E, Dubois E, Dupuis V, et al. Dynamic correlation length growth in superspin glass: bridging experiments and simulations. *Appl Phys Lett* 2012;**101**:242409.
110. Nakamae S, Crauste-Thibierge C, Komatsu K, L'Hôte D, Vincent E, Dubois E, et al. Anisotropy-axis orientation effect on the magnetization of γ-Fe_2O_3 frozen Ferrofluid. *J Phys D Appl Phys* 2010;**43**:474001.
111. Sahoo S, Petracic O, Kleemann W, Nordblad P, Cardoso S, Freitas P. Aging and memory in a superspin glass. *Phys Rev B* 2003;**67**:214422.
112. Jonsson T, Nordblad P, Svedlindh P. Dynamic study of dipole-dipole interaction effects in a magnetic nanoparticle system. *Phys Rev B* 1998;**57**:497–504.
113. De Toro JA, Lee SS, Salazar D, Cheong JL, Normile PS, Muñiz P, et al. A nanoparticle replica of the spin-glass state. *Appl Phys Lett* 2013;**102**:183104.
114. De Toro JA, Normile PS, Lee SS, Salazar D, Cheong JL, Mun P, et al. Controlled close-packing of ferrimagnetic nanoparticles: an assessment of the role of interparticle superexchange versus dipolar interactions. *J Phys Chem C* 2013;**117**:10213–9.
115. Hyeon T, Lee SS, Park J, Chung Y, Bin Na H. Synthesis of highly crystalline and monodisperse maghemite nanocrystallites without a size-selection process. *J Am Chem Soc* 2001;**123**:12798–801.
116. Katori HA, Ito A. Experimental study of the Almeida-Thouless line by using typical Ising spin-glass $Fe_xMn_{1-x}TiO_3$ with $x = 0.41$, 0.50, 0.55 and 0.57. *J Phys Soc Jpn* 1994;**63**:3122–8.
117. Dupuis V, Vincent E, Bouchaud J-P, Hammann J, Ito A, Katori H. Aging, rejuvenation, and memory effects in Ising and Heisenberg spin glasses. *Phys Rev B* 2001;**64**:174204.
118. Jonsson T, Svedlindh P, Hansen MF. Static scaling on an interacting magnetic nanoparticle system. *Phys Rev Lett* 1998;**81**:3976.
119. De Toro JA, López de la Torre MA, Riveiro JM, Beesley A, Goff JP, Thomas MF. Critical spin-glass dynamics in a heterogeneous nanogranular system. *Phys Rev B* 2004;**69**:224407.
120. Mamiya H, Nakatani I, Furubayashi T. Blocking and freezing of magnetic moments for iron nitride fine particle systems. *Phys Rev Lett* 1998;**80**:177–80.
121. Jönsson P, Jonsson T, Garcmh JL, Svedlindh P. Nonlinear dynamic susceptibilities of interacting and noninteracting magnetic nanoparticles. *J Magn Magn Mater* 2000;**222**:219–26.
122. Parisi G. Orders of parameters in spin-glasses. *Phys Rev Lett* 1983;**50**:1946–8.
123. Fisher DS, Huse DA. Equilibrium behaviour of the spin-glass ordered phase. *Phys Rev B* 1988;**38**:386–411.
124. Jönsson PE, Mathieu R, Nordblad P, Yoshino H, Katori HA, Ito A. Nonequilibrium dynamics of spin glasses: examination of the ghost domain scenario. *Phys Rev B* 2004;**70**: 174402–27.
125. Nordblad P. Spin glasses: model systems for non-equilibrium dynamics. *J Phys Condens Matter* 2004;**16**:S715–22.
126. Jonsson PE. Superparamagnetism and spin glass dynamics on interacting magnetic nanoparticle systems. *Adv Chem Phys* 2004;**128**:191–248.

127. Peddis D, Hudl M, Binns C, Fiorani D, Nordblad P. Aging experiments in a superspin glass system of Co particles in Mn matrix. *J Phys Conf Ser* 2010;**200**:72074.
128. Vasilakaki M, Trohidou KN, Peddis D, Fiorani D, Mathieu R, Hudl M, et al. Memory effects on the magnetic behavior of assemblies of nanoparticles with ferromagnetic core/antiferromagnetic shell morphology. *Phys Rev B* 2013;**88**:140402, 1–5.
129. Beloborodov I, Glatz A, Vinokur V. Electron transport in nanogranular ferromagnets. *Phys Rev Lett* 2007;**99**:066602.
130. Sankar S, Dender D, Borchers JA, Smith DJ, Erwin RW, Kline SR, et al. Magnetic correlations in non-percolated Co-SiO granular Lms. *J Magn Magn Mater* 2000;**221**:1–9.
131. Rancourt DG, Daniels JM. Influence of unequal magnetization direction probabilities on the Mossbauer spectra of superparamagnetic particles. *Phys Rev B* 1984;**18**:2410–4.
132. Glatz A, Beloborodov IS, Vinokur VM. Giant magnetoresistance in nanogranular magnets. *EPL (Europhys Lett)* 2008;**82**:47002.
133. Suber L, Imperatori P, Mari A, Marchegiani G, Mansilla MV, Fiorani D, et al. Thermal hysteresis of Morin transition in hematite particles. *Phys Chem Chem Phys* 2010;**12**:6984–9.
134. Zysler RD, Fiorani D, Testa AM, Suber L, Agostinelli E, Godinho M. Size dependence of the spin-flop transition in hematite nanoparticles. *Phys Rev B* 2003;**68**:212408.
135. Winkler E, Zysler RD, Vasquez Mansilla M, Fiorani D, Rinaldi D, Vasilakaki M, et al. Surface spin-glass freezing in interacting core-shell NiO nanoparticles. *Nanotechnology* 2008;**19**:185702.
136. Winkler E, Zysler RD, Troiani HE, Fiorani D. Magnetic properties of Co nanoparticles in a Cr_2O_3 antiferromagnetic Matrix. *Phys B Condens Matter* 2006;**384**:268–70.
137. Fiorani D, Testa AM, Suber L, Angiolini M, Montone A, Polichetti M. Size and shape effect on the canted antiferromagnetism in alpha-Fe_2O_3 particles. *Nanostruct Mater* 1999;**12**:932–42.
138. Bodker F, Morup S. Size dependence of the properties of hematite nanoparticles. *EPL (Europhys Lett)* 2000;**52**:217–23.
139. Frandsen C, Bahl CRH, Lebech B, Lefmann K, Kuhn LT, Keller L, et al. Oriented attachment and exchange coupling of alpha-Fe_2O_3 nanoparticles. *Phys Rev B* 2005;**72**:214406.
140. Madsen DE, Cervera-Gontard L, Kasama T, Dunin-Borkowski RE, Koch CB, Hansen MF, et al. Magnetic fluctuations in nanosized goethite (α-FeOOH) grains. *J Phys Condens Matter* 2009;**21**:016007.
141. Nogues J, Skumryev V, Sort J, Stoyanov S, Givord D. Shell-driven magnetic stability in core-shell nanoparticles. *Phys Rev Lett* 2006;**97**:157203–4.
142. Nunes WC, Socolovsky LM, Denardin JC, Cebollada F, Brandl AL, Knobel M. Role of magnetic interparticle coupling on the field dependence of the superparamagnetic relaxation time. *Phys Rev B* 2005;**72**:212413.
143. Loffler JF, Meier P, Doudin B, Ansermet J, Wagner W. Random and exchange anisotropy in consolidated nanostructured fe and Ni: role of grain size and trace oxides on the magnetic properties. *Phys Rev B* 1998;**57**:2915–24.
144. Bahiana M, Pereira Nunes JP, Altbir D, Vargas P, Knobel M. Ordering effects of the dipolar interaction in lattices of small magnetic particles. *J Magn Magn Mater* 2004;**281**:372–7.
145. Parisi G. Infinite number of order parameters for spin glasses. *Phys Rev Lett* 1979;**43**:1754–6.
146. Fiorani D, Del Bianco L, Testa A, Trohidou K. Glassy dynamics in the exchange bias properties of the iron/iron oxide nanogranular system. *Phys Rev B* 2006;**73**:092403.
147. Astinchap B, Moradian R, Ardu A, Cannas C, Varvaro G, Capobianchi A, et al. Bifunctional FePt @ MWCNTs/Ru nanoarchitectures: synthesis and characterization. *Chemistry of Materials* 2012;**24**:3393–400.

148. Hütten A, Reiss G, Hutten A. Applications beyond data storage. *Nat Mater* 2005;**4**:725–6.
149. Tartaj P, Del Puerto Morales M, Veintemillas-Verdaguer S, Gonzales Carreno T, Serna CJ. The preparation of magnetic nanoparticles for applications in biomedicine. *J Phys D Appl Phys* 2003;**36**:R182–97.
150. Piramanayagam SN, Srinivasan K. Recording media research for future hard disk drives. *J Magn Magn Mater* 2009;**321**:485–94.
151. Campardo G, Tiziani F, Iaculo M, editors. Memory Mass Storage. New York: Springer, 2011.
152. Parkin SSP, Hayashi M, Thomas L. Magnetic domain-wall racetrack memory. *Science* 2008;**320**:190–4.
153. Zhu J-G, Tang Y. Micromagnetics of percolated perpendicular media. *IEEE Trans Magn* 2007;**43**:687–92.
154. Varvaro G, Testa AM, Agostinelli E, Fiorani D, Laureti S, Springer F, et al. Study of microstructure and magnetization reversal mechanism in granular CoCrPt:SiO_2 films of variable thickness. *Mater Chem Phys* 2013;**141**:790–6.
155. Srinivasan K, Piramanayagam SN, Sbiaa R, Chantrell RW. Thermal stability and the magnetization process in CoCrPt–SiO_2 perpendicular recording media. *J Magn Magn Mater* 2008;**320**:3041–5.
156. Richter HJ. The transition from longitudinal to perpendicular recording. *J Phys D Appl Phys* 2007;**40**:R149–77.
157. Richter HJ, Harkness SD. Media for magnetic recording beton 100Gbis/in2. *Mater Res Bull* 2006;**31**:384–8.
158. Choe G, Zheng M, Acharya BR, Abarra EN, Zhou JN. With performance enhancement capping layer. *IEEE Trans Magn* 2005;**41**:3172–4.
159. Li WM, Shi JZ, Lim WK, Ding J. Influence of bias voltage on the switching behaviour of CoCrPt–SiO_2 perpendicular recording media. *J Phys D Appl Phys* 2013;**46**:185001.
160. Dutson JD, Hashimoto M, Inaba Y, Greaves SJ, Muraoka H. Magnetic and recording properties. *IEEE Trans Magn* 2007;**43**:814–8.
161. Inaba Y, Shimatsu T, Oikawa T, Sato H, Aoi H, Muraoka H, et al. Optimization of the SiO_2 content in CoPtCr-SiO_2 perpendicular recording media for high-density recording. *IEEE Trans Magn* 2004;**40**:2486–8.
162. Zheng M, Acharya BR, Choe G, Zhou JN, Yang ZD, Abarra EN, et al. Role of oxygen incorporation in Co-Cr-Pt-Si-O perpendicular magnetic recording media. *IEEE Trans Magn* 2004;**40**:2498–500.
163. Piramanayagam SN, Pock CK, Lu L, Ong CY, Shi JZ, Mah CS. Grain size reduction in CoCrPt:SiO_2 perpendicular recording media with oxide-based intermediate layers. *Appl Phys Lett* 2006;**89**:162504.
164. Terris BD, Albrecht M, Hu G, Thomson T, Rettner CT. Recording and reversal properties of nanofabricated magnetic islands. *IEEE Trans Magn* 2005;**41**:2822–7.
165. Krone P, Brombacher C, Makarov D, Lenz K, Ball D, Springer F, et al. Nanocap arrays of granular CoCrPt:SiO_2 films on silica particles: tailoring of the magnetic properties by Co +irradiation. *Nanotechnology* 2010;**21**:385703.
166. Ball DK, Lenz K, Fritzsche M, Varvaro G, Günther S, Krone P, et al. Magnetic properties of granular CoCrPt:SiO_2 thin films deposited on GaSb nanocones. *Nanotechnology* 2014;**25**:085703.
167. Jung HS, Kwon U, Kuo M, Velu EMT, Malhotra SS, Jiang W, et al. Effect of oxygen incorporation on microstructure and media performance in CoCrPt–SiO_2 perpendicular recording media. *IEEE Trans Magn* 2007;**43**:615–20.

168. Jung HS, Kuo M, Malhotra SS, Bertero G. Origin of incoherent magnetic switching behavior in CoCrPt–SiO_2 perpendicular magnetic recording media. *Appl Phys Lett* 2007;**91**:212502.
169. Srinivasan K, Acharya BR, Bertero G. Thermal effects on the magnetization reversal process and its interpretation in perpendicular magnetic recording media. *J Appl Phys* 2010;**107**:113912.
170. Carter A, Fernandez-Outon LE, Inaba Y, Greaves SJ, Muraoka H, O'Grady K. Anisotropy dispersion in $(CoCrPt)_{1-x}(SiO_2)_x$ perpendicular recording media. *J Magn Magn Mater* 2008;**320**:2269–72.
171. Dong Y, Victora RH. Extracting intergranular exchange coupling in perpendicular magnetic recording media. *J Magn Magn Mater* 2012;**324**:248–51.
172. Jung HS, Ghaderi A, Shi Z. Characterization of magnetic clustering and its effect on switching behavior and recording performance in perpendicular media. *J Appl Phys* 2011;**109**:07B718.
173. Morrison C, Saharan L, Ikeda Y, Takano K, Hrkac G, Thomson T. Quantifying exchange coupling in segregated granular materials. *J Phys D Appl Phys* 2013;**46**:475002.
174. Kondorsky E. On hysteresis in ferromagnetics. *J Geophys Res* 1940;**II**:161–81.
175. Coffey KR, Thomson T, Thiele J-U. Angular dependence of the switching field of thin-film longitudinal and perpendicular magnetic recording media. *J Appl Phys* 2002;**92**:4553.
176. El-hilo M, Grady KO, Chantrell RW. Fluctuation fields and reversal mechanisms in granular magnetic systems. *J Magn Magn Mater* 2002;**248**:360–73.
177. van de Veerdonk RJM, Weller D. Determination of switching field distributions for perpendicular recording media. *IEEE Trans Magn* 2003;**39**:590–3.
178. Holloway L, Laidler H, O'Grady K, Khizroev S, Howard JK, Gustafson RW, et al. Magnetic characterization of perpendicular recording media. *IEEE Trans Magn* 2002;**38**:1682–6.
179. Wu XW, van de Veerdonk RJM, Chantrell RW, Weller D. ΔM study of perpendicular recording media. *J Appl Phys* 2003;**93**:6760.
180. Lewis LH, Jiménez-Villacorta F. Perspectives on permanent magnetic materials for energy conversion and power generation. *Metall Mater Trans A* 2012;**44**:2–20.
181. Gutfleisch O, Willard MA, Brück E, Chen CH, Sankar SG, Liu JP. Magnetic materials and devices for the 21st century: stronger, lighter, and more energy efficient. *Adv Mater* 2011;**23**:821–42.
182. Jiang JS, Bader SD. Rational design of the exchange-spring permanent magnet. *J Phys Condens Matter* 2014;**26**:064214.
183. Kneller EF, Hawig R. The exchange-spring magnet: a new material principle for permanent magnets. *IEEE Trans Magn* 1991;**27**:3588–600.
184. Skomski R, Manchanda P, Kumar P, Balamurugan B, Kashyap A, Sellmyer DJ. Predicting the future of permanent-magnet materials. *IEEE Trans Magn* 2013;**49**:3215–20.
185. Betancourt I, Davies HA. Exchange coupled nanocomposite hard magnetic alloys. *Mater Sci Technol* 2010;**26**:5–19.
186. Roy D, Kumar PSA. Enhancement of $(BH)_{max}$ in a hard-soft-ferrite nanocomposite using exchange spring mechanism. *J Appl Phys* 2009;**073902**:1–5.
187. Yang C, Jia L, Wang S, Gao C, Shi D, Hou Y, et al. Single domain $SmCo_5$@Co exchange-coupled magnets prepared from core/shell Sm[Co(CN)6]·$4H_2O$@GO particles: a novel chemical approach. *Sci Rep* 2013;**3**:3542.
188. Zeng H, Li J, Wang ZL, Liu JP, Sun S. Bimagnetic core/shell FePt/Fe_3O_4 nanoparticles. *Nano Lett* 2004;**4**:187–9.
189. Fontana RE, Helzler SR, Decad G. *IEEE Trans Magn* 2012;**48**:1692.
190. Tand DD, Lee YL. *Magneti Memory: Fundamentals and Technology*. Cambridge University Press; 2010.

Chapter 5

Magnetic States and Properties of Patterned Ferromagnetic Nanostructures

Andrey A. Fraerman[*,†] and Victor L. Mironov[*,†]
*Department of Magnetic Nanostructures, Institute for Physics of Microstructures, Russian Academy of Science, Nizhny Novgorod, Russia
†Lobachevsky State University of Nizhny Novgorod, Gagarin Avenue, Nizhny Novgorod, Russia

1 INTRODUCTION

The great interest in patterned ferromagnetic nanostructures is stimulated by several factors. First, the ordered arrays of nanoscale ferromagnetic particles are regarded as a promising medium for the magnetic recording of information with ultra-high density.[1–8] Second, the ordered arrays of ferromagnetic nanoparticles (NPs) are promising sources of embedded inhomogeneous magnetic fields in hybrid structures such as ferromagnet/superconductor, ferromagnet/metal and ferromagnet/semiconductor interfaces, which can influence the orbital and spin degrees of freedom of charge carriers in such systems. In this case, the magnitude and characteristic spatial scales of the inhomogeneity of the stray fields of ferromagnetic NP arrays are determined by the material characteristics, geometric dimensions and shape of the particles, and by the spatial periods of the structure.[7–10] Third, in conducting ferromagnets the carriers are spin-polarized. Their spectrum is split into two sub-bands, which are occupied by electrons having parallel and antiparallel spin projections relative to the sample magnetization. The value of the spin "splitting" is of the same order of magnitude as the Curie temperature T_c of these materials. Thus, in ferromagnets, there is enormous "exchange" field whose strength is $H \sim kT_c/\mu \sim 10^6 - 10^7$ Oe (k is the Boltzmann constant, μ is the Bohr magneton). First, this "exchange" field was observed in transport experiments on giant magnetoresistance[11,12] and tunnel magnetoresistance[13] effects. This discovery was a powerful incentive to continue studying the spin-dependent transport effects. On the other hand, it is known[14,15] that the distribution of the magnetization in the ferromagnetic sample is determined by the competition of the magnetic anisotropy, exchange and magnetostatic

Frontiers of Nanoscience, Vol. 6. http://dx.doi.org/10.1016/B978-0-08-098353-0.00005-1

interactions. Arising from this competition, magnetic domain structure is not universal, but depends on the shape and geometry of the sample. The peculiarity of this size range for ferromagnetic materials is determined by the existence of two characteristic scales, namely, the thickness of the domain wall (DW) and the exchange length, which are a few tens of nanometres for ferromagnetic transition metals. Thus, the development in recent years of methods to control the sample geometry and hence the distribution of the magnetization in the nanometre range open up additional opportunities to find new transport and optical effects in artificial nanomaterials. It also enables the creation of new devices for storing and processing information.

2 "EXCHANGE" EFFECTS IN THE STRUCTURES WITH NON-UNIFORM MAGNETIZATION

We start the consideration of the possible transport effects of exchange nature in heterogeneous ferromagnets with a phenomenological analysis. A constant electric field $\vec{E}$ in a conductive medium generates an electric current $\vec{j}$, which is defined by the following expression:

$$j_i = \sigma_{ik}E_k + \gamma_{ijk}E_jE_k + \cdots, \tag{5.1}$$

where the tensor components of the linear and quadratic conductivity depend on the magnetic moment of the sample and its spatial derivatives. The task is to determine the form of this dependence. Considering only processes without dissipation taking place in media with an inversion centre, we conclude that the tensor of the linear and non-linear conductivities contains terms with odd powers of the magnetic moment $\vec{M}$, and the tensor σ_{ik} includes terms with even powers but the tensor γ_{ijk} includes terms with odd powers of spatial derivatives. Assuming that the expected effects are connected with the exchange interaction, we require that the expression for the current (5.1) should be invariant with respect to the coherent rotation of the sample magnetization.[16] This means that the conductivity tensor should not contain the convolution of "spatial" and "magnetic" indexes. Satisfying all these requirements, we get the following expression for the electric current in a conducting ferromagnet with an inhomogeneous distribution of magnetization[17]:

$$j_i = \sigma\left(\vec{M}\left[\frac{\partial\vec{M}}{\partial x_i}\times\frac{\partial\vec{M}}{\partial x_k}\right]\right)E_k + \gamma\left(\vec{M}\left[\frac{\partial\vec{M}}{\partial x_i}\times\frac{\partial^2\vec{M}}{\partial x_j\partial x_k}\right]\right)E_jE_k + \cdots, \tag{5.2}$$

where σ and γ are scalar constants (we consider here only the media with an isotropic or a cubic crystalline structure). The tensor of linear conductivity is antisymmetric and describes the contribution to the Hall effect, which has an exchange nature. The second term in Equation (5.2) is responsible for rectifying properties of ferromagnets. Both tensors are different from zero only in ferromagnetic samples with non-coplanar distributions of magnetization. Indeed, the expression for the conductivity tensor is easily obtained by expanding in a Taylor series the scalar triple product

$$\left(\vec{M}_1 \cdot \left[\vec{M}_2 \times \vec{M}_3\right]\right),$$

where $\vec{M}_1$, $\vec{M}_2$ and $\vec{M}_3$ are magnetic moments in the neighbouring points of the sample.

For example, we consider two non-coplanar distributions of the magnetic moment. First, let the distribution of magnetization have the following form:

$$\begin{aligned} M_x &= \sin\theta(\rho)\cos(\nu\varphi+\varphi_0),\\ M_y &= \sin\theta(\rho)\sin(\nu\varphi+\varphi_0),\\ M_z &= \cos\theta(\rho), \end{aligned} \tag{5.3}$$

where φ and ρ are cylindrical coordinates, ν is the integer number and φ_0 is the constant phase shift. Substituting these components in Equation (5.2), we obtain

$$\begin{aligned} \vec{j} &= \left[\vec{E} \times \vec{B}_{\text{eff}}\right],\\ B_{\text{eff}} &= \sigma\left(\vec{M}, \left[\frac{\partial \vec{M}}{\partial x} \times \frac{\partial \vec{M}}{\partial y}\right]\right) = \sigma\nu\frac{1}{\rho}\frac{\partial\cos\theta}{\partial\rho}. \end{aligned} \tag{5.4}$$

Thus, in structures with a non-coplanar vortex ($\nu=1$, $\varphi_0=\pm\pi/2$) or an anti-vortex ($\nu=-1$, $\varphi_0=0,\pi$) distributions of the magnetic moment, one can expect an additional contribution to the Hall effect. The number ν is so-called winding number.

Second, let us consider the distribution in the form of a conical magnetic helix

$$\begin{aligned} M_x &= m\cos(qz),\\ M_y &= m\sin(qz),\\ M_z &= m_z,\\ M^2 &= m^2 + m_z^2. \end{aligned} \tag{5.5}$$

Substituting these components in Equation (5.2), we obtain

$$j_z = \gamma q^3 m_z\left(1 - m_z^2\right)E_z^2. \tag{5.6}$$

Therefore, such a helical structure should have diode properties. The microscopic mechanism of these effects can be partially understood by the classical model proposed in Ref. 18. In this paper, it was shown that the cause of the "exchange" contribution to the Hall conductivity is the non-collinearity of the magnetic moment of the moving electron and the external magnetic field.

3 OPTICAL AND NEUTRON-OPTICAL EFFECTS IN CHIRAL MAGNETIC STRUCTURES

The simplest quantum mechanical approach to the description of the conduction electrons in ferromagnetic materials is based on the Vonsovskii–Zener s–d model, in which the problem is reduced to finding the eigenfunctions and eigenvalues of the Schrödinger equation

$$-\Delta\Psi\left(\vec{r}\right)-J\vec{M}\left(\vec{r}\right)\hat{\vec{\sigma}}\Psi\left(\vec{r}\right)=E\Psi\left(\vec{r}\right), \tag{5.7}$$

where $\vec{M}$ is the unit vector in the direction of the magnetic moment; $\vec{\sigma}$ is the vector of Pauli matrices; parameters $J=2m_e\widetilde{J}/\hbar^2$, $E=2m_e\varepsilon/\hbar^2$; $\widetilde{J}$ is the exchange constant describing the interaction of the conduction electrons (s) with localized electrons (d); m_e is the effective electron mass; ε is electron energy. Note the analogy between the description of the conduction electrons in the ferromagnet and neutrons, for which the magnetic moment vector in Equation (5.7) should be replaced by a vector of magnetic induction $\vec{B}$, and the exchange constant should be replaced by the nuclear Bohr magneton.[19] This analogy allows us to consider the transport and optical properties of conducting ferromagnets and neutron scattering in non-homogeneous magnetic systems within a unified framework.

In the case of a magnetic helix (5.5), Equation (5.7) can be solved exactly.[20] The spectrum and eigenfunctions have the following form:

$$E_{\pm}=k^2+p^2+\left(\frac{q}{2}\right)^2\pm\sqrt{q^2k^2+J^2-2m_zJqk}, \tag{5.8}$$

$$\begin{aligned}\Psi_{\pm}&=\frac{1}{\sqrt{1+(\delta_{\pm}(k))^2}}\begin{pmatrix}\delta_{\pm}(k)e^{-iqz/2}\\ e^{iqz/2}\end{pmatrix}e^{ikz}e^{i\vec{p}\vec{\rho}},\\ \delta_{\pm}&=\frac{m_zJ-qk\pm\sqrt{q^2k^2+J^2-2m_zJqk}}{J\left(1-m_z^2\right)^{1/2}},\end{aligned} \tag{5.9}$$

where k and p are the components of electron quasi-momentum along and perpendicular to the axis of the helix. From expression (5.8), one can see that the spectrum of the carriers in the cone ($m_z\neq 0$) spiral is not an even function of the quasi-momentum. Thus, the group velocities of electrons moving left and right along the axis of the helix are not equal. In a macroscopic system, this difference does not lead to the appearance of an electric current, as it is exactly compensated by the difference in the number of equilibrium electrons moving in opposite directions. However, in mesoscopic systems such as small ferromagnetic rings with a non-coplanar distribution of the magnetization, the removal of the Kramers degeneracy and the quantization of the quasi-momentum can lead to the existence of persistent currents that were predicted in Refs. 21,22. In a conical magnetic spiral, the asymmetry of the spectrum leads to the diode effect and the "easy" direction of the current flow is defined by the sign of the wave number of the spiral (the "left" or "right" spiral) and the direction of the perpendicular component of the magnetic moment m_z (see expression 5.6). In addition, as follows from Equation (5.9), the components of the wave function and hence the average value of the intrinsic magnetic moment of the electron depend on the components of the quasi-momentum along the axis of the helix. In particular, this leads to the asymmetry of the electron scattering by non-magnetic impurities, which also contributes to the diode effect.[17] Similar effects (asymmetry of the group velocity and frequency scattering by non-magnetic impurities) are responsible for the features of the spatial dispersion of the dielectric permittivity. In a magnetic conical spiral, the decomposition of the dielectric tensor can contain an additional term of the following form[23]:

$$\varepsilon_{ii} = K_{ii}\left(\vec{M}\cdot\left[\frac{\partial \vec{M}}{\partial z}\times\frac{\partial^2 \vec{M}}{\partial z^2}\right]\right)k_z, \qquad (5.10)$$

where k_z is the component of the wave vector of the electromagnetic wave directed along the helix axis.

It should be emphasized that there is a possibility of transitions between the spin sub-bands in the collinear and non-coplanar magnetic systems under the influence of the electric component of an electromagnetic wave. Using the wave function (5.9), it is easy to calculate the probability of electric dipole transitions between the spin sub-bands in the magnetic helix[15,24]:

$$W^{\pm}_{k,k'} = \frac{2\pi}{\hbar}\left(\frac{JeE_z q}{2mw^2}\right)^2 \left(1-m_z^2\right)\delta\left(\vec{k}-\vec{k'}\right)\delta(\Delta E(k_z)-\hbar w), \qquad (5.11)$$

where w is the frequency of the electromagnetic wave, ΔE is the energy separation between the spin sub-bands and $\delta(x)$ is a delta function. In non-collinear systems, these transitions lead to an additional contribution to the absorption of radiation, which is analyzed in detail in Ref. 25. In non-coplanar magnetic systems similar to the magnetic conical spiral, the electric dipole transitions (5.11) lead to the appearance of a constant electric current, that is, to the photovoltaic effect.[24]

Let us consider the features of neutron scattering in non-coplanar magnetic systems. It is known[19] that the interaction of cold neutrons with a magnetic field is small and the scattering cross section can be represented as a series in powers of the magnetic induction $\vec{B}$. The interaction between the magnetic field and the neutron spin depends only on the angle between them. Consequently, the total scattering cross section (averaged over the spin polarization of the incident and scattered neutrons) should be invariant with respect to the coherent rotation of the magnetic field at each space point. Taking into account these considerations, the scattering cross section can be written as

$$\sigma\left(\vec{k},\vec{k}',\vec{B}\left(\vec{r}\right)\right)=\sigma_0\left(\vec{k},\vec{k}'\right)+\int Q_1\left(\vec{k},\vec{k}';\vec{r}_1,\vec{r}_2\right)\left(\vec{B}\left(\vec{r}_1\right)\cdot\vec{B}\left(\vec{r}_2\right)\right)\mathrm{d}\vec{r}_1\,\mathrm{d}\vec{r}_2$$
$$+\int Q_2\left(\vec{k},\vec{k}';\vec{r}_1,\vec{r}_2,\vec{r}_3\right),\left(\vec{B}\left(\vec{r}_1\right)\cdot\left[\vec{B}\left(\vec{r}_2\right)\times\vec{B}\left(\vec{r}_3\right)\right]\right)\mathrm{d}\vec{r}_1\mathrm{d}\vec{r}_2\mathrm{d}\vec{r}_3+\cdots, \tag{5.12}$$

where $\vec{k}$ and $\vec{k}'$ are the wave vectors of the incident and scattered neutrons, respectively, Q_1 and Q_2 are scalar functions. Since the cross section should satisfy the reciprocity theorem, that is,

$$\sigma\left(\vec{k},\vec{k}',\vec{B}\left(\vec{r}\right)\right)=\sigma\left(-\vec{k}',-\vec{k},-\vec{B}\left(\vec{r}\right)\right), \tag{5.13}$$

the last term in Equation (5.12) describes the "non-reciprocal" contribution from a non-coplanar distribution of magnetic induction. Note that the calculation of this contribution goes beyond the Born approximation, which is usually used to calculate the scattering cross section of cold neutrons. The various effects of neutron scattering on the non-reciprocal non-coplanar magnetic structures are discussed in Refs. 26,27.

4 FERROMAGNETIC NANOSTRUCTURES WITH VORTEX AND ANTI-VORTEX MAGNETIZATION

The patterned ferromagnetic nanostructures were fabricated using negative e-beam lithography and ion etching processes. The initial multi-layer structure consisting of a ferromagnetic film covered by V (15 nm)/Cu (10 nm) was deposited onto a Si substrate by magnetron sputtering and then the sample was covered by e-beam resist. The initial protective mask was formed in resist by exposure in the ELPHY PLUS system (based on the scanning electron microscope (SEM) SUPRA 50VP) with subsequent chemical treatment in an organic solvent. Afterwards, the image was transferred to the Cu layer by Ar^+-ion etching and further to the V layer by plasma etching in Freon. At the final stage, the patterned system was fabricated in the ferromagnetic layer by Ar^+-ion etching.

The magnetic states and the magnetization-reversal effects in ferromagnetic nanostructures were studied using the vacuum multi-mode scanning probe microscope Solver HV. This was equipped with a dc electromagnet, producing a field of up to 0.1 Torr, incorporated in a vacuum vibration-insulating platform. The magnetic force microscope (MFM) measurements were performed in the non-contact constant height mode. The phase shift $\Delta\varphi$ of cantilever oscillations under the gradient of the magnetic force was registered to obtain the MFM contrast. All measurements were performed in a vacuum of 10^{-4} Torr, which improved the MFM signal due to an increase in the cantilever quality factor.

The vortex magnetization distribution ($v = 1$, $\varphi_0 = \pm\pi/2$ in Equation 5.3) is the ground state of the circular ferromagnetic disc under the condition that the radius and height of the disc are greater than the exchange length $l_{ex} \approx \sqrt{J/M_s^2} \sim 20\,\text{nm}$.[15,28] Due to the symmetrical shape of the magnetic vortex state, it is degenerate with respect to the direction of vorticity and usually in an array the number of discs with the "left" and "right" direction of the vortex ($\varphi_0 = \pm\pi/2$) is the same.[29]

In experimental MFM images of a circular magnetic vortex, one can see only dark or light areas in the disc centre corresponding to the vortex core.[30–33] The sign of the MFM contrast depends on the relative direction of the vortex core and the magnetization of the probe. The shell of a circular vortex creates virtually no stray fields and does not contribute to the MFM contrast. Note that if the direction of the core can be clearly identified by the MFM image, the direction of the vorticity in the circular vortex shell cannot be determined without the application of an external magnetic field.[34]

Magnetic vortex state can be controlled in various ways with the help of external homogeneous, heterogeneous and pulsed fields.[35–42] However, one of the most effective ways to control the direction of the vortex vorticity is the use of a disc shape asymmetry.[43] As an example, we consider the possibility of forming coherent magnetic vortices in an array of triangular particles (Figure 5.1A). In this case, the probability of nucleation "left" and "right" vortices after magnetization reversal in an external field directed along the base of the triangle is different[44] and depends on the direction of the field. This makes it possible to create in particle arrays a coherent vortex state with the same sign of vorticity. The MFM image of the residual state of the array of triangular particles is presented in Figure 5.1B. It is clearly seen that all particles have the same sign vorticity.

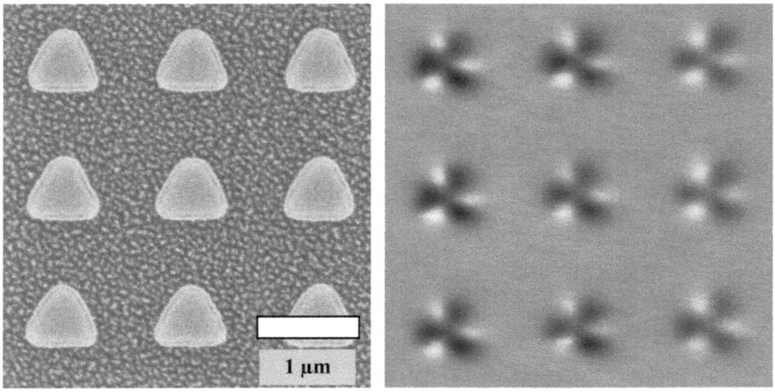

FIGURE 5.1 (A) SEM image of the triangular particles array. (B) MFM image of the residual state after magnetization in a field applied along the triangle base.

A remarkable feature of this system is an existence of a macroscopic toroidal moment

$$\vec{T} = \frac{1}{N}\sum_i \left[\vec{r} \times \vec{M}\right] \neq 0. \tag{5.14}$$

This should lead to non-reciprocal optical effects in the diffraction of light from such a magnetic lattice, which was recently confirmed experimentally.[45]

Another fundamental magnetization structure is the anti-vortex,[46–48] which is a topological counterpart of a magnetic vortex (Figure 5.2). Besides being a remarkable magnetic structure, the anti-vortex is expected to show unusual transport properties in the topological Hall effect.[49] For this distribution, the sign of the Hall voltage reverses compared to the vortex system (see Equation 5.4) since the anti-vortex has the opposite topological charge of the soliton ν.

Unlike a vortex, the realization of an anti-vortex, that is, the preparation of a nanostructure that contains only a single anti-vortex, is a challenging task. An anti-vortex has a "magnetic charge" and the implementation of such a state is difficult because of the excess (compared to the vortex state) magnetostatic energy, which is proportional to the volume of the particles.[50] To realize the anti-vortex state, we fabricated asymmetrical crosses as shown by the SEM image of the fabricated structure in Figure 5.3A. The anti-vortex state was done in several stages. At the first stage, a strong magnetic field was applied along the non-symmetrical diagonal of the cross and a quasi-uniform magnetization was formed (Figure 5.3B). Then, under the action of a specific reversed magnetic field H with a magnitude $H_1 < H < H_2$ (H_1 and H_2 are the

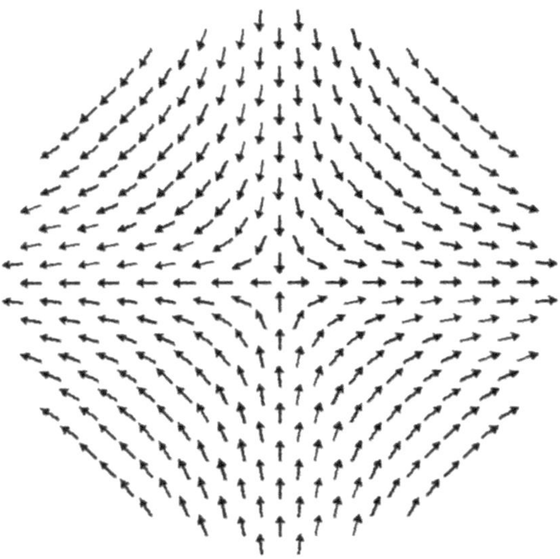

FIGURE 5.2 The distributions of magnetization for the anti-vortex states.

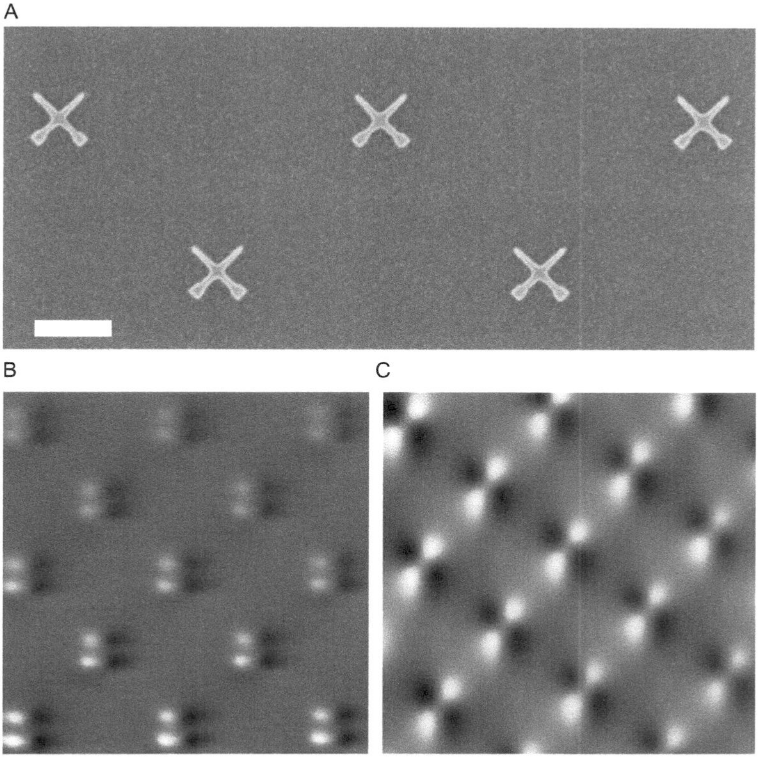

FIGURE 5.3 (A) SEM image of the asymmetric Co nanocrosses. The white scale bar is 1 μm. (B) MFM image of the crosses in the quasi-uniform state. (C) MFM image of the crosses in the anti-vortex state. MFM frame size is 12 × 12 μm.

coercivity of bulbous and tapered ends, respectively), we observed the formation of anti-vortex states in all crosses (Figure 5.3C).[50]

We also carried out the experiments on the magnetization reversal of symmetric cross-shaped particles under the action of the non-uniform field of the MFM probe to investigate the processes involved in the formation of an anti-vortex state.[38] We employed a two-stage process, in which, first, the particle was transformed, by means of the local scanning, from the quasi-homogeneous state of type A (Figure 5.4A and B) into a quasi-uniform state of type B (Figure 5.4C and D) and then into the anti-vortex state (Figure 5.4E and F).

Using the procedure of a two-stage magnetization reversal, it is possible to form anti-vortex states of different symmetries in cross-shaped particles.[38] To exemplify, Figure 5.5 shows the results of experiments on the formation of magnetic anti-vortices whose magnetization distributions are rotated by an angle of 90° relative to each other.

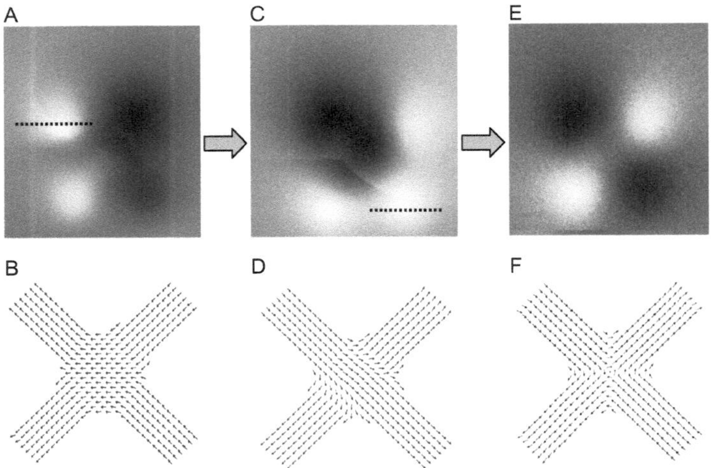

FIGURE 5.4 MFM images and corresponding magnetization distributions for probe-induced transitions in a cross-like particle. (B) Quasi-homogeneous state of type A. (D) Quasi-homogeneous state of type B. (F) Anti-vortex state. The scanning line position is indicated by the dashed line in the MFM images.

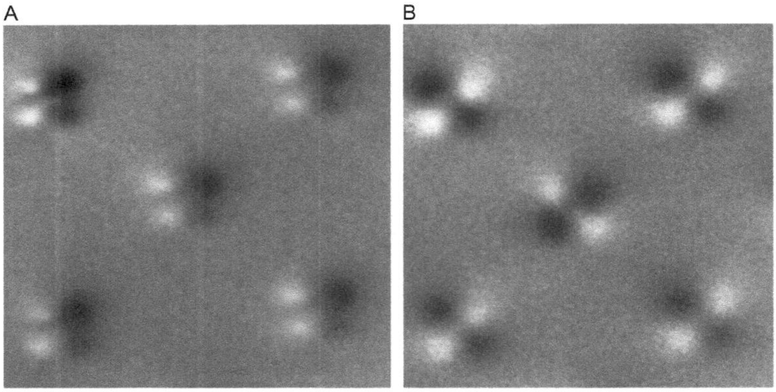

FIGURE 5.5 Formation of anti-vortex states under the effect of the field of the MFM probe. (A) Initial state. (B) Anti-vortex states formed by the MFM probe. In the central cross, a magnetic anti-vortex rotated by an angle of 90° relative to the other particles was formed.

5 FERROMAGNETIC NANOSTRUCTURES WITH A SPIRAL MAGNETIZATION DISTRIBUTION

Helical states can be realized in multi-layer nanomagnets through the interlayer magnetostatic interaction. The strength and sign of the interaction energy for two uniformly magnetized particles are determined by the mutual orientation of their total magnetic moments. For example, when two

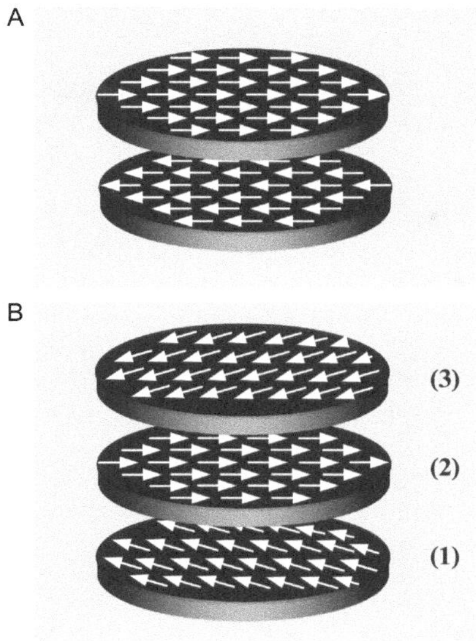

FIGURE 5.6 Magnetic states in multi-layer nanomagnets. (A) Equilibrium distribution of the magnetization in the system of two single-domain nanodiscs. (B) Non-collinear helical state in a three-layer nanomagnet.

ferromagnetic discs are separated by a non-magnetic spacer, the magnetostatic interaction leads to an antiferromagnetic orientation of their magnetization with respect to each other (Figure 5.6). The situation is significantly changed when the stack consists of three magnetic layers. The interaction between the discs leads to frustration in the magnetic moments orientation for the first and third discs. If the interaction between these discs is large enough and if the magnetic moments are confined and free to rotate in the plane, the ground state of this system is non-collinear (Figure 5.6).

Taking into account only the magnetostatic interaction, the energy of the system of three uniformly magnetized circular discs can be represented as

$$E = \varepsilon_{21}\cos\theta_{21} + \varepsilon_{23}\cos\theta_{23} + \varepsilon_{13}\cos\theta_{13}, \tag{5.15}$$

where ε_{ij} ($i,j = 1, 2, 3$) are the interaction energies between the i and j discs ($\varepsilon_{ij} > 0$) and θ_{ij} are the angles between the magnetic moment directions in the i and j discs relative to the direction of the magnetic moment in the second layer. The non-collinear state can be obtained in a stack consisting of ferromagnetic layers with unequal thicknesses. In this case, $\varepsilon_{21} \neq \varepsilon_{23} \neq \varepsilon_{13}$ and the magnitudes of ε_{ij} depend on the diameter and the thickness of the ferromagnetic discs as well as on the spacer layer thickness. The generalized phase diagram of triple nanodiscs is represented in Figure 5.7.

Depending on the ratio among ε_{12}, ε_{23} and ε_{13}, both collinear (regions 1, 2, 3 in Figure 5.7) and non-collinear (region 4) states can be obtained in triple

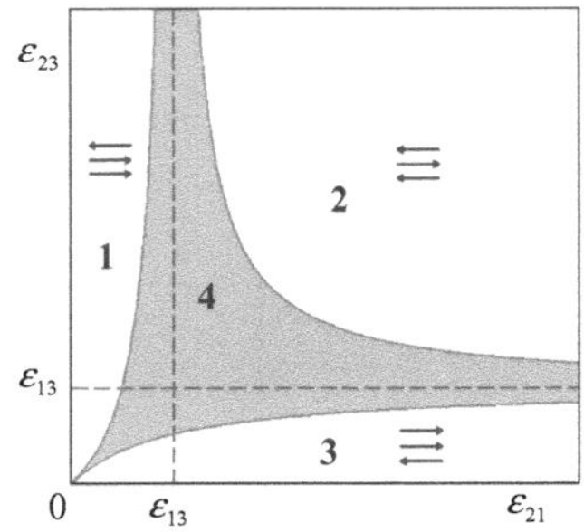

FIGURE 5.7 The diagram of states in a triple nanodisc. The central region 4 (indicated in grey colour) corresponds to the non-collinear states.

nanodiscs. The lines separating the regions with different states are defined by the following equations:

$$\varepsilon_{13} = \frac{\varepsilon_{12}\varepsilon_{23}}{\varepsilon_{12} - \varepsilon_{23}}; \quad \varepsilon_{13} = -\frac{\varepsilon_{12}\varepsilon_{23}}{\varepsilon_{12} - \varepsilon_{23}}; \quad \varepsilon_{13} = \frac{\varepsilon_{12}\varepsilon_{23}}{\varepsilon_{12} + \varepsilon_{23}}. \tag{5.16}$$

In the non-collinear state, the angles θ_{21} and θ_{23} between the magnetic moments in neighbouring layers are defined as

$$\begin{aligned} \cos\theta_{12} &= \frac{\varepsilon_{13}^2\varepsilon_{23}^2 - \varepsilon_{12}^2\varepsilon_{23}^2 - \varepsilon_{13}^2\varepsilon_{12}^2}{2\varepsilon_{12}^2\varepsilon_{23}\varepsilon_{13}}, \\ \cos\theta_{13} &= \frac{\varepsilon_{12}^2\varepsilon_{23}^2 - \varepsilon_{12}^2\varepsilon_{13}^2 - \varepsilon_{13}^2\varepsilon_{23}^2}{2\varepsilon_{12}\varepsilon_{13}^2\varepsilon_{23}}. \end{aligned} \tag{5.17}$$

In particular, for the point of asymptotic intersection ($\varepsilon_{12} = \varepsilon_{23} = \varepsilon_{13}$), the angles are $\theta_{21} = 240°$, $\theta_{23} = 120°$ for "right"-hand helicoids and $\theta_{21} = 120°$, $\theta_{23} = 240°$ for "left"-hand helicoids. Note that the spiral state is doubly degenerate, that is, the energies for left and right helicoids are identical.

Experimentally, the helical state was observed by an MFM in the [Co/Si] $\times$ 3 multi-layer stacks.[51] The model and experimental MFM pictures of a triple nanodisc are represented in Figure 5.8. The spiral symmetry of the MFM contrast distribution is clearly seen.

We investigated the dependence of electrical resistance on an applied magnetic field for a three-layer stack: Co(10 nm)/AlO_x(2 nm)/Co(5 nm)/AlO_x(2 nm)/Co(10 nm) in a helicoidal state.[52] Assuming that the magnetic field dependence of the electrical resistance for a multi-layer ferromagnetic particle is due to the effects of the exchange interaction and is determined by the scalar product of the magnetic moments of the layers,[53] we have

$$R = R_{01} + R_{02} - R_1\cos\theta_{12} - R_2\cos\theta_{23}, \tag{5.18}$$

where $R_{01,02} - R_{1,2}$ is the resistance of the first (second) tunnel junction with a parallel orientation of the magnetic moments of the discs $\theta_{12} = \theta_{23} = 0$.

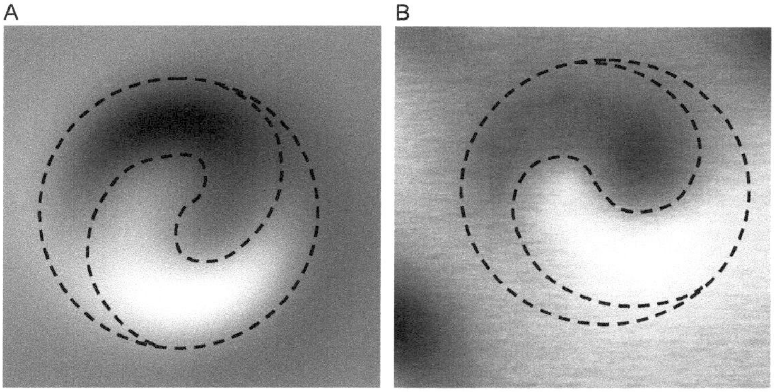

FIGURE 5.8 (A) Model MFM contrast distribution for an optimized triple nanodisc. (B) Experimental MFM image from a triple Co/Si nanodisc. Frame size 1×1 μm. The dashed lines separate the regions with dark and bright contrast to emphasize the spiral symmetry of the MFM contrast.

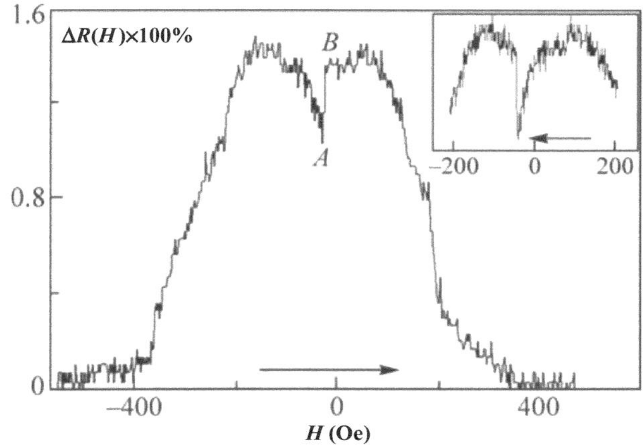

FIGURE 5.9 Measured $\Delta R(H)$ curve for a round three-layer stack 250 nm in diameter. The arrow indicates the direction of change for the applied field. The inset demonstrates that the position of the *A*–*B* resistance jump changes with a change in the field direction.

The experimental dependences of the relative change in resistance ΔR

$$\Delta R(H) = \frac{R(H) - R(H_\infty)}{R(H_\infty)} \tag{5.19}$$

on the applied magnetic field for a round particle about 250 nm in diameter are presented in Figure 5.9. At high values of the external magnetic field, the magnetic moments of all discs are parallel and, according to Equation (5.19), the resistance of the system is minimal. With a decrease in the magnetic field, the magnetic moments are disoriented, which leads to an increase in resistance. Of most interest is the region of small magnetic fields. After reaching a maximum ΔR_{max}, the resistance decreases to $\Delta R(0)$ and changes in the direction of the applied magnetic field lead to a steep growth in the resistance (A–B transition in Figure 5.9). With a further increase in the field, the resistance again reaches a peak. This magnetic field dependence of the resistance suggests a non-collinear distribution of the magnetization in the layers at zero applied magnetic field. If the resistance peak corresponds to a collinear antiferromagnetic distribution of magnetic moments, then, at zero applied magnetic field, we have the symmetric non-collinear state $\theta_{12}=\theta_{23}=\theta$. Hence, using Equation (5.19) and the experimental value $\Delta R(0)\approx 0.75\ \Delta R_{max}$ we find $\theta\approx 120°$. This scenario is also confirmed by computer simulations.

An increase in stack shape anisotropy leads to the instability of angular phases and the magnetic field dependence of the resistance demonstrates the features corresponding to the transitions between collinear states. We fabricated elliptical stacks with 100×200 nm lateral size. The corresponding field dependence of the resistance for this structure is presented in Figure 5.10.[52] The magnetic field was applied along the long axis of the stack. Indeed, we observe resistance jumps that correspond to the transitions between the collinear states indicated in the figure. Using these experimental data, we find that the magnetic field dependent parts of the resistance $R_{1,2}$ at these transitions (see Equation 5.18) differ by no more than 20%. This experiment demonstrates, first, a high degree of similarity for the connected in series tunnel junctions forming the structure under study and, second, the fundamental importance of the anisotropy for the formation of non-collinear states. Note that the existence of three stable states, which the particle has at zero external

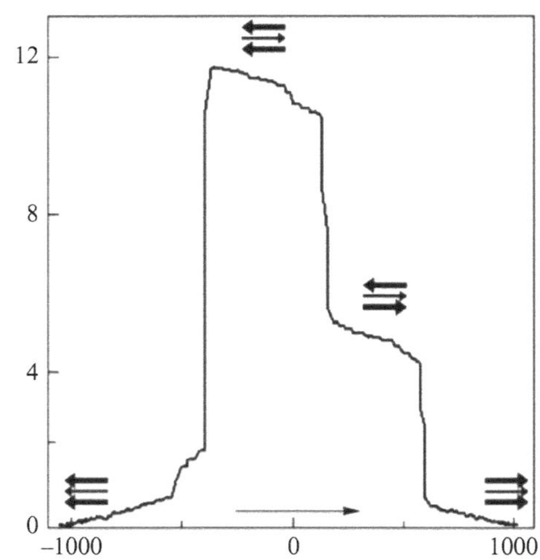

FIGURE 5.10 Measured $\Delta R(H)$ curve for an elliptical three-layer stack with a lateral size 100×200 nm. The magnetic field is applied along the long axis of the particle; the arrow indicates the direction of change in the applied field.

field, is promising for applying this system to information storage and processing devices.

6 CONTROLLABLE DW PINNING IN A FERROMAGNETIC NANOWIRE

The field-driven motion and pinning of DWs in ferromagnetic nanowires (NWs) are the subjects of intensive research motivated by promising applications for the development of magnetic logic and data storage systems.[54–57] However, the operation of DW-based magnetic logical cells and data storage systems requires controlled DW pinning for preventing the accidental erasure of data and saving the results of intermediate calculations. The simple method of DW pinning is to use geometrical features in NW topology and artificially patterned traps at the NW edge.[58–63] Recently, the non-destructive method of DW pinning based on magnetostatic interaction of a DW with a ferromagnetic nanobar has been proposed in Refs. 64,65. It was shown that the tunable system of nanobars located on one side of the NW allows one to realize the asymmetric DW potentials. A combination of trap and nanobar was used for controllable DW pinning in a nanoconstriction.[66] In this case, the nanobar was located perpendicular to the NW in the vicinity of the notch. It has been demonstrated that the nanobar stray field substantially modifies the value of the pinning energy, depending on the relative orientation of the DW and nanobar magnetic moment.

Recently, we proposed a new idea of DW pinning control based on the effective change of the stray field's configuration in a system of two single-domain ferromagnetic NPs.[67] We considered a combined system consisting of a planar ferromagnetic NW and two elongated ferromagnetic NPs placed perpendicularly on either side of the NW (Figure 5.11).

The NW has a special pad (N), which is used for the controllable nucleation of domains with opposite orientation in an external magnetic field.[68–70] The NPs are used as a magnetic gate for the field-controlled DW pinning–depinning.

The process of NW magnetization reversal is accompanied by DW propagation from the pad to the free NW end and can be stopped due to magnetostatic interaction with the NPs gate. To estimate the pinning energy and depinning fields for different configurations of magnetization in the NW–NPs system, we considered the dependence of the DW–NPs interaction energy on

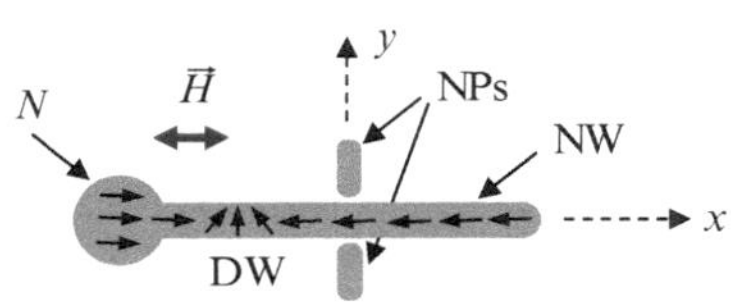

FIGURE 5.11 The schematic drawing of the field-driven NW–NPs system.

the DW position. In general, the potential energy of the NW in an external magnetic field can be calculated as follows:

$$E_{NW}(x_{DW}) = -\int_{V_{NW}} \left(\vec{M}_{NW} \cdot \vec{H}\right) dV_{NW}, \tag{5.20}$$

where $\vec{M}_{NW}\left(\vec{r}\right)$ is the magnetization distribution in the NW, $\vec{H}$ is the sum of the magnetic stray fields from the NPs and the uniform external magnetic field $\vec{H} = \vec{H}_{NP} + \vec{H}_{ex}$, x_{DW} is the position of the DW centre and the integration is performed over the NW volume.

The dependence of the interaction energy E_{NW} on the DW position allows one to estimate the pinning energy and depinning fields for different configurations of magnetization in the NW–NPs system. We carried out the numerical calculations of the energy landscape based on Equation (5.20) and the model of rigid DW, neglecting the change of exchange and demagnetizing energy as the DW propagates from the pad to the free NW end.[71]

The effects of DW pinning depend on the mutual configuration of magnetization in the NW and NPs. Different possible variations of the NW and NPs magnetization, and the corresponding dependences of the NW–NPs system energy on the DW position are represented in Figures 5.12–5.15.

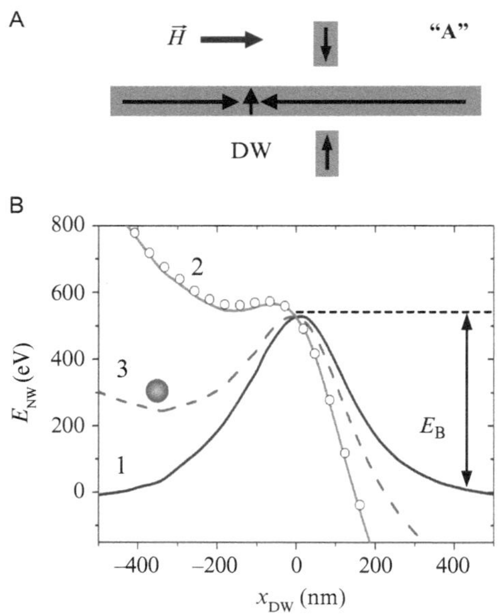

FIGURE 5.12 (A) A-type configuration of magnetization in the NW–NPs system in an external magnetic field. (B) The potential energy profiles $E_{NW}(x_{DW})$ for different external magnetic fields. The solid line (1) is the energy profile at zero field. The solid line (2) is for the critical external field $H_B = 470$ Oe. The dashed line (3) is for the intermediate field 0.5 H_B. The DW pinning position is indicated schematically by the circle in curve 3.

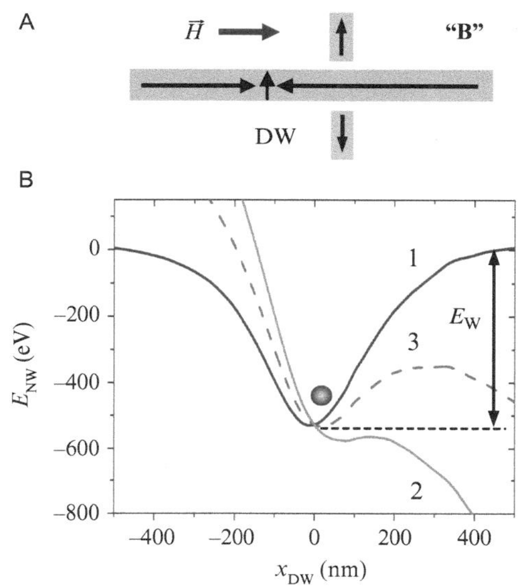

FIGURE 5.13 (A) B-type configuration of magnetization in the NW–NPs system in an external magnetic field. (B) The energy E_{NW} profile for different external magnetic fields. The solid line (1) is the energy profile at zero external field. The dot-dashed line (2) is for the critical magnetic field $H_W = 470$ Oe. The dashed line (3) is for the intermediate field 0.5 H_W. The DW pinning position is indicated schematically by the circle.

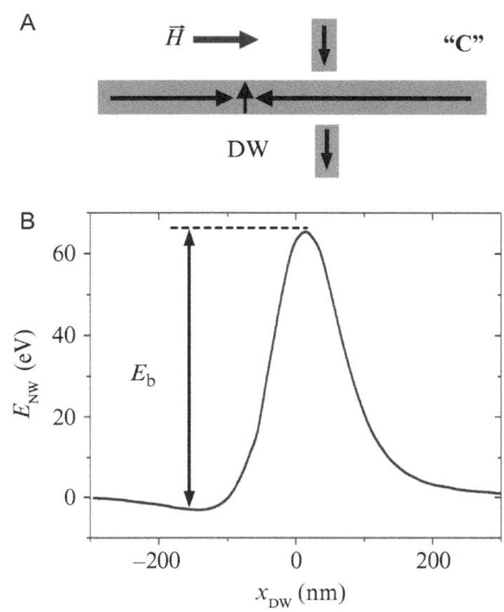

FIGURE 5.14 (A) C-type configuration of magnetization in the NW–NPs system. (B) Energy profile $E_{NW}(x_{DW})$ at zero external field.

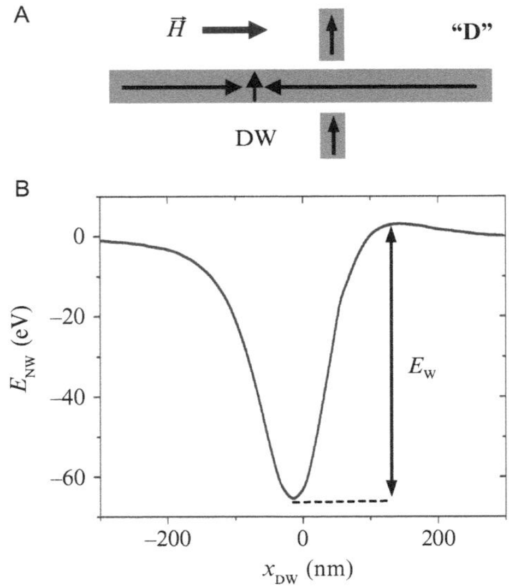

FIGURE 5.15 (A) D-type configuration of magnetization in the NW–NPs system. (B) Energy profile $E_{NW}(x_{DW})$ at zero external field.

Let us consider the configuration represented in Figure 5.12A (A-type configuration). In this case, the magnetization vectors in the NW and the NPs are directed towards each other (head-to-head or tail-to-tail configurations). The corresponding dependences of energy $E_{NW}(x_{DW})$ calculated using the rigid DW model for different external fields are presented in Figure 5.12B. The DW propagation is connected with overcoming the energy barrier, which is defined mainly by magnetostatic interaction of the quasi-uniform NW magnetization outside the DW with the x component of the NPs field. Note that the magnitude of the barrier is independent of the DW direction due to the symmetry of the NPs magnetic configuration. Hence, in a weak magnetic field, the DW will be pinned in the region before the NPs gate. The estimate of energy barrier at zero field is $E_B = 508$ eV. In an external magnetic field, the pinning barrier is decreased and at 470 Oe (depinning field H_B) it vanishes completely (curve 2 in Figure 5.12B). We performed also the comparative accurate $E_{NW}(x_{DW})$ calculations based on micromagnetic $\vec{M}_{NP}\left(\vec{r}\right)$ and $\vec{M}_{NW}\left(\vec{r}\right)$ distributions taking into account the effects of the DW and the NPs magnetization disturbance (see the circles in Figure 5.12B). It was estimated that the relative error in $E_{NW}(x_{DW})$ calculations based on the rigid DW model is less than 5% and can be neglected in qualitative considerations. The critical depinning field estimated directly from object oriented micromagnetic framework, NIST (OOMMF) modelling was $H_B = 490$ Oe. The micromagnetic simulations also confirmed that the DW is pinned before the NPs gate.

In the B-type configuration represented in Figure 5.13A, we have head-to-head (tail-to-tail) magnetization in the NW but tail-to-tail (head-to-head) magnetization in the NPs. The energy profile $E_{NW}(x_{DW})$ has a potential well (Figure 5.13B), and the DW is pinned at the region directly between the NPs. Thus, the NW remagnetization is connected with the DW escape from the energy well. The value of the energy well is defined mainly by the magnetostatic interaction of the NW magnetization outside the DW with the x component of the NPs field and does not depend on the DW direction. The estimate of the activation energy E_W at the zero field is 508 eV. In an external magnetic field, the pinning barrier connected with the potential well is decreased (see curve 3 in Figure 5.13B), and at 470 Oe (depinning field H_W) it vanishes completely (curve 2 in Figure 5.13B).

The micromagnetic modelling confirmed the DW pinning between NPs for the B-type configuration. The depinning field estimated directly from the OOMMF simulations was 490 Oe.

The third possible configuration, in which the NP moments have the same direction (head-to-tail) but the DW magnetization has the opposite direction (C-type configuration), is presented in Figure 5.14A. In this case, the x components of the stray magnetic fields from the NPs are partly compensated. The calculated energy landscape has a potential barrier caused by the interaction of the DW with the y component of the NPs field (Figure 5.14B). Hence, in the C-type configuration, a DW can be pinned in the region before the NPs. The C-type configuration has a very small pinning energy in comparison with A configurations. The estimate of the energy barrier E_b at zero external field is 65 eV. The calculated depinning field H_b (estimated as the barrier vanishing in the external field) is 90 Oe that practically coincides with the value of 100 Oe estimated directly from the OOMMF simulations.

The fourth possible configuration (D-type) is presented in Figure 5.6A. In this case, the directions of magnetization in both NPs are the same (head-to-tail) and coincide with the DW orientation. The energy landscape $E_{NW}(x_{DW})$ has a potential well caused by the interaction of the DW with the y component of the NPs field (Figure 5.6B). Hence, in this case, the DW can be pinned in the region directly between the NPs but this configuration has a small pinning energy in comparison with the B configuration. The estimate of the energy barrier E_w at zero field is 65 eV. The depinning field H_w estimated as the potential wall vanishing in the external field is 90 Oe. The depinning field estimated directly from the OOMMF simulations was 100 Oe.

The NW–NPs systems were fabricated using a negative e-beam lithography and ion etching processes. The $Co_{60}Fe_{40}$ (20 nm) structure was deposited onto a Si substrate by magnetron sputtering. The characteristic SEM image of the NW–NPs system is represented in Figure 5.16. The width of the NW was 100 nm, the NW length was about 2.8 μm, the lateral dimensions of the NPs were 100 × 200 nm, the NP–NW separation was 100 nm and the diameter of nucleating pad was 200 nm.

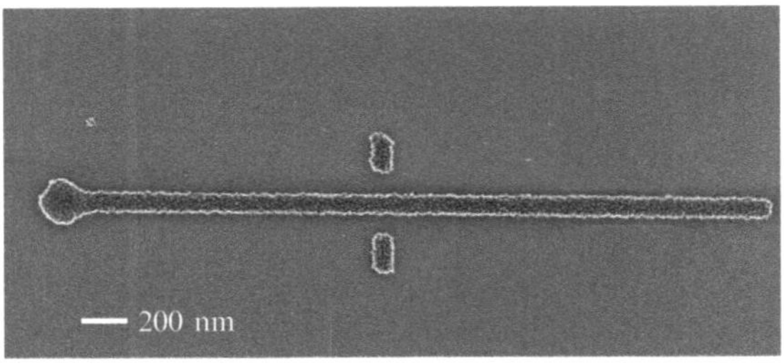

FIGURE 5.16 The SEM image of nanowire–nanoparticles system.

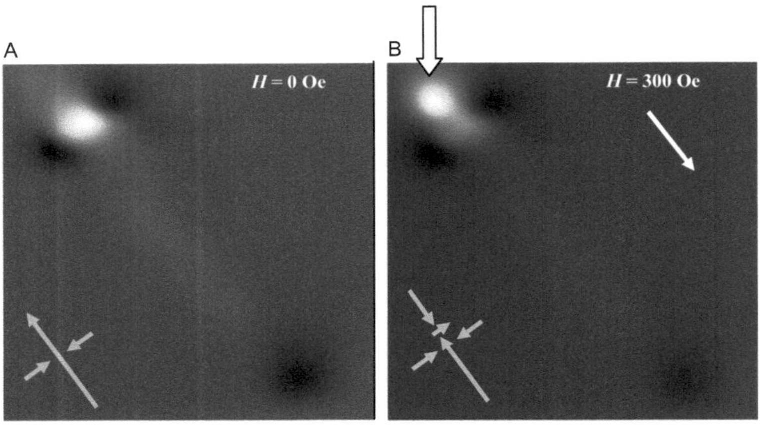

FIGURE 5.17 The MFM images of NW–NPs system after the application of an external magnetic field. (A) Initial state with the A-type configuration after magnetizing along the NW. (B) MFM image of the DW pinned at the barrier near the NPs after applying a 300-Oe external field (DW position is indicated by the white arrow). The configurations of magnetization are shown at the left bottom corners.

We investigated the dependence of the pinning–depinning processes in the NW–NPs system (Figure 5.16) on the configuration of magnetization in the NPs. The experiment was performed *in situ* in the MFM "Solver HV" vacuum chamber. First, we studied the NW magnetization reversal when magnetic moments in the NW–NPs system corresponded to the A-type configuration. The different stages of NW remagnetization experiment are represented in Figure 5.17. The initial state was prepared by magnetizing the sample along the NW. The MFM image of the initial state is presented in Figure 5.17A. It is seen that the MFM contrast pole positions in the MFM image of the initial state correspond to the type A configuration. The dark pole corresponds to the tail and bright pole corresponds to the head of the magnetization vector. Afterwards, the external field $H_{||}$ was applied in the reversed direction and the NW remagnetization effect was studied. When $H_{||}$ exceeded 300 Oe (DW nucleation field H_{Nuc}), we registered the appearance of an additional bright pole in the MFM image (indicated by the white arrow in Figure 5.17B) corresponding to the DW. So, we observed that the DW is pinned at the potential barrier before the NPs. The DW position was stable in external magnetic fields of up to $H_{\parallel}=560$ Oe, but when $H_{\parallel}$ exceeded 560 Oe (DW depinning field H_B) the remagnetization of the NW was observed.

A different situation was observed for the magnetization-reversal experiment in the B-type configuration of the NW–NPs magnetic moments. The different stages of the NW remagnetization experiment are demonstrated in Figure 5.18. The MFM image of the initial state is presented in Figure 5.18A. The MFM pole positions confirm the B-type configuration. Analogously, when the external magnetic field $H_{\parallel}$ exceeded 300 Oe, we registered the appearance of the additional bright MFM pole (indicated by white arrow in Figure 5.18B) corresponding to the DW. But it is seen that in this case the DW was pinned right in between NPs. We believe that this can be explained as the DW pinning in a potential well. The depinning field was practically the same as in the case of the A-type configuration, so when $H_{\parallel}$ exceeded 560 Oe (DW depinning field H_W) we observed the remagnetization of the NW. The similarity between the magnitudes of depinning fields H_B and H_W in A- and B- type configurations observed in the model simulations as well as in the experimental measurements allows one to suggest that the magnetostatic DW–NPs interaction in such a system is quite weak and the rigid DW approximation is valid in this case.

The magnetization reversal in the NW–NPs system with C (and D)-type configuration showed that the remagnetization effect was observed just after

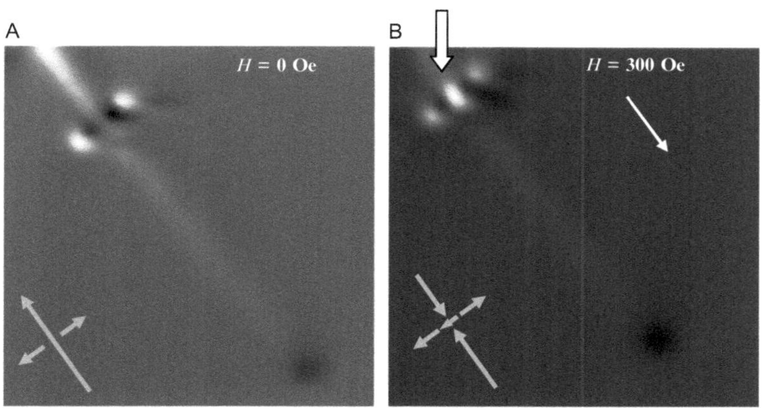

FIGURE 5.18 MFM images of the NW–NPs system after the application of an external magnetic field. (A) Initial state with a B-type configuration after magnetizing along the NW. (B) MFM image of the DW pinned at the well between NPs after applying a 300-Oe external field (DW position is indicated by the white arrow). The configurations of magnetization are shown at the left bottom corners.

applying the nucleating external magnetic field $H_{||}=H_{Nuc}=300$ Oe. This fact demonstrates that for our NW–NPs system, the DW nucleation field is larger than the depinning field connected with the DW–NPs interaction in C(D)-type configuration ($H_{Nuc}>H_b,H_w$).

The NW–NPs system described above can be considered as a prototype magnetic logic cell, which realizes the "exclusive disjunction" logical operation (so-called XOR).[67] For example, the input signals can be encoded as the direction of the magnetic moment in the NPs and the output information as the moment direction at the free end of the NW. Figure 5.19 shows a scheme of input–output and possible input–output information coding.

The algorithm of logical cell operations includes the periodic reactivation cycle and the logical computing cycle consisting of input information and the writing and reading of the result of a logical operation. The logical calculations can be organized as follows. A first stage is the initialization process when the logical "1" is written in the NW by an external magnetic field $H_{||}$ ($H_{||}>H_B,H_W$) applied in the x-direction. Afterwards, the input information is written in the NPs by local magnetic fields, which can be created for instance with current buses B1–B2 and B3–B4 (see Figure 5.19). At the final stage, the reversed external magnetic field $H_{||}$ with amplitude $H_B,H_W>H_{||}>H_{Nuc}$ ($H_{Nuc}>H_b,H_w$) is applied in the x-direction and the output information is read. The magnetic state of the NW free end can be analyzed using the local magneto optical Kerr effect or by means of a tunnelling magnetoresistance element. Afterwards, the cycle of operations is repeated. The correspondence between input and output information in such a magnetic logical cell is represented in Table 5.1.

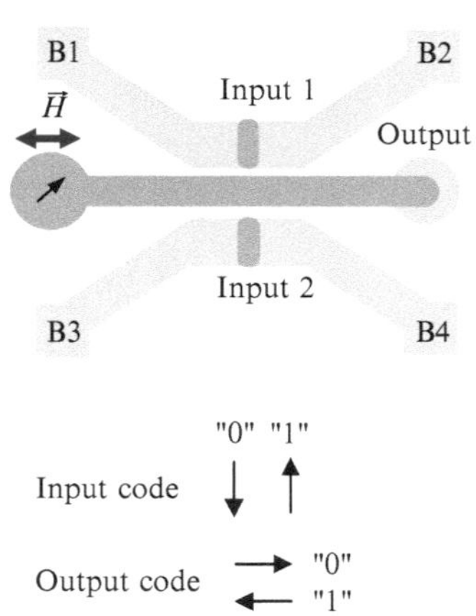

FIGURE 5.19 The schematic NW–NPs logical cell morphology and the coding of input–output states.

TABLE 5.1 The Logical Output States for All Input States (Truth Table)

Input 1	Input 2	Output
0	0	0
0	1	1
1	0	1
1	1	0

7 CONCLUSION

Thus, in this chapter, we discussed some properties of patterned ferromagnetic nanostructures:

- New transport, optical and neutron-optical effects in ferromagnetic nanosystems with non-coplanar magnetization distribution were predicted.
- Methods of nanolithography and scanning probe microscopy allowed us to realize vortex, anti-vortex and spiral distributions of magnetization in ferromagnetic nanostructures.
- It was established experimentally that in multi-layer anisotropic ferromagnetic particles with tunnel barriers, there are stable collinear states with different magnitudes of tunnel resistance that makes these systems promising for use in storage devices and processing.
- The possibility of controlled pinning of DWs in ferromagnetic NW under the influence of local stray fields of ferromagnetic NPs was shown. Based on this system, we proposed that the logic cell implements a logical XOR function.

ACKNOWLEDGEMENTS

This work was supported by Russian Foundation for Basic Research; Programs of Russian Academy of Science ("Fundamentals of technology of nanostructures and nanomaterials", "Spin phenomena in solid nanostructures and spintronics", "Quantum macro-physics") and The Ministry of Education and Science of The Russian Federation.

REFERENCES

1. Chappert C, Bernas H, Ferré J, Kottler V, Jamet J-P, Chen Y, et al. Planar patterned magnetic media obtained by ion irradiation. *Science* 1998;**280**:1919–22.
2. Albrecht M, Moser A, Rettner CT, Anders S, Thomson T, Terris BD. Writing of high-density patterned perpendicular media with a conventional longitudinal recording head. *Appl Phys Lett* 2002;**80**:3409–11.

3. Kryder MH, Gustafson RW. High-density perpendicular recording—advances, issues, and extensibility. *J Magn Magn Mater* 2005;**287**:449–58.
4. Richter HJ, Dobin AY, Heinonen O, Gao KZ, Veerdonk RJMvd, Lynch RT, et al. Recording on bit-patterned media at densities of 1 Tb/in^2 and beyond. *IEEE Trans Magn* 2006;**42**:2255–60.
5. Moser A, Hellwig O, Kercher D, Dobisz E. Off-track margin in bit patterned media. *Appl Phys Lett* 2007;**91**(16):162502.
6. Albrecht M, Rettner CT, Moser A, Best ME, Terris BD. Recording performance of high-density patterned perpendicular magnetic media. *Appl Phys Lett* 2002;**81**:2875–7.
7. Martn JI, Nogues J, Liu K, Vicent JL, Schuller IK. Ordered magnetic nanostructures: fabrication and properties. *J Magn Magn Mater* 2003;**256**:449–501.
8. Lau JW, Shaw JM. Magnetic nanostructures for advanced technologies: fabrication, metrology and challenges. *J Phys D Appl Phys* 2011;**44**:303001.
9. Kodama RH. Magnetic nanoparticles. *J Magn Magn Mater* 1999;**200**:359–72.
10. Sapozhnikov MV, Fraerman AA, Vdovichev SN, Gribkov BA, Gusev SA, Klimov AY, et al. Effect of ferromagnetic nanoparticles on the transport properties of a GaMnAs microbridge. *Appl Phys Lett* 2007;**91**(6):062513.
11. Baibich MN, Broto JM, Fert A, Nguyen van Dau AF, Petroff F, Etienne P, et al. Giant magnetoresistance of (001)Fe/(001)Cr magnetic superlattices. *Phys Rev Lett* 1988;**61**:2472.
12. Binasch G, Grunberg P, Saurenbach F, Zinn W. Enhanced magnetoresistance in layered magnetic structures with antiferromagnetic interlayer exchange. *Phys Rev B* 1989;**39**:4828.
13. Julliere M. Tunneling between ferromagnetic films. *Phys Lett A* 1975;**54**:225.
14. Prejbeanu IL, Natali M, Buda LD, Ebels U, Lebib A, Chen Y, et al. In-plane reversal mechanisms in circular Co dots. *J Appl Phys* 2002;**91**:7343–5.
15. Metlov KL, Lee Y. Map of states for thin circular nanocylinders. *Appl Phys Lett* 2008;**92**:112506.
16. Landau LD, Lifshitz EM. *Electrodynamics of continuous media*. New York: Pergamon; 1984.
17. Fraerman AA, Udalov OG. Diode effect in a medium with helical magnetic structure. *Phys Rev B* 2008;**77**:094401.
18. Aharonov Ya, Stern A. Origin of the geometric forces accompanying Berry's geometric potentials. *Phys Rev Lett* 1992;**69**:3593.
19. Izumov YA, Naish VE, Ozerov RP. *Neutron diffraction of magnetic materials*. New York: National Consulting Bureau; 1991.
20. Matveev VM, Nagaev EL. Effect of indirect exchange via conduction electrons on magnetic ordering of the layer type. *Sov Phys JETP* 1975;**42**:1094.
21. Loss D, Goldbart P, Balatsky AV. Berry's phase and persistent charge and spin currents in textured mesoscopic rings. *Phys Rev Lett* 1990;**65**:1655.
22. Tatara G, Kohno H. Permanent current from noncommutative spin algebra. *Phys Rev B* 2003;**67**:113316.
23. Karashtin EA, Udalov OG, Fraerman AA. Optical activity in media with noncoplanar magnetization distribution. *JETP* 2009;**109**:973.
24. Fraerman AA, Udalov OG. Photogalvanic effect in ferromagnets with a noncoplanar magnetization distribution. *JETP Lett* 2008;**87**:159.
25. Karashtin EA, Udalov OG. Absorption of electromagnetic radiation by systems with a noncollinear magnetization distribution. *J Exp Theor Phys* 2011;**113**:992–9.
26. Ignatovich VK, Nikitenko Yu V, Fraerman AA. *J Exp Theor Phys* 2010;**110**:5.

27. Tatarsky DA, Udalov OG, Fraerman AA. Nonreciprocal elastic scattering of unpolarized neutrons by magnetic systems with the noncoplanar magnetization distribution. *J Exp Theor Phys* 2012;**115**:626–63.
28. Cowburn RP, Koltsov DK, Adeyeye AO, Welland ME. *Phys Rev Lett* 1999;**83**:1042–5.
29. Vdovichev SN, Gribkov BA, Gusev SA, Mironov VL, Nikitushkin DS, Fraerman AA, et al. Possibility of observing chiral-symmetry effects in ferromagnetic nanoparticles. *Phys Solid State* 2006;**48**:1902–5.
30. Natali M, Prejbeanu IL, Lebib A, Buda LD, Ounadjela K, Chen Y. Correlated magnetic vortex chains in mesoscopic cobalt dot arrays. *Phys Rev Lett* 2002;**88**(15):157203.
31. Okuno T, Shigeto K, Ono T, Mibu K, Shinjo T. MFM study of magnetic vortex cores in circular permalloy dots: behavior in external field. *J Magn Magn Mater* 2002;**240**:1–6.
32. Shima H, Novosad V, Otani Y, Fukamichi K, Kikuchi N, Kitakamai O, et al. Pinning of magnetic vortices in microfabricated permalloy dot arrays. *J Appl Phys* 2002;**92**:1473–6.
33. Pokhil T, Song D, Nowak J. Spin vortex states and hysteretic properties of submicron size NiFe elements. *J Appl Phys* 2000;**87**:6319–21.
34. Fernandez A, Cerjan CJ. Nucleation and annihilation of magnetic vortices in submicron-scale Co dots. *J Appl Phys* 2000;**87**:1395–401.
35. Taniuchi T, Oshima M, Akinaga H, Ono K. Vortex-chirality control in mesoscopic disk magnets observed by photoelectron emission microscopy. *J Appl Phys* 2005;**97**:10J904.
36. Chang J, Mironov VL, Gribkov BA, Fraerman AA, Gusev SA, Vdovichev SN. Magnetic state control of ferromagnetic nanodots by magnetic force microscopy probe. *J Appl Phys* 2006;**100**(10):104304.
37. Mironov VL, Gribkov BA, Fraerman AA, Gusev SA, Vdovichev SN, Karetnikova IR, et al. MFM probe control of magnetic vortex chirality in elliptical Co nanoparticles. *J Magn Magn Mater* 2007;**312**:153–7.
38. Mironov VL, Fraerman AA, Gribkov BA, Ermolayeva OL, Klimov AY, Gusev SA, et al. Control of the magnetic state of arrays of ferromagnetic nanoparticles with the aid of the inhomogeneous field of a magnetic-force microscope probe. *Phys Met Metallogr* 2010;**110**:708–34.
39. Jain S, Ren Y, Adeyeye AO, Singh N. Configurational anisotropy and control of magnetic vortex chirality in arrays of circular Ni80Fe20nanoscale dots. *Phys Rev B* 2009;**80**:132401.
40. Van Waeyenberge B, Puzic A, Stoll H, Chou KW, Tyliszczak T, Hertel R, et al. Magnetic vortex core reversal by excitation with short bursts of an alternating field. *Nature* 2006;**444**:461.
41. Hertel R, Gliga S, F′ahnle M, Schneider CM. Ultrafast nanomagnetic toggle switching of vortex cores. *Phys Rev Lett* 2007;**98**:117201.
42. Yamada K, Kasai S, Nakatani Y, Kobayashi K, Ono T. Switching magnetic vortex core by a single nanosecond current pulse. *Appl Phys Lett* 2008;**93**:152502.
43. Shneider M, Hofman H, Zweck J. Magnetic switching of single vortex permalloy elements. *Appl Phys Lett* 2001;**79**:3113.
44. Yakata S, Miyata M, Nanoguchi S, et al. Control of vortex chirality in regular polygonal nanomagnets using in-plane magnetic field. *Appl Phys Lett* 2010;**97**:222503.
45. Udalov OG, Sapozhnikov MV, Karashtin EA, Gribkov BA, Gusev SA, Skorohodov EV, et al. Nonreciprocal light diffraction by a lattice of magnetic vortices. *Phys Rev B* 2012;**86**:094416.
46. Gliga S, Yan M, Hertel R, Schneider CM. Ultrafast dynamics of a magnetic antivortex: Micromagnetic simulations. *Phys Rev B* 2008;**77**:060404R.
47. Wang H, Campbell CE. Spin dynamics of a magnetic antivortex: Micromagnetic simulations. *Phys Rev B* 2007;**76**:220407.

48. Shigeto K, Okuno T, Mibu K, Shinjo T, Ono T. Magnetic force microscopy observation of antivortex core with perpendicular magnetization in patterned thin film of permalloy. *Appl Phys Lett* 2002;**80**:4190.
49. Neubauer A, Pfleiderer C, Binz B, Rosch A, Ritz R, Niklowitz PG, et al. Topological Hall Effect in the A Phase of MnSi. *Phys Rev Lett* 2009;**102**:186602.
50. Mironov VL, Ermolaeva OL, Gusev SA, Klimov AY, Rogov VV, Gribkov BA, et al. Antivortex state in crosslike nanomagnets. *Phys Rev B* 2010;**81**(9):094436.
51. Fraerman AA, Gribkov BA, Gusev SA, Klimov AY, Mironov VL, Nikitushkin DS, et al. Magnetic force microscopy of helical states in multilayer nanomagnets. *J Appl Phys* 2008;**103**(7):073916.
52. Vdovichev SN, Gribkov BA, Gusev SA, Klimov AY, Mironov VL, Nefedov IM, et al. Magnetoresistance and noncollinear structures of multilayer ferromagnetic nanoparticles. *JETP Lett* 2011;**94**(5):386–9.
53. Slonczewski JC. Conductance and exchange coupling of two ferromagnets separated by a tunneling barrier. *Phys Rev B* 1989;**39**:6995.
54. Allwood DA, Xiong G, Cooke MD, Faulkner CC, Atkinson D, Vernier N, et al. Submicron ferromagnetic NOT gate and shift register. *Science* 2002;**296**:2003.
55. Allwood DA, Xiong G, Faulkner CC, Atkinson D, Petit D, Cowburn RP. Magnetic Domain-Wall Logic. *Science* 2005;**309**:1688.
56. O'Brien L, Read DE, Zeng HT, Lewis ER, Petit D, Cowburn RP. Bidirectional magnetic nanowire shift register. *Appl Phys Lett* 2009;**95**:232502.
57. Zeng HT, Read DE, O'Brien L, Sampaio J, Lewis ER, Petit D, et al. Asymmetric magnetic NOT gate and shift registers for high density data storage. *Appl Phys Lett* 2010;**96**:262510.
58. Zhu X, Allwood DA, Xiong G, Cowburn RP, Grütter P. Spatially resolved observation of domain-wall propagation in a submicron ferromagnetic NOT-gate. *Appl Phys Lett* 2005;**87**:062503.
59. Lewis ER, Petit D, Thevenard L, Jausovec AV, O'Brien L, Read DE, et al. Magnetic domain wall pinning by a curved conduit. *Appl Phys Lett* 2009;**95**:152505.
60. Petit D, Jausovec AV, Read DE, Cowburn RP. Domain wall pinning and potential landscapes created by constrictions and protrusions in ferromagnetic nanowires. *J Appl Phys* 2008;**103**:114307.
61. Petit D, Jausovec AV, Zeng HT, Lewis ER, O'Brien L, Read DE, et al. Mechanism for domain wall pinning and potential landscape modification by artificially patterned traps in ferromagnetic nanowires. *Phys Rev B* 2009;**79**:214405.
62. O'Shea K, McVitie S, Chapman JN, Weaver JMR. Direct observation of changes to domain wall structures in magnetic nanowires of varying width. Appl. *Phys Lett* 2008;**93**:202505.
63. Bogart LK, Atkinson D, O'Shea K, McGrouther D, McVitie S. Dependence of domain wall pinning potential landscapes on domain wall chirality and pinning site geometry in planar nanowires. *Phys Rev B* 2009;**79**:054414.
64. Zeng HT, Petit D, O'Brien L, Read D, Lewis ER, Cowburn RP. The influence of wire width on the charge distribution of transverse domain walls and their stray field interactions. *J Magn Magn Mater* 2010;**322**:2010.
65. O'Brien L, Petit D, Lewis ER, Cowburn RP, Read DE, Sampaio J, et al. Tunable Remote Pinning of Domain Walls in Magnetic Nanowires. *Phys Rev Lett* 2011;**106**:087204.
66. Ahn SM, Moon KW, Cho CG, Choe SB. Control of domain wall pinning in ferromagnetic nanowires by magnetic stray fields. *Nanotechnology* 2011;**22**:085201.

67. Mironov VL, Ermolaeva OL, Skorohodov EV, Klimov AY. Field-controlled domain wall pinning-depinning effects in ferromagnetic nanowire-nanoislands system. *Phys Rev B* 2012;**85**(1–9):144418.
68. Allwood DA, Xiong G, Cowburn RP. Writing and erasing data in magnetic domain wall logic systems. *J Appl Phys* 2006;**100**:123908.
69. Kunz A, Reiff SC. Dependence of domain wall structure for low field injection into magnetic nanowires. *Appl Phys Lett* 2009;**94**:192504.
70. Glathe S, Zeisberger M, Mattheis R, Hesse D. Influence of slanted nanostripe edges on the dynamics of magnetic domain walls. *Appl Phys Lett* 2010;**97**:112508.
71. O'Brien L, Petit D, Zeng HT, Lewis ER, Sampaio J, Jausovec AV, et al. Near-field interaction between domain walls in adjacent permalloy nanowires. *Phys Rev Lett* 2009;**103**:077206.

Chapter 6

Medical Applications of Magnetic Nanoparticles

Chris Binns
Department of Physics and Astronomy, University of Leicester, Leicester, United Kingdom

1 INTRODUCTION

In the last two decades, a number of treatment and diagnosis methods based on magnetic nanoparticles have been suggested and in some cases developed to clinical use.[1] The magnetism in nanoparticles is used in four basic modes, that is: (a) An external magnetic field gradient is applied to produce a force on magnetic nanoparticles within the body so that they can be steered to and held at a specific place, which is used to concentrate attached drugs where they are required thus minimising systemic exposure and side effects[2]; (b) Their ability to heat up in response to an alternating magnetic field (AMF) applied from outside the body is utilised to focus heat at tumour sites causing cancerous cells to die by hyperthermia[3]; (c) Their effect on nuclear magnetic relaxation is used to produce contrast enhancement in magnetic resonance imaging (MRI) images[4];(d) Their response to an AMF is used to map their concentration in the body in a new technique called magnetic particle imaging (MPI).[5] In some cases, more than one magnetic response can be combined, for example, Brusentsov *et al.*[6] combined magnetic nanoparticle hyperthermia (MNH) of tumours in mice with the application of a magnetic field gradient to initially concentrate the particles in the tumour, which resulted in a more effective treatment. Another example is where magnetic nanoparticles used for drug delivery can also be stimulated by an oscillating field to heat up and release the drug at the target site.[7]

This chapter describes each of these medical applications of magnetic nanoparticles in detail, outlining any theory not already described in Chapter 1 and discusses the current state-of-the-art in addition to future perspectives. Following introductory sections 1–3, the role of magnetic nanoparticles as MRI contrast enhancers, which have been in clinical use for a number of years, is described (Section 4). Section 5 presents the newly developed medical diagnostic method of MPI, which so far has used similar nanoparticles to those required for MRI. MNH, which has reached phase 2 clinical trials (Section 6), is

Frontiers of Nanoscience, Vol. 6. http://dx.doi.org/10.1016/B978-0-08-098353-0.00006-3

described next followed by magnetic drug delivery, which is still at the fundamental research stage (Section 7). All the magnetic nanoparticles in clinical use to date are Fe oxides (maghemite, Fe_2O_3 or magnetite, Fe_3O_4) due to the ease of manufacture in suspension by chemical methods[8] and also their relative biocompatibility.[8] We begin in Section 2 with a description of how magnetic nanoparticles interact with living organisms, followed in Section 3 with a discussion on the coatings required on magnetic nanoparticles for clinical use.

2 INTERACTION OF MAGNETIC NANOPARTICLES WITHIN LIVING ORGANISMS

2.1 Size of Nanoparticles Relative to Relevant Biological Structures

Magnetic nanoparticles are used for many treatment and diagnostic methods described in the rest of the chapter and the optimum particle sizes varies according to the application. The size range encompasses particles that are required to be superparamagnetic (<5 nm) up to the largest size that will maintain a single-domain magnetic configuration (~50 nm). The size range is illustrated in Figure 6.1 and it is observed that even the largest nanoparticles used are tiny in comparison to the smallest blood vessels (capillaries) and biological cells (erythrocytes). On the other hand, antibodies used to target the nanoparticles to specific cell sites are about the same size as the largest nanoparticles (Figure 6.1C) but dwarf superparamagnetic nanoparticles. For these, it may be more appropriate to use smaller targeting molecules such as folic acid (Figure 6.1D).

2.2 Uptake of Nanomaterials by the Body

When nanoparticles enter the body, they trigger a response from the *mononuclear phagocyte system*, also known as the *reticuloendothelial system* (RES), which consists of a population of different types of cells distributed throughout the body. These are specialised to ingest and process foreign invaders and also transport then to the liver and kidneys. From there they can be further processed before being ejected from the body in the faeces and urine. The system has evolved to deal with a range of intruders including bacteria, viruses, toxins and passive nanoparticles, which are abundant in the environment.

In this chapter, we will assume that the nanoparticles used for diagnosis or therapy are passive (i.e. non-toxic) and that they enter the bloodstream directly (say by intravenous injection). As discussed in Section 2.1, it is safe to assume that even the 'large' ones (~50 nm) are too small to block even the narrowest blood vessels (capillaries), which have diameters of at least 5 μm, though large agglomerates of nanoparticles are a potential problem.

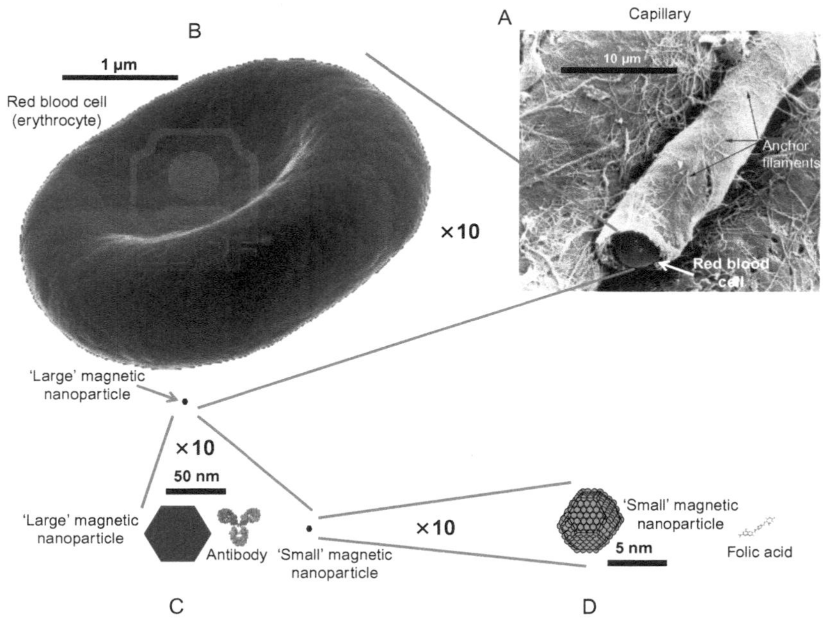

FIGURE 6.1 Comparison of sizes of magnetic nanoparticles used in medicine with biological and chemical structures. (A) Scanning electron microscope (SEM) image of smallest blood vessel (capillary) showing erythrocyte (red blood vessel) emerging from the broken end. (B) comparison of erythrocyte with the largest nanoparticles used in medical applications (~50 nm). (C) Comparison of 'large' magnetic nanoparticle with antibody (used for targeting) and the size of the smallest nanoparticles used (<5 nm). (D) 'Small' nanoparticle compared to a folic acid molecule, which is one of the smallest targeting molecules used. *SEM image reproduced with permission from Dr. Thomas Cececi, Virginia-Maryland College of Veterinary Medicine.*

Once in the bloodstream, the particles encounter monocytes that are a subpopulation of leukocytes or white blood cells, which make up about 1% of whole blood by volume. They are designed to engulf foreign particles by *phagocytosis* but it is difficult for this to happen directly because the monocytes and particles normally both have a negative charge in the form of an electrical double layer (EDL) and tend to repel each other. This is overcome by an intermediate process known as *opsonisation* in which circulating proteins, mostly antibodies, bind to the particle. The proteins bound to the particle are attracted to receptors expressed on the monocyte surface. Once attached, the membrane of the monocyte surrounds the particle and brings it into the cell wrapped in a piece of membrane. The package consisting of the particle surrounded by cytoplasm and wrapped in membrane is referred to as a *phagosome*. The phagosome is transported through the cell to organelles (*lysosomes*) that process the particle further and attempt to break it down to basic constituents. Any organic coatings on the nanovectors will be broken down by the lysosomes but solid metal or oxide nanoparticles remain intact and are kept internalised to be transported to the liver, spleen and bone marrow where specialist cells capture the nanoparticles and attempt to break them down further or get them ready for expulsion from the body. For example, if they end up in the liver, they are incorporated into the bile and removed by faeces. Alternatively, depending on the size, they can be filtered by the kidneys and removed in the urine. In the conventional model of the RES,[9] the circulating monocytes are resident in the blood for 1–3 days before entering tissues throughout the body where they mature into different types of macrophages, which are also capable of phagocytosis.

In general, the residence time of particles within the bloodstream increases as the particle size decreases[10] but can still be quite short. The blood residence time for a small (~10 nm) nanoparticle is typically hours and for large nanoparticles (~100 nm) can be as short as a few minutes. This can present a significant problem for treatment since nanoparticles with a very high-performance *in vitro* are useless if *in vivo* they are removed before their targeting can work or before they can be activated. It is important therefore to develop 'stealth' technologies for the particles that prevent them from being recognised by monocytes and this means preventing the initial step of opsonisation. Clearly such a thing is possible since the red blood cells evade the monocytes and have a lifespan of about 120 days while they fulfil their task of transporting oxygen. One factor in determining the relatively long life span of erythrocytes is a coating of sugar polymers that prevent the adsorption of opsonins and the resulting recognition by monocytes and macrophages. Similar coatings can be used to protect therapeutic nanoparticles as discussed in Section 3.2.

3 COATING MAGNETIC NANOPARTICLES FOR MEDICAL APPLICATIONS

Coatings are required on nanoparticles to carry out four basic functions, that is, to produce stability in suspension, to render them biocompatible, to provide longevity of survival against the immune system *in vivo* and in some cases to provide active biological targeting to specific sites on cells.

3.1 Producing Stable Suspensions

In the vast majority of clinical applications, the nanoparticles are required to be dispersed in a liquid suspension. There are notable exceptions, for example, research on steering magnetic nanoparticles with attached drugs in the form of an aerosol into a specific lung (see Section 7) but in this section, we will focus on hydrosols. An important characteristic of a suspension is that it should be stable against particle agglomeration in the presence of various attractive and repulsive interactions. The attractive van der Waals force arises ultimately from quantum vacuum fluctuations and so is present regardless of any other interactions in the system. In the case of a suspension of blocked magnetic nanoparticles, the magnetic dipolar force will be an additional attractive interaction. In the absence of any repulsive forces, Brownian motion will always bring the nanoparticles sufficiently close for the van der Waals and magnetic (if present) attractions to irreversibly bind the particles and the process will accelerate as the agglomerates increase in size. Repulsive forces between the particles arise from the formation of an EDL if there are ions in the solution (see below) or by steric repulsion for particles coated in chain-like molecules or polymers.[11] Suspensions are stabilised by using one or both of these repulsive forces.

The formal theory of nanoparticles interacting through liquid via the van der Waals and EDL mechanisms was originally developed in the 1940s by Derjaguin, Verwey, Landau and Overbeek[12,13] and is hence known as DVLO theory. The attractive part is formulated by summing the pairwise van der Waals attraction between all the individual atoms, of the form, $-C/r^6$, where C is a constant dependent on the type of atom and yields, for identical spherical particles:

$$U_A(D) = -\frac{AR}{12D} \tag{6.1}$$

where A is the Hamaker constant given by $A = \pi^2 C\rho^2$ for the particle surfaces with atomic densities (atoms/unit area) ρ, R is the radius of the particles and D is the distance between their surfaces.

The EDL repulsive force arises from charges that accumulate on particle surfaces either as a result of ions in the liquid attracted onto the particle by the image charge or by the dissociation of surface groups (e.g. Fe and O for Fe oxide nanoparticles). The surface potential attracts oppositely charged ions from the liquid and the surface charge is balanced by a sphere of counterion-rich fluid surrounding the particle. It is this excess charge that produces a repulsive interaction with similar particles surrounded by their own EDL (Figure 6.2).

The charge bound firmly to the particle surface is the Stern layer and surrounding this is the counterion-rich *diffuse* layer, which extends to a notional boundary known as the slipping plane. Within this boundary, the ions in solution are considered to move with the particle, while outside it the ions remain with the bulk liquid. The potential drops with distance away from the surface and at the slipping plane reaches a value referred to as the zeta potential, which can be considered to be the potential experienced by neighbouring particles. If the repulsive force is too low, then Brownian motion will bring particles sufficiently close for the attractive force to take over and produce coalescence. As a general rule of thumb a zeta potential of $>\pm 30$mV is required to produce a stable suspension if the van der Waals force is the only attractive interaction.

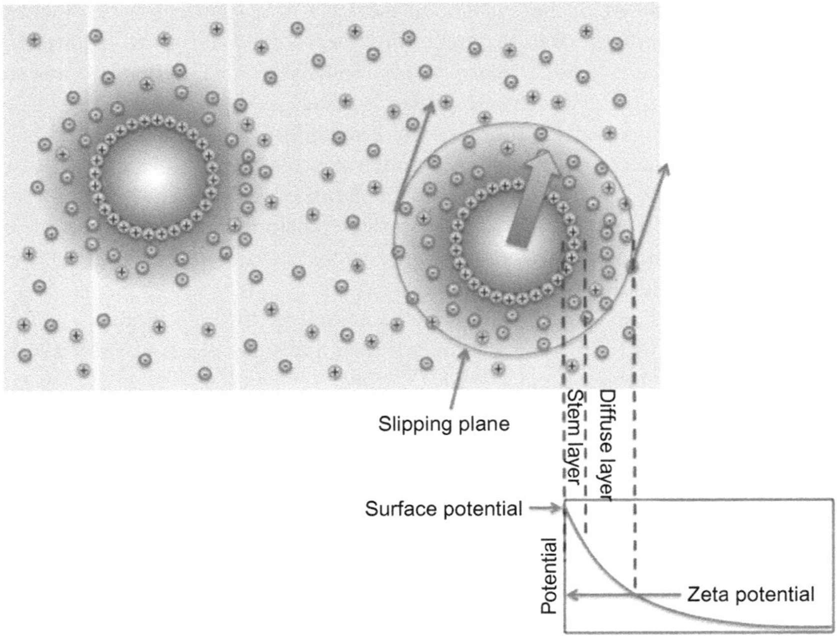

FIGURE 6.2 Formation of electrical double layer (EDL) around nanoparticles in suspension. The charge bound strongly to the surface is the Stern layer and the counterion-rich region around the particle is the diffuse layer. This extends to a boundary referred to as the slipping plane with ions beyond the slipping plane regarded as being within the bulk fluid. The potential drops on moving away from the surface and reaches the zeta potential at the slipping plane.

According to the DVLO theory the repulsive force gives an interaction energy between two spheres of radius R, which can be written in terms of the zeta potential, ζ as[14]:

$$U_R(D) = 2\pi\varepsilon\varepsilon_0 R\zeta^2 e^{-\kappa D} \tag{6.2}$$

where

$$\kappa = \sqrt{\frac{\rho e^2 z^2}{\varepsilon\varepsilon_0 k_B T}} \tag{6.3}$$

is the inverse of the Debye screening length for a fluid with a relative permittivity of ε containing ions with a valency, z and a density ρ. Thus the total (attractive + repulsive) interaction potential between particles is given by:

$$U_A(r) = -\frac{AR}{12D} + 2\pi\varepsilon\varepsilon_0 R\zeta^2 e^{-\kappa D} \tag{6.4}$$

The Hamaker constant for magnetite, maghemite and hematite Fe oxide nanoparticles interacting through water has been calculated to be in the range

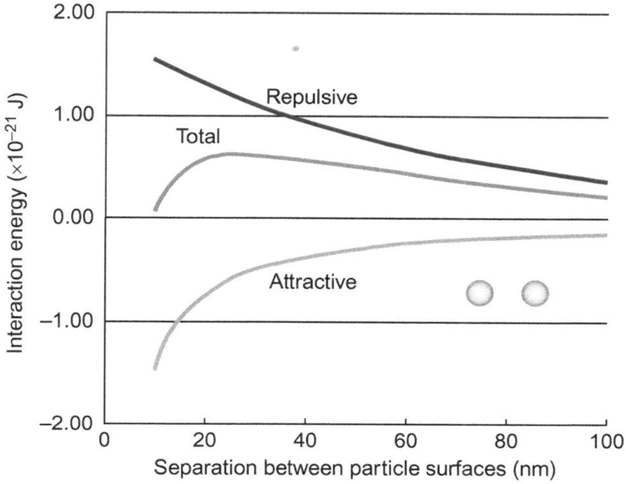

FIGURE 6.3 Attractive, repulsive and total interaction energies between two Fe oxide nanoparticles of radius 5 nm through water for a stable suspension at 300 K. The Hamaker constant, $A = 3.5 \times 10^{-20}$ J, the zeta potential = 9 mV, the monovalent ion concentration is $3 \times 10^{22}/\text{m}^3$ and $\varepsilon = 80$. The interaction is repulsive down to a separation of 10 nm, illustrated by the particles drawn to scale.

3.3–3.9 × 10^{-21} J[15] so the van der Waals part can be determined absolutely. For the repulsive part, assuming a zeta potential of 9 mV, which would not normally give a stable suspension and a density of monovalent ions of $3 \times 10^{22}/\text{m}^3$ the variation of interaction energy with surface separation for two Fe oxide nanoparticles of radius 5 nm is shown in Figure 6.3. It is evident that the interaction is repulsive down to a separation of 10 nm, illustrated to scale by the particles drawn in the figure.

Fe oxide nanoparticles with a radius of 5 nm will be superparamagnetic and so the magnetic dipolar force between them should be zero, though this will not be the case if they are magnetised by an external field. The blocking temperature of a nanoparticle with a uniaxial anisotropy is given by (see Chapter 1, Section 4):

$$T_b = \frac{KV}{k_B \ln\left(\frac{\tau}{\tau_0}\right)} \tag{6.5}$$

where V is the particle volume, K is the anisotropy constant, τ is the characteristic timescale at which the particle is blocked and τ_0 is the timescale of the moment fluctuation at the high-temperature limit. An anisotropy constant of 4.7×10^4 Jm^{-3} was measured for maghemite nanoparticles[16] and typical values for τ and τ_0 are 1 s and 10^{-9} s, respectively, so a blocking temperature of 300 K corresponds to a particle radius of about 8 nm. Above this size at room temperature, the nanoparticles will exhibit a static magnetic moment and there will be an additional attractive interaction given by:

$$\frac{\mu_0 \mu^2}{4\pi r^3} \tag{6.6}$$

where μ is the magnetic moment of the nanoparticles (assumed identical) and r is the distance between their centres. Figure 6.4 compares the attractive,

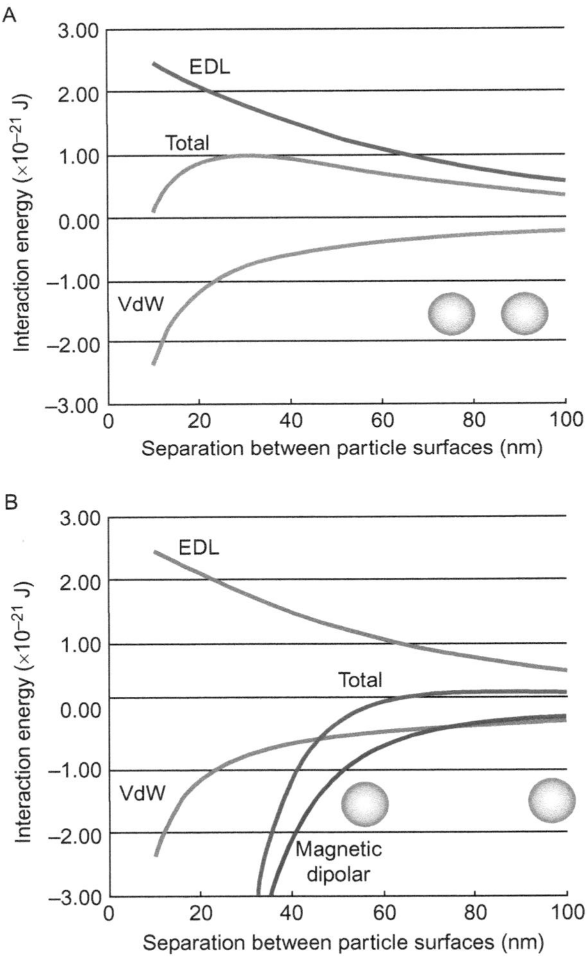

FIGURE 6.4 Attractive, repulsive and total interaction energies between two Fe oxide nanoparticles of radius 8 nm through water at 300 K with all parameters as in Figure 6.3. (A) Without the magnetic dipolar force included the total interaction becomes attractive at a separation of 10 nm as illustrated by the particles drawn to scale. (B) Including the magnetic dipolar interaction between blocked particles gives a total interaction that becomes attractive at 54 nm illustrated by the particles drawn to scale.

repulsive and total interaction energies for two 8 nm diameter maghemite particles with and without the magnetic force included. It is evident that including the magnetic force results in an attractive force becoming dominant at a much larger separation (54 nm compared to 10 nm) and is a strong destabilising influence. Additional sources of stabilisation, for example, steric repulsion would be required for blocked magnetic nanoparticle suspensions. The magnetic dipolar force in blocked particles becomes even stronger when they contain a pure Fe core as in suspensions synthesised in attempts to produce higher performance in hyperthermia and MRI contrast enhancement (see Section 6.5).

A problem with relying on EDL repulsion to stabilise suspensions is that at high ionic concentrations, the Debye screening length reduces leading to a higher rate of change of the force with distance and this can produce a second minimum in the total interaction leading to a weak binding of nanoparticles referred to as flocculation. Issues such as this and the need to strengthen repulsion in some suspensions have led to the use of *steric stabilisation* produced by coating the nanoparticles in chain molecules and polymers (Figure 6.5).[11] Steric repulsion is a complex interplay of forces but there are two main mechanisms.[17] The inter-digitation of the molecular chains surrounding approaching nanoparticles produces a higher local concentration and a higher osmotic pressure in the overlap region leading to repulsion. The other main mechanism is the mechanical compression of the chains and if the molecules are charged this results in a further repulsive term. One advantage of using steric stabilisation is that the repulsive force is independent of ion concentration, which is important in some systems. Many polymers have been used to coat magnetic nanoparticles including dextran,[18–20] polyethylene glycol (PEG),[21] polyvinyl alcohol,[22,23] polyvinylpyrrolidone,[24] alginate[25] and chitosan.[26] Some of these have

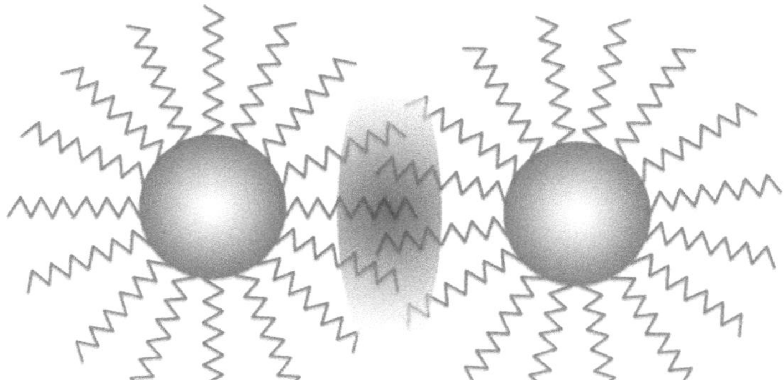

FIGURE 6.5 Steric repulsion of nanoparticles coated in polymers. The higher local concentration of molecules in the overlap region produces an increase in osmotic pressure leading to repulsion. The mechanical compression of the molecules and electrostatic repulsion if they are charged are further mechanisms.

secondary functions including rendering the nanoparticles biocompatible and screening them from the immune system (see Section 3.2).

3.2 Biocompatibility of Magnetic Nanoparticles

In addition to stability against agglomeration, magnetic nanoparticles *in vivo* should not be toxic and must avoid being removed by the immune system for long enough to play their role in diagnosis or therapy. These are complex issues and mainly beyond the scope of the chapter but are covered very briefly here in order to instil some feel for how particles need to be functionalised to perform *in vivo*. The blood circulation time of injected nanoparticles depends on their size, charge and thickness of their coating[27] and can vary from minutes to hours. In general, to maximise the lifetime within the blood of an artificially introduced particle it should be as small as possible, electrically neutral and have a hydrophilic surface. Commonly used coatings to increase longevity are PEG[21,28] and dextran,[29] which are thought to keep changes in nanoparticles surface charge to a minimum as proteins accumulate. As shown in Figure 6.6, the uptake of 10 nm diameter magnetite nanoparticles by mouse macrophage (RAW 264.7)[21] is significantly lower when they are coated with PEG indicating that the coating would increase the survival time against the macrophages of the RES.

The polymers PEG and dextran both improve nanoparticle longevity against clearance by the RES, and as shown in Section 3.1, they also are effective at preventing agglomeration by steric repulsion but another issue is toxicity. It has been shown that the nanoparticle coating has a significant influence on the viability of cells that have taken up the nanoparticles.[29,30] For example Figure 6.7 shows the proliferation of a population of human derma fibroblasts after 24 h in a culture medium containing 8–15 nm diameter iron oxide nanoparticles that were uncoated (U), coated with dextran (D) or

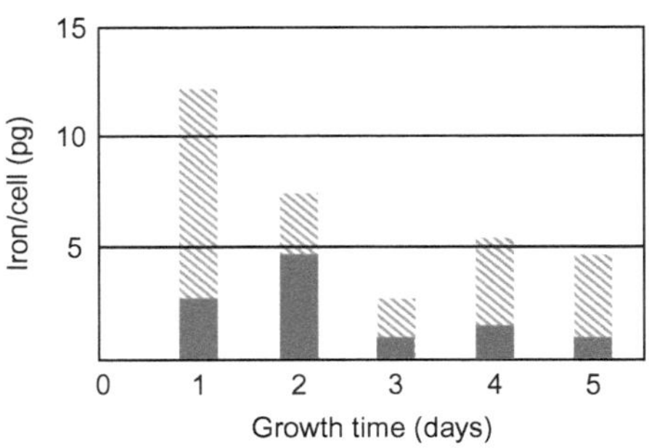

FIGURE 6.6 Uptake of 10 nm diameter magnetite nanoparticles by mouse macrophage RAW 264.7 as a function of growth time for a culture grown in a medium containing 0.2 mg ml^{-1} of nanoparticles. The shaded bars are the uptake of the uncoated nanoparticles and the solid bars the uptake of the PEG-coated nanoparticles. *Reproduced with permission from Ref. 21.*

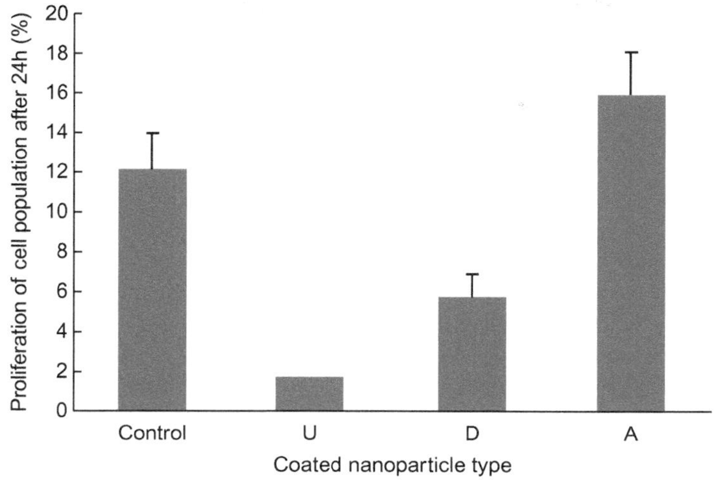

FIGURE 6.7 Proliferation of human dermal fibroblasts after 24 h in a culture medium containing 8–15 nm diameter Fe oxide nanoparticles that are uncoated (U), coated with dextran (D) or albumin (A) in comparison to the unexposed control group. *Reproduced with permission from Ref. 29.*

coated with albumin (A). It is clear that the uncoated nanoparticles have a profound detrimental effect on cell growth. Coating with dextran improves the cell growth but still not to the level of the control group whereas, within the error, the albumin coating allows the cells to proliferate without hindrance. To illustrate the complexity of the issue, however, it has been shown that in the case of polyethylene oxide-coated nanoparticles, the length of the polymer correlates inversely with toxicity,[30] that is the shorter the chain length the higher the toxicity.

4 MAGNETIC NANOPARTICLES AS MRI CONTRAST ENHANCERS

MRI has become an important diagnostic tool in medicine due to its ability to show contrast between different soft tissues and its good spatial resolution. In an applied magnetic field, the magnetic moments of the hydrogen nuclei can populate two states with different energy levels but, since the energy difference is much less than kT, the populations in the higher and lower levels are almost equal. Nevertheless, the slight imbalance produces a residual longitudinal magnetisation. A secondary oscillating field, applied at the resonance frequency (i.e. at the Larmor frequency corresponding to the DC field) in the transverse direction equalises the two populations and creates coherence in the precessing nuclear magnetization. This simultaneously reduces the longitudinal magnetization and creates transverse magnetization.

The secondary field is in the radiofrequency range, and is applied as a short burst, hence it is termed an 'RF pulse'. The characteristic times for the longitudinal magnetization to return to the pre-pulse value (T_1) and for the transverse magnetization to return to zero (T_2) are highly sensitive to the local chemical environment and vary for different tissues but T_1 is typically hundreds of ms, while T_2 is tens of ms. Spatial discrimination is achieved by applying spatial gradients in the DC field so that the Larmor frequency varies with position, while image contrast is produced by introducing delays between applying the RF pulse and measuring. This allows the transverse or longitudinal magnetization to relax by an amount that depends on the tissue type. Since 1988, artificially introduced magnetic fluids have been used clinically to enhance the contrast in MRI by locally reducing relaxation times and two currently available agents are Gd chelates and superparamagnetic Fe oxide nanoparticles.

The performance of a magnetic contrast agent is characterised by its relaxivity, which is defined as the increase in the relaxation rate that is produced by adding 1 mmol l^{-1} of the active ion (e.g. Gd or Fe). It is given by the equation:

$$R_{i(obs)} = \frac{1}{T_{i(obs)}} = \frac{1}{T_{i(\mathrm{diamag})}} + r_i C \tag{6.7}$$

where $i = 1$ or 2 (relaxation of longitudinal or transverse magnetisation respectively), $R_{i(\mathrm{obs})}$ is the observed relaxation rate of the system with the contrast agent added, $T_{i(\mathrm{diamag})}$ is the relaxation time of the system prior to adding the agent, r_i is the relaxivity of the agent and C is the concentration of its active ions in mmol l^{-1}. The units of r_i are thus s^{-1}mmol^{-1}l.

Longitudinal relaxation is brought about by transverse magnetic fields that have an oscillatory component at the Larmor frequency, while transverse relaxation is caused by any transverse fields, with an effect that increases with decreasing frequency. These fluctuating fields are produced by the stochastic motion of the fields generated by nearby magnetic dipoles (other magnetic nuclei or unpaired electrons) or by motion relative to an introduced magnetic contrast-enhancing agent. In the case of a superparamagnetic nanoparticle, under certain conditions, the fluctuating magnetic moment also provides a random time varying transverse field that can produce relaxation. Thus, in general, the relaxation times are dependent on the diffusion coefficient of the surrounding tissue fluids, and of the nanoparticle and its size in a complicated relationship.

One of the most important factors in calculating relaxation time is the spectral density function, $J(\omega)$. This is the Fourier transform of the spin correlation function, $G(\tau)$, which is the average of the normalised field $F(t)$ experienced by the spins at time t multiplied by that at time $t+\tau$, ($F(t+\tau,)$) that is, for N spins it is given by:

$$G(\tau) = \frac{1}{N}\sum F(t)F(t+\tau) \tag{6.8}$$

$G(\tau)$ has a maximum value of 1 when $\tau=0$ and decays away at positive and negative τ. For example, if one assumes that in a particular system $G(\tau)$ has the simple form:

$$G(\tau)=G(0)\exp\left(-\frac{|\tau|}{\tau_c}\right) \tag{6.9}$$

where τ_c is the correlation time, then $J(\omega)$, the Fourier transform, is a Lorentzian given by:

$$J(\omega)=\frac{2\tau_c}{1+\omega^2\tau_c^2} \tag{6.10}$$

These model functions are illustrated in Figure 6.8, and although in a real system the functional form is much more complicated, the general shape of $J(\omega)$ is similar to that plotted in Figure 6.8.

The relaxation mechanisms are generally divided into innersphere and outersphere effects. Innersphere relaxation deals with the direct exchange of energy between protons and electrons located in the first hydration sphere of the paramagnetic ion, while outersphere relaxation is due to the movement of water protons within the magnetic field gradients around each ion. For superparamagnetic nanoparticles, the innersphere relaxation is a minor contribution and the model presented below is for outersphere relaxation only.

In a solvent containing solute nanoparticles, the outersphere spectral density can be given in closed form[31]:

$$J_n(\omega,\tau_D,\tau_{Sn})=Re\left(\frac{1+\left(\frac{1}{4}\right)\left(i\omega\tau_D+\frac{\tau_D}{\tau_{Sn}}\right)^{\frac{1}{2}}}{1+\left(i\omega\tau_D+\frac{\tau_D}{\tau_{Sn}}\right)^{\frac{1}{2}}+\left(\frac{4}{9}\right)\left(i\omega\tau_D+\frac{\tau_D}{\tau_{Sn}}\right)+\left(\frac{1}{9}\right)\left(i\omega\tau_D+\frac{\tau_D}{\tau_{Sn}}\right)^{\frac{3}{2}}}\right) \tag{6.11}$$

where τ_D is the timescale associated with fluctuations in the magnetic interaction between the water protons and the nanoparticle arising from relative diffusion. This term can be written as $\tau_D=r^2/D$, where D is the diffusion

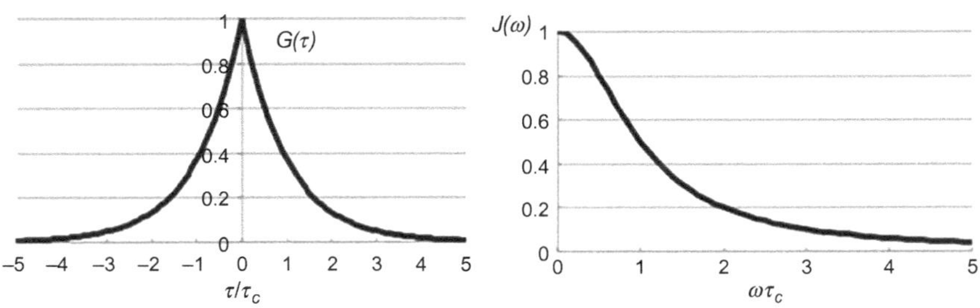

FIGURE 6.8 Model correlation and spectral density functions.

coefficient of water and r is the radius of the nanoparticle. The other time constant, τ_{Sn}, is the lifetime of the magnetic moment of the nanoparticle within the applied field and, in general, can have different values in the longitudinal (i.e. along the field) and transverse directions, distinguished by $n=1$ and 2, respectively. In a saturating field however, the fluctuations within the nanoparticles are quenched and both components tend to infinity leaving a single spectral density function. Figure 6.9 shows the variation in the spectral density function of nanoparticles in water as τ_s varies between 10^{-8} s, which is the value measured for unconstrained Fe nanoparticle moments at the high-temperature limit[32] and infinity as would occur in a saturated array of nanoparticles.

For typical fields used in clinical MRI ($B=0.2-3$ T) and typical contrast agent nanoparticle sizes (10–50 nm), the ratio $\mu_B/kT \gg 1$, where μ is the nanoparticle magnetic moment and the nanoparticles are magnetically saturated in the applied field. Thus, the assumption that both τ_{Sn} components tend to infinity is reasonable, and in this limit, the R_1 and R_2 relaxation rates per nanoparticle are[8,33]:

$$R_1 = \frac{1}{T_1} = \left(\frac{6400\pi}{81}\right)\left(\frac{\mu_0}{4\pi}\right)^2 \gamma_I^2 \mu^2 \frac{1}{rD} 9J_1(\omega_I, \tau, \tau_{S1} \to \infty) \tag{6.12}$$

and

$$R_2 = \frac{1}{T_2} = \left(\frac{6400\pi}{81}\right)\left(\frac{\mu_0}{4\pi}\right)^2 \gamma_I^2 \mu^2 \frac{1}{rD}[4.5J_1(\omega_I, \tau, \tau_{S1} \to \infty) + 3J_1(0)] \tag{6.13}$$

where γ_i is the gyromagnetic ratio of the proton and μ is the magnetic moment of the nanoparticles. The relaxivity is normally quoted per mmol l^{-1} of the

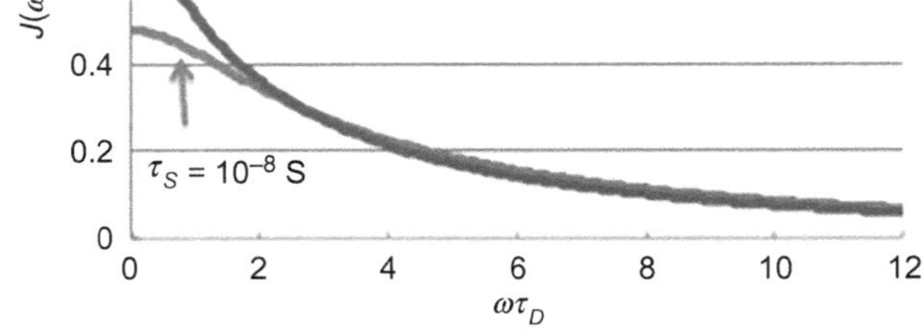

FIGURE 6.9 Variation in the spectral density function of nanoparticles in water as τ_s varies between 10^{-8} s (unconstrained nanoparticle moments) and infinity (saturated array).

magnetically active atoms (e.g. Fe), so the relaxivities calculated from Equations (6.12) and (6.13) should be multiplied by $6 \times 10^{20}/N$, where N is the number of atoms in each nanoparticle to get the values per mmol l^{-1}.

Within the high-field limit where the array is saturated and fluctuations of the nanoparticle moments do not play a role, the only characteristics of the nanoparticle that affect R_1 and R_2 according to the model are the diameter, d, and the saturation magnetisation. For a given material, the moment scales with the number of atoms, that is, d^3 and the number of nanoparticles, for a given number of mmol of Fe, decreases as d^3. The moment is squared in Equations (6.12) and (6.13), so the relaxivities R_1 and R_2 for a given material should be proportional to d^3/r, where r is the closest approach of water molecules. At the lower limit, this can also be taken to be d since, irrespective of the coating, the first hydration sphere can reach the nanoparticle surface resulting in a d^2 dependence of R_1 and R_2.

For a given particle size, Equations (6.12) and (6.13) predict that R_1 and R_2 should be proportional to the square of the saturation magnetisation, so given the assumption of saturated nanoparticles with infinite relaxation times, the size/material dependence of the relaxivities condenses to:

$$R_i \propto M_s^2 d^2 \tag{6.14}$$

Thus for maximum relaxivity, the particles should be composed of a material with a high saturation magnetisation and be as large as possible while still remaining a single domain ($d < \sim 50$ nm).

The requirement for high saturation magnetisation has led to the search for nanoparticles with a higher magnetisation than the Fe oxide nanoparticle contrast agents clinically available (e.g. Resovist®). Table 6.1 shows the relaxivity obtained from suspensions of FeCo nanoparticles with graphite shells,[36] a series of spinel metal ferrite nanoparticles coated in DSMA[35] and Co nanoparticles coated in polymer,[34] all of which have a higher saturation magnetisation than Fe oxide and show a higher relaxivity than Resovist.[37] Figure 6.10 shows the relaxivity of all the nanomaterials listed in Table 6.1 plotted versus $d^2 \times M_s^2$, which according to Equation (6.14) should be a linear relationship. It is evident that, despite the different nature of the materials being compared and not taking any account of factors such as shape (see below), the Universal scaling given by Equation (6.14) is reasonable.

Recently, the relaxivity of suspensions of Fe@Fe oxide core-shell nanoparticles synthesised by the gas-phase method described in Chapter 3 was measured.[38] These have a metallic Fe core with a high M_s value so they are expected to show high relaxivities. In addition, a feature of the gas-phase method is that the nanoparticles can be annealed in vacuum by passing the beam through a hot tube, which modifies their shape enabling the effect of this parameter on relaxivity to be observed. As can be seen in Figure 6.11, the unheated nanoparticle beam consists of a mixture of shapes, as determined by observing TEM images of the vacuum-deposited nanoparticles, including

TABLE 6.1 R_2 Relaxivities for Different Magnetic Nanoparticle Contrast Agents

Material in Nanoparticles (Coating)	Core Nanoparticle Diameter (nm)	Hydrodynamic Diameter (nm)	Saturation Magnetisation (emu g^{-1})	R_2 Relaxivity at 1.5 T	Reference
Fe_2O_3 (Resovist®) (dextran)	5–25	45–60	55	82	34
$Ni^{2+}Fe_2O_4$ (DSMA)	12	18	85	152	35
$Co^{2+}Fe_2O_4$ (DSMA)	12	18	99	172	35
$Fe^{2+}Fe_2O_4$ (DSMA)	12	18	101	218	35
$Mn^{2+}Fe_2O_4$ (DSMA)	12	18	110	358	35
FeCo (graphite/PEG)	4	~30	215	185	36
FeCo (graphite/PEG)	7	~30	215	644	36
Co (polymer)	3.3	13	128	88	34
Co (polymer)	3.9	28	128	88	34

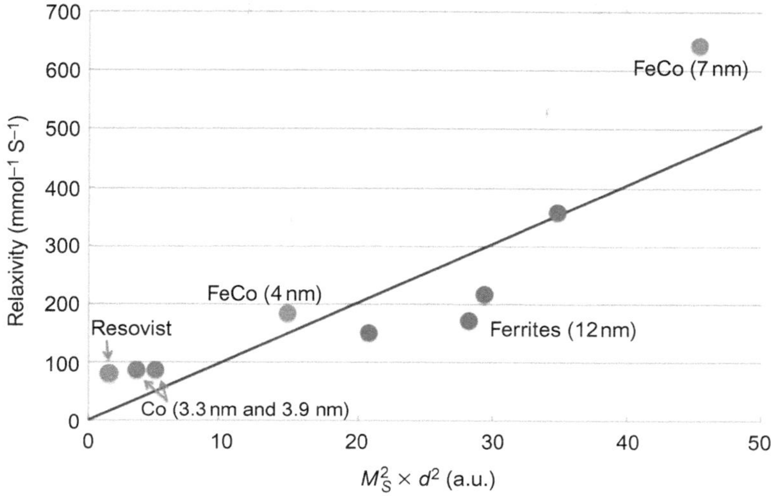

FIGURE 6.10 Relaxivity of all the nanomaterials listed in Table 6.1 plotted versus $d^2 \times M_s^2$ illustrating the general agreement with the universal scaling predicted by Equation (6.14).

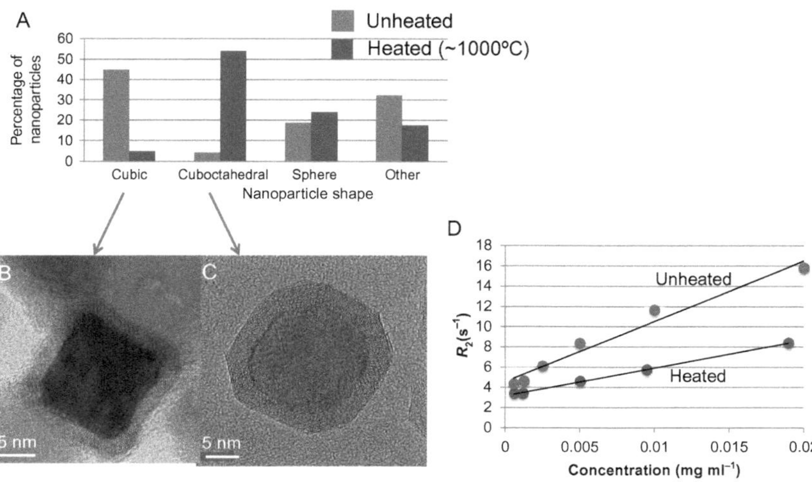

FIGURE 6.11 (A) Change in populations of gas-phase Fe nanoparticles with different shapes after heating the free beam in vacuum by passing it through a hot tube. (B) and (C) Cubic and cuboctahedral shapes observed by TEM after depositing the nanoparticles in vacuum. The nanoparticles are produced and deposited in UHV and the oxide layer forms after removing the substrate from vacuum. (D) Relaxivity observed after depositing the heated and unheated nanoparticles into water (see Chapter 3) showing a significantly lower value for the heated nanoparticles, which contain a much lower proportion of cubes and a correspondingly higher proportion of cuboctahedra.

cubes (Figure 6.11B), cuboctahedra (Figure 6.11C), spheres and more complex shapes. The largest population however is in cubes and the main effect of heating the free nanoparticle beam to ~1000 °C is to convert the cubes to cuboctahedra. As observed in Figure 6.11D, this significantly lowers the relaxivity of the final suspension when the nanoparticles are deposited into water (see Chapter 3). The reason for this is not fully understood but may be due to canted spins at the surface of the cuboctahedra reducing the effective magnetic moment.

5 MAGNETIC PARTICLE IMAGING

MPI is emerging as a new diagnostic method for providing *in vivo* images and is a tracer-based method. That is, the magnetic nanoparticles are injected into the organism and a scanner then provides an image of their density in space thus following, for example, blood vessels. Alternatively, if there is a process that concentrates them at a site by active biological targeting or a natural process such as the enhanced permeability and retention[39] mechanism that tends to concentrate nanoparticles at tumour sites, the location of the site will be revealed. The detection of the nanoparticle tracer is at very high contrast as the background tissue produces no signal[40] and images can be obtained with a very high temporal resolution combined with a spatial resolution of around 1 mm. The method was invented in 2001 by Bernhard Gleich working at the Philips Research Laboratories in Hamburg in 2001[41] and has developed rapidly since. From taking static two-dimensional images of an array of Resovist-filled holes in a plate in 2005,[42] it evolved to *in vivo* three-dimensional imaging of a beating mouse heart at 46 frames a second by 2009[5] (see Figure 6.12). Since then the size of the scanned volume has been increased to 12 cm[43] taking an important step towards human body sized scanners and the idea of acoustic MPI, which measures acoustic emissions caused by magnetic fluctuations of nanoparticles has been investigated.[44] This offers the possibility of extending MPI to distinguish different soft tissues by the changes in the velocity and attenuation of sound waves they produce.

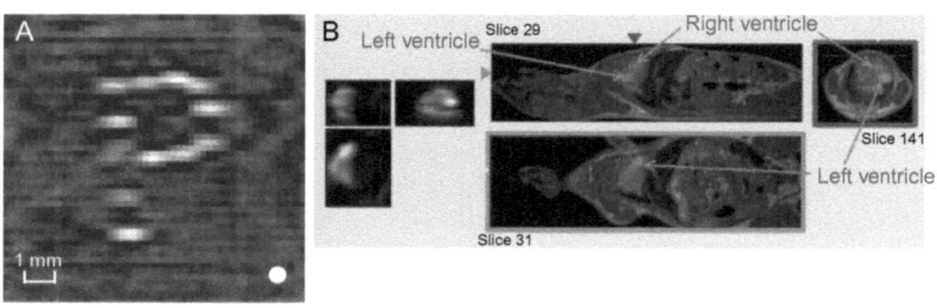

FIGURE 6.12 Development of MPI between 2005 and 2009. (A) Static image of Resovist-filled holes. (B) Dynamic image of blood circulation in a mouse taken at 46 frames a second. *(A) Reproduced with permission from Ref. 42; (B) reproduced with permission from Ref. 5.*

Another important development towards human clinical use has been the development of a single sided scanner[45] circumventing the problem of building a scanner large enough to contain a human body. In addition, new medical diagnostic capabilities have been demonstrated including stem cell tracking.[46]

The basic principle of the method is to excite an assembly of magnetic nanoparticles by an external magnetic excitation field oscillating in the tens to hundreds of kHz range and detect the oscillating magnetic moment of the nanoparticles inductively by pick-up coils (Figure 6.13A). The magnetization, M, of the nanoparticle assembly in an applied magnetic field, H, is described by the Langevin function (see Chapter 1, Section 5):

$$M(H,T) = c\mu\left[\cos h\left(\frac{\mu_0\mu H}{kT}\right) - \frac{k_B T}{\mu_0\mu \mathrm{H}}\right] \tag{6.15}$$

where c is the concentration of nanoparticles (number per m^3) and μ is their magnetic moment. The nonlinear dependence of M on H enables the voltage

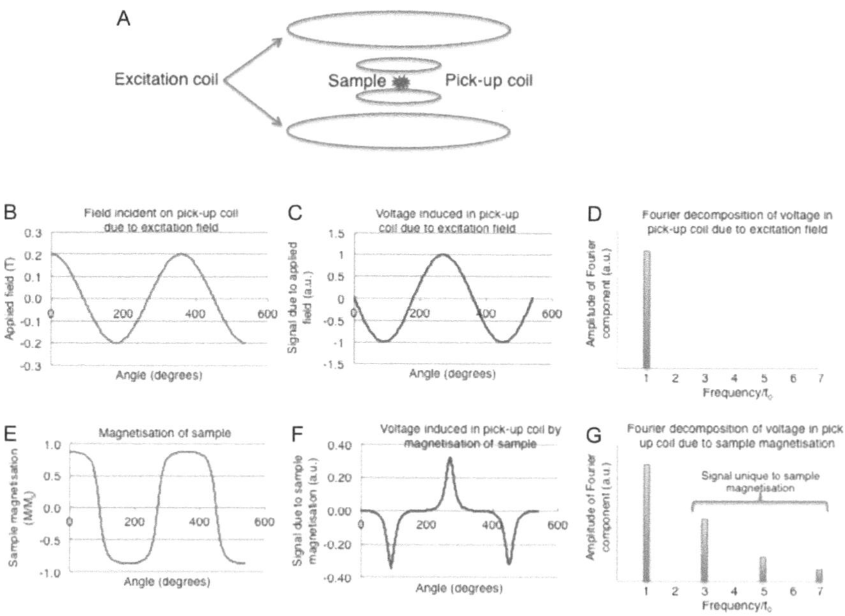

FIGURE 6.13 (A) Schematic of the geometry of the excitation coil and the pick-up coil to detect the sample magnetisation in MPI. (B) Field produced by the excitation coil. (C) Voltage on pick-up coil produced by direct induction from the excitation coil. (D) Fourier transform of the direct pick-up voltage signal showing a single Fourier component. (E) Magnetisation of a sample composed of 10 nm diameter maghemite (Fe_2O_3) nanoparticles produced by the excitation field shown in (B). (F) Voltage generated in the pick-up coil by the magnetisation of the sample. (G) Fourier transform of the voltage signal in the pick-up by the magnetisation of the sample showing components at all odd harmonics. Measuring only the harmonics isolates the nanoparticle signal from the direct pick-up signal.

on the pick-up coil induced by the nanoparticles, which is orders of magnitude weaker than the direct voltage induced by the excitation field, to be measured by detecting only harmonics of the applied field frequency. Figure 6.13C shows the waveform induced in the pick-up coil ($\propto \mathrm{d}B/\mathrm{d}t$) by a cosine wave in the excitation coil (Figure 6.13B) and as illustrated in Figure 6.13D, only the fundamental Fourier component is present in the signal. Figure 6.13E and F show respectively the magnetization of an assembly of 10 nm diameter maghemite (Fe_2O_3) nanoparticles in the same excitation field obtained using Equation (6.15) and the corresponding signal in the pick-up coil. The nonlinear response of the sample to the field produces a waveform heavily distorted relative to the cosinusoidal excitation field with a significant proportion of the signal in higher harmonics and if only these are measured, the magnetization of the sample is separated from the directly induced signal in the pick-up coil. The presence of only odd harmonics in the pick-up coil signal in the simple set up shown in Figure 6.13A is a result of the waveform crossing zero at the origin (i.e. purely odd). However, it was demonstrated by Weaver *et al.*[47] that adding an offset static magnetic field to the excitation field introduces even harmonics, which significantly increases the total signal arising from the sample magnetisation.

The signal can be picked up at a particular point in space by superimposing a field gradient on the excitation field such that the nanoparticle assembly is magnetically saturated at all points, thus producing no signal, except close to where the bias field crosses zero. This field-free point (FFP) can then be scanned either by physically moving the specimen to build an image or in more sophisticated scanners, moving the FFP through the sample by scanning a homogenous drive field.[48] Scanning techniques have steadily become more sophisticated and in recent designs it is possible to carry out a three-dimensional scan in ~20 ms enabling movies of the nanoparticle distribution to be made.

6 MAGNETIC NANOPARTICLE HYPERTHERMIA

6.1 Background to Cancer and Treatment by Hyperthermia

The success in treating cancer has steadily improved in the last few decades and this has in turn been a factor in increasing the life expectancy of the population in developed nations. However, if the death rates are adjusted to reflect the changes in the population age profile, as shown in Figure 6.14 for the United States, it is seen that the deaths per 100,000 people have only improved marginally since 1950. In other words for somebody alive today, the chances of dying of cancer are almost the same as in 1950, contrary to other major illnesses, which when normalised in the same way, show significant reductions in death rate. Recently, Paul Davies writing in Physics World has suggested that finding a cure is such a difficult problem because cancer is a

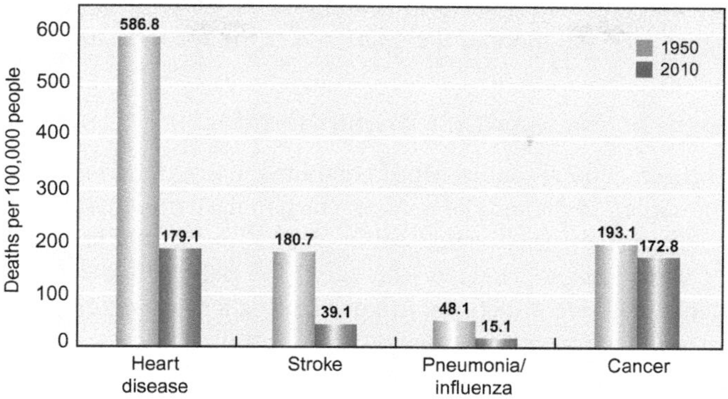

FIGURE 6.14 Comparison of death rates from major illnesses in 1950 and 2010 in the United States population adjusted for changes in the age profile. While there have been significant reductions in the proportion of the population dying from heart disease, stroke and influenza, the reduction for cancer is marginal. *Reproduced with permission from Physics World.*[49]

fundamental genetic programme within cells that dates back to the point when multicellular organisms first evolved over a billion years ago.[49] The tenacious nature of the disease has encouraged the search for radical new therapies including MNH.

The observation that elevated temperatures can regress tumours has been known since ancient times and Hippocrates (479–377 BC) was a keen advocate of using hyperthermia as a treatment.[50] Raising the temperature of cells to around 43 °C produces a number of reactions including denaturing of proteins, making cell membranes pervious and the release of heat shock proteins. Cancers tend to be more sensitive to heating than healthy tissue since *in vivo* the abnormal vasculature around a tumour limits heat loss by blood perfusion from the region so it becomes a hot spot. In addition, the disturbed vasculature produces a poor nutrient supply so acidosis and hypoxia further increases the sensitivity of the tumour to rises in temperature.[51]

In order for hyperthermia to become an effective modern tool for cancer treatment, it needs to utilise technologies that elevate just the temperature of the tumour without heating surrounding tissue. Various means have been devised to do this by focusing external sources of energy, including high-intensity focused ultrasound[52] and focused microwave thermotherapy.[53] In addition, interstitial methods have been developed in which a 'thermoseed' is implanted in the cancer and heats in response to a relatively weak external stimulus that normal tissue does not respond to. Early attempts to do this involved surgically implanting small ferrite cores encased in metal, which were inductively heated[54] but it proved difficult to heat sufficiently between the seeds with a tolerable number of implants. A more recent development in interstitial heating is MNH in which a ferrofluid containing magnetic nanoparticles is injected into the tumour.[55] Thus, the thermoseeds are shrunk to the nanoscale, which provides uniform heating of the dosed region. MNH is the most promising of the thermal therapies and is the focus of this section.

6.2 Magnetic Nanoparticle Hyperthermia

The first human clinical trial of MNH combined with external beam radiotherapy on a group of 14 patients with recurrent glioblastoma multiforme (GBM), a severe brain cancer, was carried out in 2007.[55] In this Phase I trial, the aim was not to treat but to evaluate the safety of the method and the results demonstrated that side effects are minor and it is a safe treatment. The ferrofluid used was an aqueous suspension of 12 nm diameter magnetite (Fe_3O_4) nanoparticles with an aminosilane coating and a Fe concentration of 112 mg ml^{-1}. After injecting the fluid at various sites within the tumour, the nanoparticles were heated interstitially by applying a 100 kHz oscillating magnetic field whose amplitude could be varied in the range 2–15 kA m^{-1}. The clinical trials on recurrent GBM treated by combining MNH and radiotherapy were extended to a Phase II study on 59 patients in 2010[56] with the aim of evaluating the effectiveness of the treatment. It was shown that the life expectancy after a second occurrence of GBM was extended to 13.4 months compared to 6.2 months reported for conventional treatment consisting of combined radio and chemotherapy, despite a reduced radiotherapy dose used in the MNH study.

The Phase I clinical trials have been extended to prostate cancer[57] in which the complex morphology around the prostate presents a significant challenge for MNH as the boundaries separating regions of different dielectric constant produces internal reflections of the applied field and hot spots. Thus, the applied field amplitude has to be kept much lower than for the treatment of brain tumours. Again, it was found that the treatment is well tolerated by patients but the difficulty of producing a sufficient temperature rise throughout the whole tumour has steered the on-going Phase II trial to combining MNH with brachytherapy. The first results from this trial are expected around 2015.[57]

In the last 5 years, there has been growing clinical experience with treating tumours using MNH and significant hardware and software resources have been developed. These include the Magforce Nanotechnologies NanoActivator, which can apply magnetic fields at a frequency of 100 kHz and field amplitudes in the range 2–15 kA m^{-1} to large areas of the body (Figure 6.15) and the NanoPlan® planning software. This enables physicians to determine the temperature distribution within a tumour for a specific applied field amplitude once the density distribution of nanoparticles injected into the tumour is determined by a computerised tomography scan. There has also been developed a standard measure of therapeutic heat dose, which is expressed as the cumulative equivalent minutes at 43 °C for 90% of the tumour volume (CEM 43 T_{90}).[58]

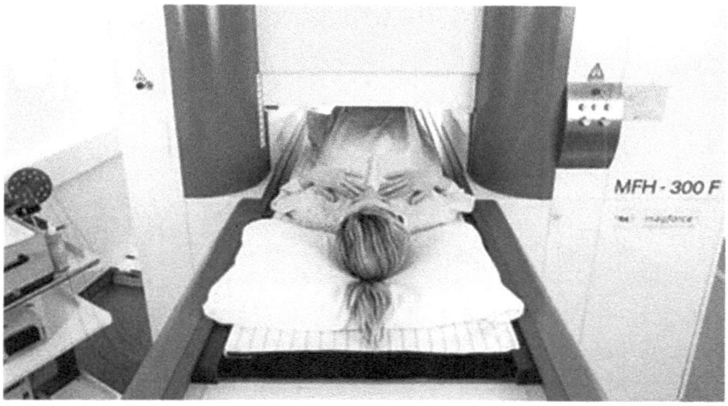

FIGURE 6.15 Magforce Nanotechnologies NanoActivator, which can apply magnetic fields at a frequency of 100 kHz and field amplitudes in the range 2–15 kA m^{-1} to large areas of the body. *Reproduced with permission from Magforce AG, Berlin.*

Despite the rapid move towards clinical use, the real potential of MNH as a generic stand-alone treatment almost free of side effects has yet to be realised, and it is clear that the main restriction lies right at the heart of the method, that is, the nanoparticles themselves, which need to produce more heat per gramme of injected material without increasing the power of the applied field. MNH applied to mice, in which, because of the small volume, higher field-frequency values can be used, have demonstrated complete tumour regression[6] indicating that MNH as a stand-alone therapy is possible, at least in principle.

6.3 Limits on Applied Magnetic Field

In the idealised conceptualisation of MNH, the applied AMF has no effect anywhere apart from the location of the magnetic nanoparticles where heat is generated. Any oscillating magnetic field however will generate electric field gradients given by Faradays law:

$$\nabla \times \mathbf{E} = -\frac{\partial \mathbf{B}}{\partial t} \tag{6.16}$$

and these field gradients occurring in tissue can produce eddy current heating, peripheral nerve stimulation and, if sufficiently strong, direct muscle stimulation. The physiological properties of tissue are well characterised and it is possible to calculate with reasonable confidence what should be safe limits to the applied AMF. Starting with eddy current heating, this is the direct Joule heating produced by current flowing through resistive tissue in response to the field gradient predicted by Equation (6.16). In general, the calculation of eddy currents is complex due to the spatial distribution of current throughout the heated volume but making certain assumptions it is possible to derive the following analytical expression for the power deposited per unit volume (Wm^{-3}) in a cylinder of uniform tissue by an AMF.[59]

$$P = \pi^2 \frac{\mu^2 \mu_0^2 \sigma}{2} f^2 r^2 H_0^2 \tag{6.17}$$

where μ is the relative permeability of the tissue (≈ 1), σ is its conductivity (Sm^{-1}), f is the frequency of the AMF (Hz), r is the radius of the specimen (m) and H_0 is the amplitude of the AMF (Am^{-1}). It has been proposed that the upper safe limit for heating tissue is 0.025 Wcm^{-3},[1] which is about 25% of the typical heating power density generated by magnetic nanoparticles during MNH. The power deposited in tissue by an AMF of a given amplitude and frequency depends on the conductivity σ, which varies between 0.025 Sm^{-1} for fat and 0.64 Sm^{-1} for blood.[60] Figure 6.16 shows the regions of AMF amplitude and frequency that would generate in excess of 0.025 Wcm^{-3} in these two types of tissue in a sample with a radius of 15 cm (thin torso sized). Equation (6.17) shows that it is the product $H_0 \times f$ that is important in determining whether the applied AMF is safe and a nominal value of 4.85×10^8 $Am^{-1}s^{-1}$ referred to as the Atkinson–Brezovich limit is often quoted.[61]

Also shown in Figure 6.16 are the regions of amplitude and frequency of the AMF that would stimulate cardiac tissue or peripheral nerves.[1] At low frequencies, for increasing frequency the field amplitude required for threshold stimulation of peripheral nerves decreases and this trend continue up to a 'corner' frequency that depends on the physiology and size of the nerve. Beyond this corner frequency, the field amplitude required for stimulation becomes almost independent of frequency. Bearing in mind that the typical frequency of the AMF during MNH treatment is 100 kHz, it is evident that peripheral nerve stimulation is not likely to be a major issue if the AMF is kept within

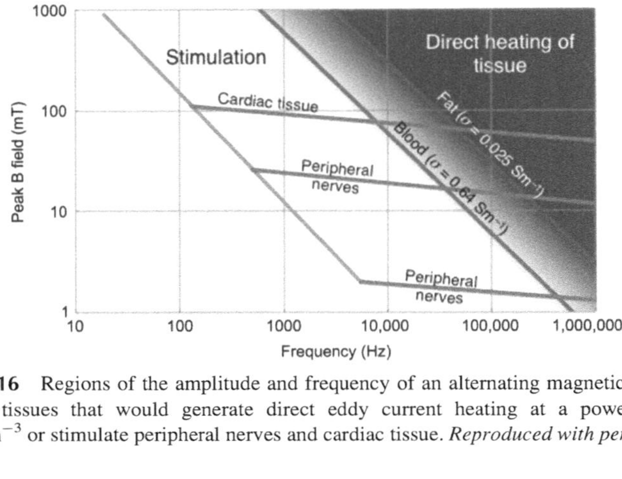

FIGURE 6.16 Regions of the amplitude and frequency of an alternating magnetic field (AMF) applied to tissues that would generate direct eddy current heating at a power density of 0.025 W cm^{-3} or stimulate peripheral nerves and cardiac tissue. *Reproduced with permission from Ref. 1.*

the safe zone to avoid inductive heating of tissue. This is backed up by the clinical trials[55–57] in which a minority of patients reported any discomfort during treatment.

For sufficiently large amplitudes of the AMF, the electric field generated can directly stimulate muscles, which can be extremely dangerous around the heart. Fortunately, this occurs at much higher amplitudes than required for peripheral nerve stimulation at all frequencies so the patients themselves can provide a warning that the field amplitude is unsafe. In summary, the amplitude of a 100 kHz AMF needs to be kept below around 20 mT (16 kAm^{-1}) and so the magnetic fluid must provide sufficient heating to raise the temperature of its host tumour tissue to 43 °C under these conditions. The limits are calculated for a torso-sized specimen but the r^2 dependence of the power generated (Equation 6.17) shows that for smaller parts of the body or small animals it would be safe to apply higher amplitudes of the AMF without directly heating tissue. Thus, Brusentsov *et al.* applied fields an order of magnitude stronger to mice than could be used in humans and demonstrated complete regression of tumours using MNH alone.[6] This in turn suggests that if the heat generated per gramme of magnetic fluid could be increased by an order of magnitude, the method would work as a safe stand-alone treatment in humans.

In the discussion above, it was assumed that the tissue subject to the AMF is homogenous but in reality there may be a complex structure of different tissues, and so for a given amplitude and frequency, the actual power deposited depends on where the field is applied. This was shown in the clinical studies[55–57] in which it was found that much higher field amplitudes could be used in the brain, which is homogenous tissue, than in the prostate region in which there is a complex morphology of tissue and bone. Given that there is a maximum safe limit of the amplitude of the applied field at a given frequency and this is already used during treatment, a priority is to develop nanoparticle suspensions that produce more heat per gramme of injected material in order to further develop the therapy. To do this requires an understanding of the mechanisms by which nanoparticles generate heat in response to an AMF.

6.4 Heating Mechanisms

Heat is generated by a colloidal suspension of magnetic nanoparticles in response to an AMF by more than one mechanism and the relative contribution from each depends on the size distribution of the nanoparticles. The most commonly used theory to date assumes that the nanoparticles are superparamagnetic[62] but suspensions used in practise encompass a size range that spans the boundary between superparamagnetic and blocked particles. Various extensions to the SPM model have been developed to include larger particle sizes.[63,64]

Recently the processes by which heat is produced have been placed in a coherent framework by Vallejo-Fernandez *et al.* that covers the entire size range of particles used.[65] The theory is detailed below but the basic physics underlying the heating mechanisms is illustrated in Figure 6.17. The smallest particles are superparamagnetic and in zero applied field their moment fluctuates on the timescale of nanoseconds so they present no permanent magnetic

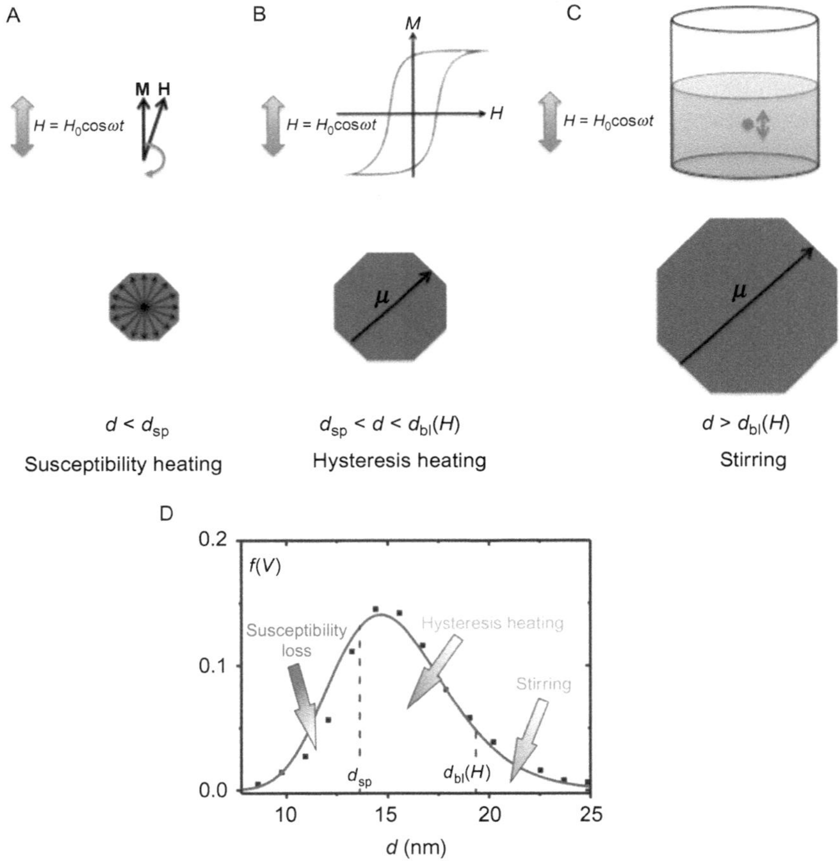

FIGURE 6.17 Mechanisms by which magnetic nanoparticles produce heat in response to AMF. (A) Assemblies of superparamagnetic nanoparticles ($d < d_{sp}$) have a phase lag between the applied field and sample magnetisation, represented by an imaginary susceptibility (χ'') and the energy generated per cycle is proportional to χ''. (B) For $d > d_{sp}$ the particles are blocked and the assembly magnetisation follows a hysteresis loop. The energy generated per cycle is proportional to the area of the loop. (C) Above a (field-dependent) critical size ($d_{bl}(H)$) the applied field is unable to switch the particles but they will move in response to field gradients and generate heat by stirring. (D) A typical nanoparticle size distribution of Fe_2O_3 nanoparticles used to generate heat showing the range of sizes that produce heat by the different mechanisms. *Reproduced with permission from Ref. 65.*

moment (Figure 6.17A). When a magnetic field is applied, the assembly as a whole develops a magnetisation along the applied field and if the field oscillates the magnetisation will follow. The finite lifetime of the nanoparticle magnetic moments (see below) however introduces a phase lag between the magnetisation and the applied field, which is represented by an imaginary (loss) component of the susceptibility (χ''), the magnitude of which determines how much power is produced. This mechanism is dubbed susceptibility heating in Ref. 65.

Above a critical size, labelled d_{sp} in Figure 6.17, the nanoparticles are blocked and susceptibility heating no longer plays a role as each particle will switch its magnetisation in response to the AMF. The magnetisation of the entire assembly follows a hysteresis loop and the heat generated per cycle of the AMF is proportional to the area enclosed by the loop. The coercivity of the nanoparticles increases with size and above a second critical size-labelled $d_{bl}(H)$ the AMF amplitude (H_0) will be insufficient to switch the particle magnetic moments and hysteresis heating will vanish also. The labelling as a function of H emphasises that this critical size is dependent on the strength of the applied field. This largest group however can still produce heat by stirring within the AMF, which will inevitably include field gradients that apply forces to the nanoparticles. Figure 6.17D shows a log-normal size distribution of Fe_2O_3 nanoparticles that is typical of an assembly used for hyperthermia indicating the range of sizes that produce heat by the different mechanisms[65] for which the detailed theory is presented below.

Starting with susceptibility heating generated by superparamagnetic nanoparticles, this will occur for particle diameters up to the blocking temperature, that is up to (see Chapter 1, Equation 1.18):

$$d_{sp} = \left(\frac{6k_B T \ln(tf_0)}{\pi K}\right)^{\frac{1}{3}} \tag{6.18}$$

where t is the measurement time (or the inverse of the frequency of the AMF), f_0 is the high-temperature limit of the nanoparticle magnetic moment fluctuation frequency (usually taken to be $\sim 10^9\ s^{-1}$) and K is the anisotropy constant (assuming uniaxial anisotropy). Within this regime, the power generated per unit volume by an applied AMF of the form $H = H_0\cos(\omega t)$ is given by[62]:

$$P_{sus} = f\mu_0\pi\chi'' H_0^2 \tag{6.19}$$

That is, for a given field amplitude, the power is directly proportional to the loss component of the susceptibility, χ''. This can be written explicitly in terms of the lifetime of the nanoparticle magnetic moment, τ and the equilibrium susceptibility of the assembly, χ_0, in an applied field of H_0 as:

$$\chi'' = \frac{\omega\tau}{1+(\omega\tau)^2}\chi_0 \tag{6.20}$$

For a superparamagnetic assembly whose magnetisation follows the Langevin function (see Chapter 1, Section 5), χ_0 in turn can be written explicitly in terms of the initial susceptibility, χ_i, of the assembly as:

$$\chi_0 = \frac{3\chi_i}{\xi}\left(\cot h\xi - \frac{1}{\xi}\right) \quad (6.21)$$

where ξ is the factor $\mu_0 M_s HV/kT$ with V being the nanoparticle volume and M_s the nanoparticle saturation magnetisation. Finally χ_i can be determined by differentiating the Langevin function at the low field limit to give:

$$\chi_i = \frac{\mu_0 \phi M_s^2 V}{3k_B T} \quad (6.22)$$

where ϕ is the nanoparticle volume fraction. This can be set to 1 to obtain the heating intrinsic to the material and calculate the power generated per m^3 using Equation (6.19). Dividing the result by the density of the material gives the power generated per kg or the specific absorption rate (SAR). It has become normal practise to quote the SAR in units of W g^{-1}.

The only remaining unknown factor to provide a rigorous prediction of heating is the nanoparticle magnetic moment time constant, τ in Equation (6.20) but this can also be given in closed form. The lifetime associated with the nanoparticle magnetic moment arises from the rotation of the moment within the particle (Néel relaxation) and the rotation of the particle in the fluid carrying the moment with it (Brownian relaxation). The lifetime associated with Néel relaxation is given by[65]:

$$\tau_N = \frac{1}{f_0}\exp\left[\frac{KV(1 - H_0/H_K)^2}{k_B T}\right] \quad (6.23)$$

where K is the anisotropy constant of the nanoparticles (assumed uniaxial) and H_K is the anisotropy field given by $H_K = 2K/\mu_0 M_s$. The lifetime associated with Brownian relaxation can be written as[62]:

$$\tau_B = \frac{3V_H \eta}{k_B T} \quad (6.24)$$

where η is the viscosity of the fluid and V_H is the hydrodynamic volume, that is, the effective diameter of the core magnetic nanoparticle plus the coating added for stabilisation of the suspension, targeting, etc. (see Section 3). The two lifetimes are combined using:

$$\frac{1}{\tau} = \frac{1}{\tau_N} + \frac{1}{\tau_B} \quad (6.25)$$

For a size distribution, τ_N, τ_B, χ_0, χ_i (and probably K) all depend on the nanoparticle volume so to determine the susceptibility heating an integration needs to be carried out for diameters up to d_{sp} weighted by the probability for each size. Generally, nanoparticle size distributions are reasonably well represented by a log-normal distribution:

$$g(d) = \frac{1}{\sqrt{2\pi}\sigma d}\exp\left[\frac{-(\ln(d/d_0))^2}{2\sigma^2}\right] \tag{6.26}$$

where $\ln(d_0)$ is the median and σ the standard deviation. The function $g(d)$ is normalised so that:

$$\int_0^\infty g(d)\mathrm{d}d = 1 \tag{6.27}$$

Thus, the power generated per unit volume by susceptibility heating is, from Equation (6.19):

$$P_{\mathrm{sus}} = f\mu_0\pi H_0^2 \int_0^{d_{\mathrm{sp}}} \chi''(d)g(d)\mathrm{d}d \tag{6.28}$$

For diameters above d_{sp}, the susceptibility heating vanishes and is replaced by hysteresis heating for which the power generated is proportional to the frequency of the AMF multiplied by the area of the hysteresis loop. This mechanism will occur up to a size $d_{\mathrm{bl}}(H)$ above which the applied field amplitude is too weak to switch the particle magnetisation. The nanoparticle coercivity, H_c, increases with size according to[65]:

$$H_c = H_K\left[1 - \left(\frac{6k_BT\ln(tf_0)}{\pi Kd^3}\right)^{1/2}\right] \tag{6.29}$$

and it can be shown that for an assembly of nanoparticles with randomly oriented anisotropy axes the critical size above which the field is unable to switch the particle magnetisation is[65]:

$$d_{\mathrm{bl}}(H) = \left[1 - \frac{\mu_0 HM_s}{0.96K}\right]^{-2/3} d_{\mathrm{sp}} \qquad 6.30)$$

For particle sizes in the range d_{sp} to $d_{\mathrm{bl}}(H)$, assuming uniaxial anisotropy, so that the remanence of the assembly is half the saturation magnetisation (see Chapter 1, Section 7), the power generated per unit volume by hysteresis heating is[65]:

$$P_{\mathrm{hyst}} = 2f\mu_0 M_s \int_{d_{\mathrm{sp}}}^{d_{\mathrm{bl}}(H)} H_c(d)g(d)\mathrm{d}d \tag{6.31}$$

For particles with diameters larger than $d_{\mathrm{bl}}(H)$, P_{hyst} vanishes but heat can still be generated by frictional losses due to mechanical stirring. It is not possible to write explicit equations for this mechanism but it is clearly a significant effect as demonstrated by Vallejo-Fernandez *et al.*[65] by comparing the SAR of similar nanoparticle assemblies dispersed in low-viscosity paraffin and wax. It is unclear how large this contribution will be *in vivo* as in some environments particle motion will be more restricted than in others. Restricting particle motion can also remove the Brownian contribution to the nanoparticle magnetic moment (Equation 6.24), which reduces the P_{sus} contribution.[66]

It is clear that the heat generated by an ensemble of nanoparticles depends on parameters such as particle size, M_s, K, H_0 and f in a complicated relationship; however, it is possible to use the model described above to determine some general characteristics that optimise the heat produced. Figure 6.18 shows the SAR calculated using Equations (6.28) and (6.31) for ensembles of nanoparticles with a log-normal size distribution (Equation 6.26) as a function of d_0 with σ set to a constant value of 0.25. Representative size distributions for $d_0 = 10$, 20 and 30 nm are plotted in Figure 6.18A. In all the SAR calculations, a value of 3×10^{-3} Pa s (or 3cP) was used for the fluid viscosity (approximately the value for blood) and the applied AMF had

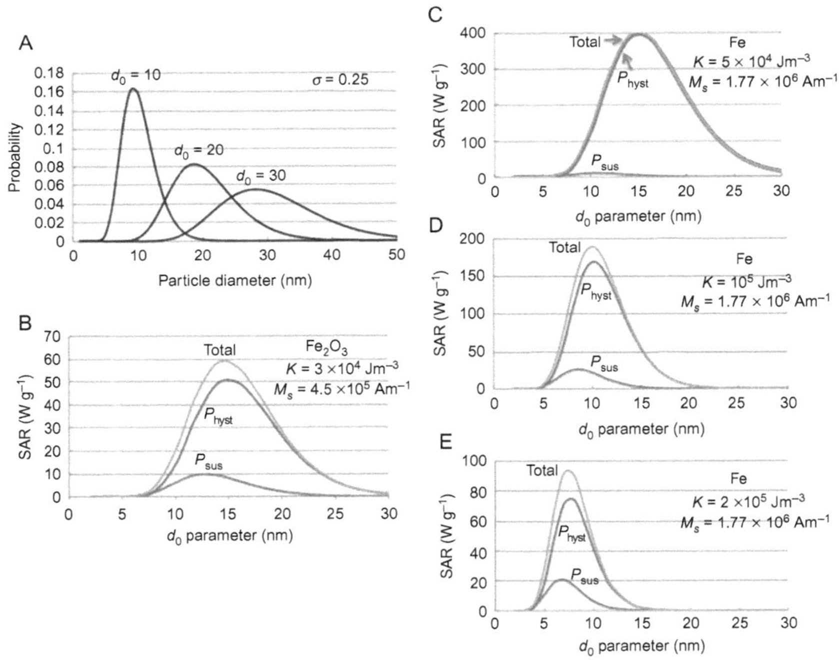

FIGURE 6.18 Calculation of SAR as a function of the d_0 parameter in a log-normal size distribution with $\sigma = 0.25$ for Fe_2O_3 and Fe for different values of the anisotropy constant, K in an applied AMF with $H_0 = 10{,}000$ Am^{-1} and $f = 10$ kHz (other parameters given in the text). (a) Log-normal size distributions plotted for $d_0 = 10$, 20 and 30 nm and $\sigma = 0.25$. (B) SAR for Fe_2O_3 with the M_s and K values shown. (C)–(E) SAR for Fe assuming $M_s = 1.77 \times 10^6$ Am^{-1} (the bulk value) and K values shown, varying from the bulk value to the upper limit expected for Fe nanoparticles.

$H_0 = 10{,}000\ \mathrm{Am^{-1}}$ (12.6 mT) and $f = 100$ kHz. According to Figure 6.16, this is on the edge of the safe region but would be suitable for brain tumours. Figure 6.18(B) shows P_{sus}, P_{hyst} and the total heating power in W g^{-1} for Fe_2O_3 maghemite particles with $M_s = 4.5 \times 10^5\ \mathrm{Am^{-1}}$ and a K value of $3 \times 10^4\ \mathrm{Jm^{-3}}$ taken from Ref. 65. It is evident that most of the heat is generated by P_{hyst}, which is also true for the other samples discussed below. The absolute values obtained are highly sensitive to the value chosen for the anisotropy since this determines the range of integration over the size distribution for P_{hyst} and the anisotropy is in turn sensitive to the particle shape, coating material stoichiometry, etc. Whatever value is chosen however does not change the general shape of the SAR versus d_0 curve, which shows a maximum at a specific value of d_0. It is clear that good size control is important in obtaining high SAR values.

Another general result emerging from the modelling is that large increases in SAR are obtained by increasing the saturation magnetization of the nanoparticle material. Figure 6.18C shows the calculation repeated for Fe nanoparticles with $M_s = 1.77 \times 10^6$ and $K = 5 \times 10^4\ \mathrm{Jm^{-3}}$ (the values for bulk Fe) and at the peak the SAR value is seven times that of Fe_2O_3 nanoparticles. Generally, nanoparticles will have a higher anisotropy than the bulk and Figure 6.18D and E show the SAR vs diameter for Fe nanoparticles with $K = 10^5\ \mathrm{Jm^{-3}}$ and $K = 2 \times 10^5 \mathrm{Jm^{-3}}$. This latter value is that measured for very small 2 nm Fe nanoparticles[67] and can be taken to be the upper limit of Fe nanoparticles in the size of interest for hyperthermia. It is observed that the peak power occurs at smaller particle sizes and reduces in magnitude as the anisotropy increases but even at the upper limit of K, the power available is still significantly higher than for the oxide. Clearly, to be used for hyperthermia, Fe nanoparticles need to be protected by a biocompatible shell that prevents the metal core from oxidising but suspensions of core-shell nanoparticles with pure Fe cores are routinely available using gas-phase synthesis[68] (see Chapter 3).

6.5 New Nanoparticle Materials

Up to now, Fe oxide (mostly Fe_2O_3) is the only nanoparticle material used for *in vivo* hyperthermia and improvements in performance have focused around achieving better particle size control and increasing the particle density within suspensions. As shown in the previous section however, large improvements in SAR values are predicted by using higher moment materials. This has led to the investigation of a number of new types of nanoparticles including Fe@Fe oxide core-shell nanoparticles produced by gas-phase synthesis[68] (see Chapter 3), Co@CoO core-shell nanoparticles prepared by thermolysis of $Co_2(CO)_8$[69,70] and chemically synthesised Fe(0) nanocubes[71] all of which have higher published SAR values than maghemite.

For example, Figure 6.19 shows the measured SAR values (open circles) from Fe@Fe oxide core-shell nanoparticles as a function of frequency with

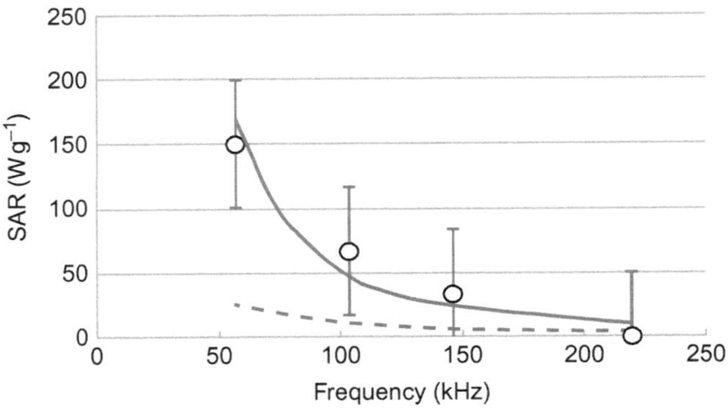

FIGURE 6.19 Measured SAR (open circles) of Fe@Fe oxide core-shell nanoparticles produced by gas-phase synthesis.[68] The applied AMF maintains $H_0 \times f = 4.85 \times 10^8$ Am^{-1}s^{-1} (safe for application in humans) thus, for example, at 100 kHz the field amplitude is 4850 Am^{-1}. The line through the data points is a calculation using the model described by Vallejo-Fernandez *et al.* assuming the nanoparticles have an anisotropy $K = 4.9 \times 10^4$ Jm^{-3}, a saturation magnetisation $M_s = 1.77 \times 10^6$ Am^{-1} and a log-normal size distribution (measured) with $d_0 = 16.5$ nm and $\sigma = 0.2$. The dotted line shows a calculation using the same model for nanoparticles of the same size but with $K = 3 \times 10^4$ Jm^{-3} and a saturation magnetisation $M_s = 4.5 \times 10^5$ Am^{-1}, representative of Fe_2O_3. *Data reproduced with permission from Ref. 68.*

the AMF amplitude adjusted so that at each frequency, the $H_0 \times f$ product is 4.85×10^8 Am^{-1}s^{-1}, that is, the Atkinson–Brezovich limit. The total SAR resulting from the P_{sus} and P_{hyst} terms calculated using the Vallejo-Frenandez model is also displayed as a grey line and shows a good fit with the nanoparticle magnetic anisotropy set to $K = 4.9 \times 10^4$ Jm^{-3}, which is slightly above the bulk Fe value. Any heating by the stirring mechanism is not included in this calculation so the $P_{sus} + P_{hyst}$ heating could be lower than the measured values indicating a higher anisotropy of the nanoparticles but the increased SAR relative to Fe_2O_3, shown by the red line, is clear.

As shown in Chapter 3, magnetosomes manufactured by bacteria are an attractive way to produce nanoparticles and in the case of medical applications have the advantage that the particles are naturally biocompatible.[72] They can also display impressive SAR values despite the fact that they are low-moment magnetic minerals rather than the high-moment metallic core nanoparticles described in this section. In addition the large size of these particles (36.5 nm) indicates that, according to the heating model presented above, the contribution from both P_{sus} and P_{hyst} will be very small so that all heat generation should be by stirring.

A problem with comparing these novel nanoparticle materials with each other is that all reported SAR values are measured with different field amplitudes and frequencies. An attempt to provide a scaling method that gives an intrinsic material figure of merit (the intrinsic loss parameter or ILP) that is independent of the AMF parameters used was published by Kallumadil *et al.*[73] This normalises the reported SAR values to the product $H_0^2 \times f$, which is appropriate for the P_{sus} heating mechanism (Equation 6.19) but as shown in the previous section, this is a minor contribution in most cases. The scaling of SAR with H_0 is not simple for the P_{hyst} heating mechanism however for both mechanisms the power is proportional to the frequency used. A simpler and perhaps more appropriate normalisation is simply to scale the frequency so that the $H_0 \times f$ product is the same for all samples and an appropriate value to use is $4.85 \times 10^8 \mathrm{Am^{-1}s^{-1}}$, that is, the Atkinson–Brezovich limit.

Table 6.2 shows the reported SAR values for various nanoparticle materials and the applied AMF frequency and amplitude used. Also shown is the value scaled to $H_0 \times f = 4.85 \times 10^8\ \mathrm{Am^{-1}s^{-1}}$ and it is clear that all the new materials have higher SAR values than maghemite. Note that the value for maghemite obtained by Fortin *et al.*[74] was obtained for size-selected nanoparticles and is close to the best that can be obtained for this material. For ease of comparison, the scaled SAR values in Table 6.2 are plotted as a function of particle size in Figure 6.20. Thus, if other problems (biocompatibility stability of suspensions, etc.) can be solved, new nanoparticle materials may produce significant improvements in the effectiveness of hyperthermia.

6.6 Targeting Magnetic Nanoparticles to Tumours

Currently all clinical studies of MNH have injected the ferrofluid directly into the tumour at typical concentrations of about $10\ \mathrm{mg\ ml^{-1}}$ and relied on natural mechanisms to keep it in place during therapy. In research laboratories, however, there has been considerable progress in conjugating the nanoparticles with targeting agents, including antibodies, aptamers, peptides and folic acid that bind specifically to tumour cells. Such active targeting is likely to increase the efficiency of filling the tumour with ferrofluid with minimal leakage out into healthy tissue. As shown by animal studies,[76] it is even be possible to accumulate nanoparticles in the tumour after injecting them at a different site.

In the last few years, MNH has been tested in mice bearing breast cancers using magnetic nanoparticles conjugated with antibodies suitable for specific targeting to breast cancer and shows encouraging results. The first such study,[76] employed Fe oxide nanoparticles embedded in a dextran matrix, forming a 20 nm bead, coated with PEG and conjugated with L6 (ChL6) antibodies attached (one or two per nanoparticle) labelled with ^{111}In as illustrated in Figure 6.21. The purpose of the ^{111}In radioactive tracer was to confirm the uptake of the nanoparticles in the tumour.

To test the effectiveness of the targeting, the nanoparticles were injected into the tail vein of mice bearing human breast cancer HBT3477 and 3 days later the mice were exposed to an AMF with an amplitude of 56, 80 or $103\ \mathrm{kA\ m^{-1}}$ at a frequency of 153 kHz. These are much higher doses than could be used for human sized bodies but as discussed in Section 6.3, in

TABLE 6.2 Reported SAR Values for a Selection of High-Performance Nanoparticles with the AMF Parameters Used and the Scaled SAR Values to Allow a Comparison

Material in Nanoparticle Core	Diameter of Core (nm)	H_0 (kAm^{-1})	f (kHz)	Reported SAR (Wg^{-1})	Scaled SAR (Wg^{-1})	Reference
Fe_2O_3 maghemite	16.5	24.8	700	1650	46	74
Fe@Fe oxide	16.5	6.5	57	150	150	68
Fe magnetosome	36.5	10	410	960	114	72
Co@Co oxide	6	10	410	720	85	69,70
Fe(0) nanocubes	16	52.5	300	1690	52	71
$CoFe_2O_4$@$MnFe_2O_4$ soft/hard core-shell particles	15	37.3	500	4000	104	75

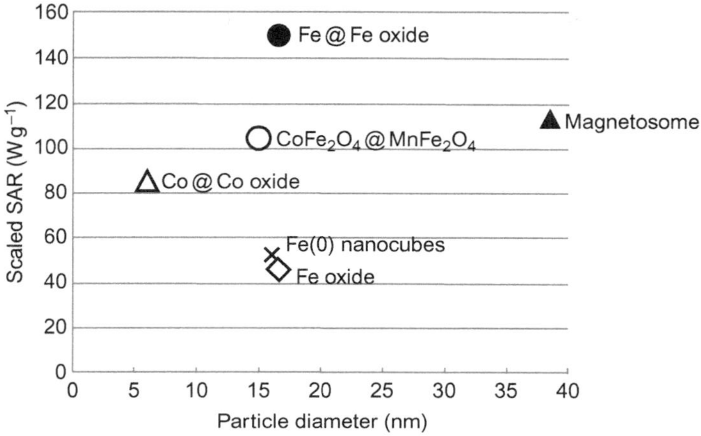

FIGURE 6.20 Best SAR values (scaled) for comparison for a selection of high-performance nanomaterials compared with Fe oxide.

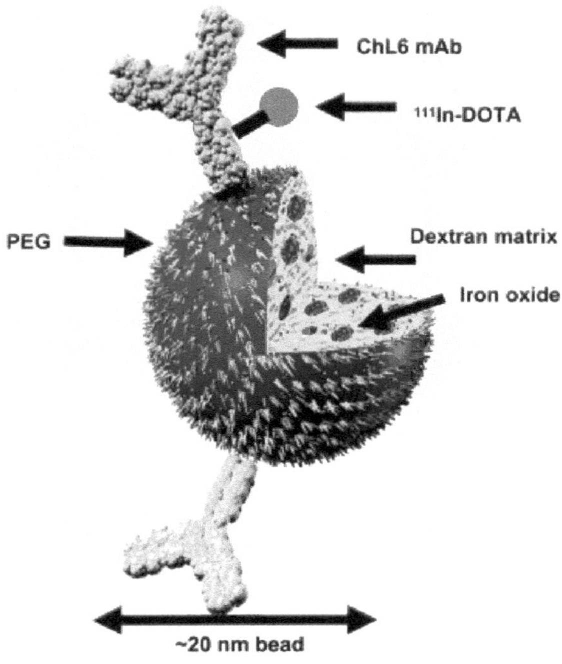

FIGURE 6.21 PEG- and dextran-coated Fe oxide nanoparticle conjugated to ChL6 antibody for studies of MNH using targeted nanoparticles. *Reproduced with permission from Ref. 77.*

smaller volumes the direct heating of tissue by the AMF is reduced. Nonetheless, some toxic side effects were observed for the group exposed to the highest field amplitude. The tumours in all the groups subjected to the AMF showed a significant decrease in the tumour growth rate relative to untreated animals.

In a later study,[77] using the same nanoparticles and breast cancer type, the group showed, via the radioactive tracer, that around 14% of the injected nanoparticles per gramme of tumour were concentrated at the tumour site despite their introduction into the tail. In addition, the therapeutic effect was observed at a concentration of nanoparticles within the tumour 30 times lower than used in human clinical trials. This is partly offset by the applied AMF being about an order of magnitude higher than could be used on humans but is still an encouraging result indicating that targeting does increase the efficiency of heating.

7 MAGNETIC NANOPARTICLES FOR TARGETED DRUG DELIVERY

In its simplest form, magnetic targeting of drugs to a specific area involves attaching magnetic nanoparticles to the drug to be delivered and confining it within the target region using a magnetic field gradient. Thus, the concentration can be greatly enhanced relative to systemic administration, which is particularly important for toxic drugs, such as anticancer agents. This simple idea however is proving difficult to put into practise in clinical applications. The force, F, on a magnetic nanoparticle with a total magnetic moment μ in a magnetic field $\mathbf{B}$ is given by:

$$F = (\mu \cdot \nabla)\mathbf{B} \tag{6.32}$$

or in one dimension:

$$F = \mu \frac{\partial \mathbf{B}}{\partial z} \tag{6.33}$$

Within a single-domain particle, μ is proportional to the volume and so the force increases with particle diameter. Above the critical size, for the particle to remain a single domain, however, μ will reduce to the average magnetisation of the different domains along the applied field. So unless the magnetic field is strong enough to saturate the particles, their magnetisation will be reduced relative to single-domain particles in which the magnetisation is permanently saturated. The optimum size is thus the largest particle that is a single domain. This critical size depends on the anisotropy but generally is in the range 50–100 nm.

The technical difficulties involved can be illustrated by a 'back of the envelope' estimate of the force on a 100 nm diameter Fe particle containing $\sim 4 \times 10^7$ atoms. Assuming the bulk value for the magnetic moment per atom, that is, 2.2 μ_B, the total moment of the particle would be $\mu = 4 \times 10^7 \times 2.2 \times 9.274 \times 10^{-24} = 8 \times 10^{-16}$ Am^2. In practise, the largest field gradient that can be achieved within the body by applying a magnetic field externally is about 50 Tm^{-1} so the maximum force that can be applied is 4×10^{-14}N. The drag force for a particle with a radius r in a fluid with a viscosity η moving with a velocity v is:

$$F = 6\pi\eta vr \quad (6.34)$$

The viscosity of blood is $\eta = 0.0027$ Ns m^{-2} so for an Fe particle with a diameter of 100 nm the magnetic force would be overcome by a blood flow faster than about 15 μms^{-1}. This is slower than in even small blood vessels so in practise the technique relies on the particle being pulled to a tissue wall and sticking rather than being stopped in its tracks by pure magnetic force. The above calculation is for pure Fe nanoparticles, which are not available for medical applications and repeating the calculation for the commonly used oxide nanoparticles produces smaller forces.

The numbers become much more favourable for applications close to the surface of the body and recently Lueshen *et al.*[78] demonstrated a ninefold increase in the concentration of Au-coated superparamagnetic Fe oxide magnetic nanoparticles in the intrathecal space of a spine model when the external magnetic field was applied. Thus, for regions where a high-field gradient can be applied due to their proximity to the surface, such as the spine, targeting using pure magnetic forces is feasible.

For applications deep in the body, it is possible to introduce a local high gradient magnetic field by implanting magnetic needles or meshes at the target and theoretical modelling demonstrates that this approach can produce a sufficiently high magnetic force to capture drug-loaded magnetic nanoparticles.[79] In addition, it has been demonstrated by modelling that periodic magnetised arrays can be much more efficient at trapping and holding magnetic nanoparticles than conventional magnetic sources that produce field gradients.[80,81]

An alternative approach to using the magnetic force therapeutically is to apply it to enhance a natural targeting process and this mode was demonstrated by Muthana *et al.*[82] Attempts have been made to utilise monocytes, which naturally migrate into tumours from the bloodstream, to carry therapeutic agents to the heart of a tumour. Generally, the efficiency of targeting tumours this way has been found to be quite low but a significant increase was demonstrated in mice bearing tumours when the monocytes were loaded with magnetic nanoparticles and a magnetic force was applied close to the target site.[82] The advantage of using the magnetic force in this way is that the loaded cells do not need to be anchored by the magnetic field but simply slowed down sufficiently so that chemical messengers from the tumour are able to bring a higher proportion of monocytes into the core than would normally be the case.

In aerosols, the strong drag force from blood is absent and magnetic nanoparticles respond much more sensitively to a magnetic gradient, which has encouraged studies on the use of magnetic targeting to guide aerosol into the lungs. Many pulmonary diseases, including asthma, cystic fibrosis and lung cancer are treated by inhaled drugs and with conventional inhalers only about 4% of the drug makes it through to the windpipe so in order to get sufficient drug to where it is needed, some tissues have to be over-exposed leading to unwanted side effects. A study by Dames *et al.*[83] tested aerosols in which liquid droplets contained iron oxide nanoparticles and a drug. By using a magnetic field, gradient next to one lung of a mouse they found that this lung received eight times more drug than its neighbour. Applying this to humans will be more difficult because of the larger lung volume and the difficulty already discussed above with getting a significant magnetic field gradient deep in the body but with improved magnetic field generators such as those described in Refs. 80,81 and magnetic particles with a higher magnetisation than iron oxide, this may become a viable technology.

Finally, although it is beyond the scope of this chapter, it is worth pointing out that there is also a burgeoning area of *in vitro* research using magnetic nanoparticles (where applying large field gradients is not an issue) such as the manipulation and control of cell function using magnetic nanoparticles. The use of magnetic nanoparticles to probe the mechanical properties of cells dates back to the 1920s[84] but recently these methods have become highly sophisticated and include the ability to actuate/deactuate ion channels in cell membranes. Some of the recent advances in these techniques are reviewed in Ref. 1.

ACKNOWLEDGEMENTS

Discussions with Dr. Mark Horsfield in the Department of Cardiovascular Sciences, University of Leicester on MRI and the model presented in Section 4 led to significant improvement in the description.

REFERENCES

1. Pankhurst QA, Thanh NTK, Jones SK, Dobson J. Progress in applications of magnetic nanoparticles in biomedicine. *J Phys D Appl Phys* 2009;**42**:224001.
2. Dobson J. Magnetic nanoparticles for drug delivery. *Drug Dev Res* 2006;**67**:55–60.
3. Binns C. Magnetic nanoparticle hyperthermia treatment of tumours. In: Akas B, Mikailzade F, editors. *Nanostructured materials for magnetoelectronics*. Heidelburg, New York, Dordrecht, London: Springer; 2013 [Chapter 8].
4. Mornet S, Vasseur S, Grasset F, Duguet E. Magnetic nanoparticle design for medical diagnosis and therapy. *J Mater Chem* 2004;**14**:2161–75.
5. Weizenecker J, Gleich B, Rahmer J, Dahnke H, Borgert J. Three-dimensional real time in vivo magnetic particle imaging. *Phys Med Biol* 2009;**54**:L1–10.
6. Brusentsov NA, Nikitin LV, Brusentsova TN, Kuznetsov AA, Bayburtskiy FS, Shumakov LI, et al. Magnetic fluid hyperthermia of the mouse experimental tumor. *J Magn Magn Mater* 2002;**252**:378–80.

7. Brazel CS. Magnetothermally responsive nanomaterials: combining magnetic nanostructures and thermally-sensitive polymers for triggered drug release. *Pharm Res* 2009;**26**:644–56.
8. Laurent S, Forge D, Port M, Roch A, Robic C, Elst LV, et al. Magnetic iron oxide nanoparticles: synthesis, stabilization, vectorization, physicochemical characterizations, and biological applications. *Chem Rev* 2008;**108**:2064–110.
9. van Furth R. Production and migration of monocytes and kinetics of macrophages. In: van Furth R, editor. *Mononuclear phagocytes*. Dordrecht: Kluwer; 1992 [Chapter 1].
10. Arruebo A, Fernández-Pacheco R, Ibarra MR, Santamaria J. Magnetic nanoparticles for drug delivery. *Nanotoday* 2007;**2**:22–32.
11. Napper DH. Steric stabilization. *J Colloid Interface Sci* 1977;**58**:390–407.
12. Derjaguin B, Landau L. Theory of the stability of strongly charged lyophobic sols and of the adhesion of strongly charged particles in solutions of electrolytes. *Acta Physico Chemica URSS* 1941;**14**:633–62.
13. Verwey EJW, Overbeek JThG. *Theory of the stability of lyophobic colloids*. Amsterdam: Elsevier; 1946.
14. Eastman J. Stability of charge-stabilised colloids. In: Cosgrove T, editor. *Colloid science: principles, methods and applications*. Chichester, UK: Wiley; 2010 [Chapter 3].
15. Faure B, Salazar-Alvarez G, Bergström L. Hamaker constants of iron oxide nanoparticles. *Langmuir* 2011;**27**:8659–64.
16. Fiorani D, Testa AM, Lucari F, D'Orazio F, Romero H. Magnetic properties of maghemite nanoparticle systems: surface anisotropy and interparticle interaction effects. *Phys B* 2002;**320**:122–6.
17. Fritz G, Schädler V, Willenbacher N, Wagner NJ. Electrosteric stabilization of colloidal dispersions. *Langmuir* 2002;**18**:6381–90.
18. Berry CC, Wells S, Charles S, Curtis ASG. Dextran and albumin derivatised iron oxide nanoparticles: influence on fibroblasts in vitro. *Biomaterials* 2003;**24**:4551–7.
19. Laurent S, Nicotra C, Gossuin Y, Roch A, Ouakssim A, Van der Elst L, et al. Influence of the length of the coating molecules on the nuclear magnetic relaxivity of superparamagnetic colloids. *Phys Status Solidi C* 2004;**1**:3644–50.
20. Gamarra LF, Brito GES, Pontuschka WM, Amaro E, Parma AHC, Goya GF. Biocompatible superparamagnetic iron oxide nanoparticles used for contrast agents: a structural and magnetic study. *J Magn Magn Mater* 2005;**289**:439–41.
21. Zhang Y, Kohler N, Zhang M. Surface modification of superparamagnetic magnetite nanoparticles and their intracellular uptake. *Biomaterials* 2002;**23**:1553–61.
22. Lee J, Isobe T, Senna M. Preparation of ultrafine Fe_3O_4particles by precipitation in the presence of PVA at high pH. *J Colloid Interface Sci* 1996;**177**:490–4.
23. Schöpf B, Neuberger T, Schulze K, Petri A, Chastellain M, Hofmann M, et al. Methodology description for detection of cellular uptake of PVA coated superparamagnetic iron oxide nanoparticles (SPION) in synovial cells of sheep. *J Magn Magn Mater* 2005;**293**:411–6.
24. Liu H-L, Ko SP, Wu J-H, Jung M-H, Min JH, Lee JH, et al. One-pot polyol synthesis of monosize PVP-coated sub-5 nm Fe_3O_4 nanoparticles for biomedical applications. *J Magn Magn Mater* 2007;**310**:e815–7.
25. Nishio Y, Yamada A, Exaki K, Miyashita Y, Furukawa H, Horie K. Preparation and magnetometric characterization of iron oxide-containing alginate/poly(vinyl alcohol) networks. *Polymer* 2004;**45**:7129–36.
26. Kim EH, Lee HS, Kwak BK, Kim B-K. Synthesis of ferrofluid with magnetic nanoparticles by sonochemical method for MRI contrast agent. *J Magn Magn Mater* 2005;**289**:328–30.
27. Moghimi SM, Hunter AC, Murray JC. Long-circulating and target-specific nanoparticles: theory to practice. *Pharm Rev* 2001;**53**:283–316.

28. Zahr AS, Davis CA, Pishko MV. Macrophage uptake of core-shell nanoparticles surface modified with poly(ethylene glycol). *Langmuir* 2006;**22**:8178–85.
29. Berry CC, Wells S, Charles S, Curtis ASG. Detran and albumin derivatised iron oxide nanoparticles: influence on fibroblasts in vitro. *Biomaterials* 2003;**24**:4551–7.
30. Häfeli UO, Riffle JS, Harris-Shekhawat L, Carmichael-Baranauskas A, Mark F, Dailey JP, et al. Cell uptake and *in vitro* toxicity of magnetic nanoparticles suitable for drug delivery. *Mol Pharm* 2009;**6**:1417–26.
31. Freed JH. Dynamic effects of pair correlation functions on spin relaxation by translational diffusion in liquids. II. Finite jumps and independent T1 processes. *J Chem Phys* 1978;**68**:4034–7.
32. Jackson TJ, Binns C, Forgan EM, Morenzoni E, Niedermayer Ch, Glückler H, et al. Superparamagnetic relaxation in Fe nanoclusters measured by Low energy muon spin rotation. *J Phys Condens Matter* 2000;**12**:1399–411.
33. Koenig SH, Keller EK. Theory of $1/T_1$ and $1/T_2$ NMRD profiles of solutions of magnetic nanoparticles. *Magn Reson Med* 1995;**34**:227–33.
34. Parkes LM, Hodgson R, Tu LT, Tung LD, Robinson I, Fernig DG, et al. Cobalt nanoparticles as a novel magnetic resonance contrast agent—relaxivities at 1.5 and 3 Tesla. *Contrast Media Mol Imaging* 2008;**3**:150–6.
35. Lee J-H, Huh Y-M, Jun Y-W, Seo J-W, Jang J-T, Song H-T, et al. Artificially engineered magnetic nanoparticles for ultra-sensitive molecular imaging. *Nat Med* 2007;**13**:95–9.
36. Seo W-S, Lee J-H, Sun X, Suzuki Y, Mann D, Liu Z, et al. FeCo/graphitic-shell nanocrystals as advanced magnetic resonance-imaging and near-infrared agents. *Nat Mater* 2006;**5**:971–6.
37. Reimer P, Balzer T. Ferucarbotran (Resovist): a new clinically approved RES-specific contrast agent for contrast-enhanced MRI of the liver: properties, clinical development, and applications. *Eur Radiol* 2003;**13**:1266–76.
38. Aktas S., Horsfield M., Binns C., Thornton S.C. University of Leicester; 2013, unpublished.
39. Torchilin V. Tumor delivery of macromolecular drugs based on the EPR effect. *Adv Drug Deliv Rev* 2011;**63**:131–5.
40. Goodwill P, Krishnan KM, Conolly SM. Magnetic particle imaging for angiography, stem cell tracking, cancer imaging and inflammation imaging. In: Thanh NTK, editor. *Magnetic nanoparticles from fabrication ot clinical applications*. Boca Raton, Florida: CRC Press; 2012 [Chapter 20].
41. Gleich B. Verfahren zur ermittlung der räumlichen verteilung magnetischer partikel. German Patent No. DE-10151778-A1; 2001.
42. Gleich B, Weizenecker J. Tomographic imaging using the nonlinear response of magnetic particles. *Nature* 2005;**435**:1214–7.
43. Gleich B, Weizenecker J, Timminger H, Bontus C, Schmale I, Rahmer J, et al. Fast MPI demonstrator with enlarged field of view. In: *Proceedings of ISMRM*. vol. 18; 2010. p. 216, Stockholm (May 2010).
44. Gleich B, Weizendecker J, Borgert J. Theory, simulation and experimental results of the acoustic detection of magnetization changes in superparamagnetic iron oxide. *BMC Med Imaging* 2011;**11**:16.
45. Sattel TF, Knopp T, Biederer S, Gleich B, Weizenecker J, Borgert J, et al. Single-sided device for magnetic particle imaging. *J Phys D Appl Phys* 2009;**42**:022001.
46. Saritas EU, Goodwill PW, Croft LR, Konkle JJ, Lu K, Zheng B, et al. Magnetic particle imaging (MPI) for NMR and MRI researchers. *J Magn Res* 2013;**229**:116–26.
47. Weaver JB, Rauwerdink AM, Sullivan CR, Baker I. Frequency distribution of the nanoparticle magnetization in the presence of a static as well as a harmonic magnetic field. *Med Phys* 2008;**35**:1988–94.

48. Knopp T, Buzug TM. *Magnetic particle imaging*. Springer; 2012.
49. Davies P. Exposing cancer's deep evolutionary roots. *Phys World* 2013;**26**(7):37–40.
50. DeNardo GL, DeNardo SJ. Update: turning the heat on cancer. *Cancer Biother Radiopharm* 2008;**23**:671–9.
51. Vaupel P, Kallinowski F. Physiological effects of hyperthermia. In: Streffer C, editor. *Hyperthermia and the therapy of malignant tumors*. Berlin/Heidelberg, Germany: Springer-Verlag; 1987.
52. Kennedy JE, ter Haar GR, Cranston D. High intensity focused ultrasound: surgery of the future? *Br J Radiol* 2003;**76**:590–9.
53. Dooley WC. Focused microwave thermotherapy for preoperative treatment of nnvasive breast cancer: a review of clinical studies. *Ann Surg Oncol* 2009;**17**(4):1076–93.
54. Cetas TC, Gross EJ, Contractor Y. A ferrite core metallic sheath thermoseed for interstitial thermal therapies. *IEEE Trans Biomed Eng* 1998;**45**:68–77.
55. Maier-Hauff K, Rothe R, Scholz R, Gneveckow U, Wust P, Thiesen B, et al. Intracranial thermotherapy using magnetic nanoparticles combined with external beam radiotherapy: results of a feasibility study on patients with glioblastoma multiforme. *J Neuro-Oncol* 2007;**81**:53–60.
56. Maier-Hauff K, Ulrich F, Nestler D, Niehoff H, Wust P, Thiesen B, et al. Efficacy and safety of intratumoral thermotherapy using magnetic iron-oxide nanoparticles combined with external beam radiotherapy on patients with recurrent glioblastoma multiforme. *J Neurooncol* 2011;**103**:317–24.
57. Johannsen M, Thiesen B, Wust P, Jordan A. Magnetic nanoparticle hyperthermia for prostate cancer. *Int J Hyperther* 2010;**26**:790–5.
58. Sapareto SA, Dewey WC. Thermal dose determination in cancer therapy. *Int J Radiat Oncol Biol Phys* 1984;**10**:787–800.
59. Jordan A, Wust P, Fähling H, John W, Hinz A, Felix R. Inductive heating of ferromagnetic particles and magnetic fluids: physical evaluation of their potential for hyperthermia. *Int J Hyperthermia* 1993;**9**:51–66.
60. Faes TJC, van der Meij HA, de Munck JC, Heethaar RM. The electrical resistivity of human tissue (100 Hz – 10 MHz): a meta-analysis of review studies. *Physiol Meas* 1999;**20**:R1–R10.
61. Atkinson WJ, Brezovich IA, Chakraborty DP. Usable frequencies in hyperthermia with thermal seeds. *IEEE (T) Bio-Med Eng* 1984;**31**:70.
62. Rosensweig RE. Heating magnetic fluid with alternating magnetic field. *J Magn Magn Mater* 2002;**252**:370–4.
63. Hergt R, Meffre S, Röder M. Effects of size distribution on hysteresis losses of magnetic nanoparticles for hyperthermia. *J Phys Condens Matter* 2008;**20**:385214.
64. Hergt R, Dutz S, Zeisberger M. Validity limits of the Néel relaxation model of magnetic nanoparticles for hyperthermia. *Nanotechnology* 2010;**21**:015706.
65. Vallejo-Fernandez G, Whear O, Roca AG, Hussain S, Timmis J, Patel V, et al. Mechanisms of hyperthermia in magnetic nanoparticles. *J Phys D Appl Phys* 2013;**46**:312001.
66. Suto M, Hirota Y, Mamiya H, Fujita A, Kasuya R, Tohji K, et al. Heat dissipation mechanism of magnetite nanoparticles in magnetic fluid hyperthermia. *J Magn Magn Mater* 2009;**321**:1493–6.
67. Binns C, Maher MJ, Pankhurst QA, Kechrakos D, Trohidou KN. Magnetic behaviour of nanostructured films assembled from pre-formed Fe clusters embedded in Ag. *Phys Rev B* 2002;**66**:184413.
68. Binns C, Prieto P, Baker SH, Howes PB, Dondi R, Burley G, et al. Preparation of hydrosol suspensions of elemental and core-shell nanoparticles by co-deposition with water vapour from the gas-phase in ultra-high vacuum conditions. *J Nanoparticle Res* 2012;**14**:1136.

69. Bönnemann H, Brijoux W, Brinkmann R, Matoussevitch N, Waldöfner N, Palina N, et al. A size-selective synthesis of air stable colloidal magnetic cobalt nanoparticles. *Inorg Chim Acta* 2003;**350**:617–24.
70. Hergt R, Dutz S, Müller R, Zeisberger M. Magnetic particle hyperthermia: nanoparticle magnetism and materials development for cancer therapy. *J Phys Condens Matter* 2006;**18**: S2919–34.
71. Mehdaoui B, Meffre A, Lacroix L-M, Carrey J, Lachaize S, Gougeon M, et al. Large specific absorption rates in the magnetic hyperthermia properties of metallic iron nanocubes. *J Magn Magn Mater* 2010;**322**:L49–52.
72. Hergt R, Hiergeist R, Zeisberger M, Schüler D, Heyen U, Hilger I, et al. Magnetic properties of bacterial magnetosomes as potential diagnostic and therapeutic tools. *J Magn Magn Mater* 2005;**293**:80–6.
73. Kallumadil M, Tada M, Nakagawa T, Abe M, Southern P, Pankhurst QA. Suitability of commercial colloids for magnetic hyperthermia. *J Magn Magn Mater* 2009;**321**:1509–13.
74. Fortin J-P, Wilhelm C, Servais J, Ménager C, Bacri J-C, Gazeau F. Size-sorted anionic iron oxide nanomagnets as colloidal mediators for magnetic hyperthermia. *J Am Chem Soc* 2007;**129**:2628–35.
75. Lee J-H, Jang J-T, Choi J-S, Moon SH, Noh S-H, Kim J-W, et al. Exchange-coupled magnetic nanoparticles for efficient heat induction. *Nat Nanotechnol* 2011;**6**:418–22.
76. DeNardo S, DeNardo GL, Miers LA, Natarajan A, Foreman AR, Gruettner C, et al. Development of tumour targeting bioprobes (^{111}In-Chimeric L6 monoclonal antibody nanoparticles) for alternating magnetic field cancer therapy. *Clin Cancer Res* 2005;**11**:7087s–92s.
77. DeNardo S, DeNardo GL, Natarajan A, Miers LA, Foreman AR, Gruettner C, et al. Thermal dosimetry predictive of efficacy of ^{111}In-ChL6 nanoparticle AMF–induced thermoablative therapy for human breast cancer in mice. *J Nucl Med* 2007;**48**:437–44.
78. Lueshen E, Venugopal I, Kanikunnel J, Soni T, Alaraj A, Linninger A. Intrathecal magnetic drug targeting using gold-coated magnetite nanoparticles in a human spine model. *Nanomedicine* 2013:1–15, published, online July 18th. http://dx.doi.org/10.2217/nnm.13.69.
79. Iacob Gh, Rotariu O, Strachan NJC, Häfeli UO. Magnetizable needles and wires—modeling an efficient way to target magnetic microspheres *in vivo*. *Biorheology* 2004;**41**:599–612.
80. Hayden ME, Häfeli UO. 'Magnetic bandages' for targeted delivery of therapeutic agents. *J Phys Condens Matter* 2006;**18**:S2877–S2891.
81. Häfeli UO, Gilmour K, Zhou A, Lee S, Hayden ME. Modeling of magnetic bandages for drug targeting: button vs. Halbach arrays. *J Magn Magn Mater* 2007;**311**:323–9.
82. Muthana M, Scott SD, Farrow N, Morrow F, Murdoch C, Grubb S, et al. A novel magnetic approach to enhance the efficacy of cell-based gene therapies. *Gene Ther* 2008;**15**:902–10.
83. Dames P, Gleich B, Flemmer A, Hajek K, Seidl N, Wiekhorst F, et al. Targeted delivery of magnetic aerosol droplets to the lung. *Nat Nanotechnol* 2007;**2**:495–9.
84. Heilbronn A. A new method for the estimation of viscosity in living protoplasts. *Jahrb Wiss Bot* 1922;**61**:284–338.

Chapter 7

Environmental Applications of Magnetic Nanoparticles

Andrew Pratt
Department of Physics, University of York, York, United Kingdom

1 INTRODUCTION

Many of the most pressing issues facing the world over the coming decades and beyond relate to the quality of the environment in which we live. Global warming, energy, material and food security, and water cleanliness have all been identified as environmental grand challenges that need addressing now. Amongst the many technologies and solutions currently being investigated by scientists and engineers, the use of magnetic nanoparticles, either directly or indirectly, plays an increasingly important role in all of these fields.

Perhaps the area in which magnetic nanoparticles has made the biggest impact to date is in the treatment of aqueous environments, for example through the removal of industrial contaminants from groundwater and marine environments or by improving the quality of drinking water supplies. The potential benefits are enormous—approximately one billion people around the globe lack access to a safe water supply contributing to millions of deaths each year from diarrhoeal disease and millions of cases of cholera and schistosomiasis. Heavy metals and organic molecules are the most common pollutants making their way into various water systems as a result of anthropogenic activities (industry, farming, urban runoff, etc.; Figure 7.1) and even low concentrations at the ng L^{-1} to μg L^{-1} level can lead to a variety of cancers and respiratory and skin diseases. Mercury poisoning is a well-known example of heavy metal contamination with epidemics occurring at an alarming rate throughout the twentieth century.[1–3] A notorious case that occurred in Japan in 1956 involved the release of industrial wastewater containing an organomercury compound into Minamata Bay.[3] Bioaccumulation in fish and shellfish, subsequently consumed by humans and animals, led to over 2200 victims of mercury poisoning, most of whom died. Chemical release, either sudden or gradual, can also have a dramatic effect on local environments as in the well-documented feminization of fish and alligators that has been reported at multiple sites where natural waters are located close to industrial

Frontiers of Nanoscience, Vol. 6. http://dx.doi.org/10.1016/B978-0-08-098353-0.00007-5

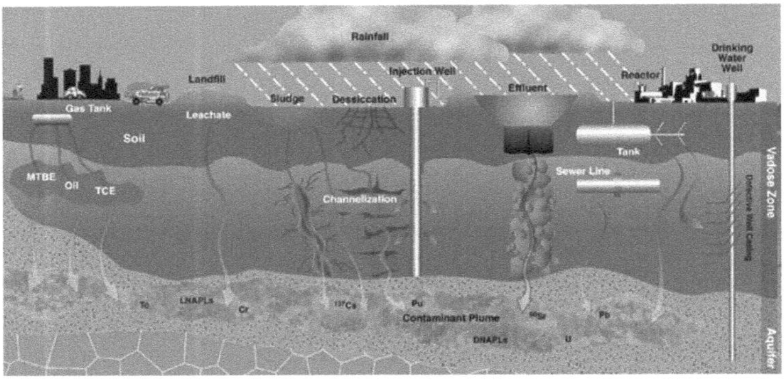

FIGURE 7.1 Schematic diagram showing potential anthropogenic sources of water, groundwater and soil contamination. *Figure reproduced from Ref. 6.*

activity.[4,5] Successful removal of pollutants and microorganisms (bacteria, fungi, algae, viruses) allows soil to be regenerated, marine and freshwater environments to be cleaned and wastewater to be hygienically processed for reuse leading to massive benefits to human health and aquatic biota.

Magnetic nanoparticles couple the modified behaviour uniquely associated with the nanoscale, for example enhanced reactivities arising from enormous surface-to-volume ratios, with inherent magnetic phenomena such as superparamagnetism. This powerful combination enables magnetic nanoparticles to not only act as efficient sorbents for pollutants but also to be magnetically separated from the surrounding aqueous media for further processing and removal. Coating and functionalization of a magnetic nanoparticle creates a core–shell structure that enables specific chemical and metallic contaminants to be targeted with impressive selectivity and ultrahigh sensitivity. Attracting increased attention recently are environmental chemosensors using magnetic nanoparticles functionalized with chemogenic or fluorogenic molecules. As well as selectively capturing a target pollutant, these molecules also undergo some physicochemical change as a result of that capture producing a readable signal that enables rapid *in situ* detection. The inherent reactivity of iron, in the form of nanoscale zero-valent iron (nZVI), may also be exploited in the treatment of soil and groundwater contamination by reducing pollutants to less harmful compounds.

Hence, the environmental application of magnetic nanoparticles to contaminant removal can be divided into three main categories: removal through chemical reaction, removal through physical adsorption, and detection through chemosensing. Following an overview of the properties of magnetic nanoparticles relevant to these applications (Section 2) and of the many methods used for nanoparticle production, coating and functionalization (Section 3), this chapter focuses on each of these applications in the form of soil and groundwater remediation (Section 4), magnetic separation (Section 5) and environmental sensing (Section 6). Other applications of magnetic nanoparticles are also briefly discussed (Section 7) before the chapter closes with a summary and discussion of future prospects (Section 8).

2 RELEVANT PROPERTIES OF MAGNETIC NANOPARTICLES

The general properties of magnetic nanoparticles have been reviewed in detail in Chapter 1 of this volume and here the intention is to focus only on those properties that are especially relevant to environmental applications. The inherent magnetism of these particles is an obvious reason for their use, however, the physicochemical properties of, for example, Fe-based nanoparticles are at least equally important in many settings. Drawbacks associated with the magnetic and nanoscale behaviour also require attention as agglomeration and deposition can easily occur due to attractive magnetic interactions between particles. Attempts to prevent this through passivation and the use of surfactants itself change the fundamental properties of the particles. The same applies to coating of nanoparticles to prevent corrosion or to add functionality.

2.1 Physical Properties

As the size of a particle reduces to the nanoscale, its specific surface area drastically increases, particularly below ~10 nm, as demonstrated in Figure 7.2.[7] Hence, reducing particle dimensions from the microscale to the nanoscale can increase the capacity for contaminant adsorption by several orders of magnitude. This was demonstrated by Yean *et al.* who showed that

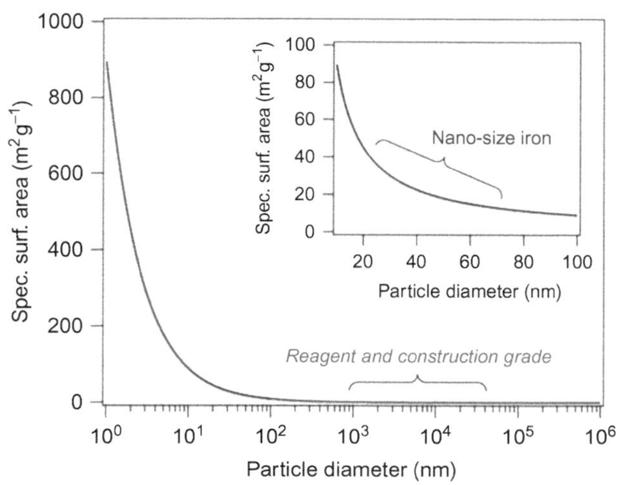

FIGURE 7.2 Plot showing the specific surface area of a spherical particle as a function of its diameter calculated based on the average densities of Fe^0 and Fe_3O_4. The drastic increase in particle–surface area associated with smaller particles can lead to enhanced efficiency in metal ion removal from water, as seen in Table 7.1 which shows the maximum adsorption capacities, q_{max}, for arsenic ions. *Reproduced with permission from Ref. 7.*

TABLE 7.1 Maximum adsorption capacities, q_m, for arsenite (As^{3+}) and arsenate (As^{5+}) removal from water

Arsenic	Diameter (nm)	Surface area ($m^2\ g^{-1}$)	q_m ($\mu mol\ g^{-1}$)
As^{3+}	300	3.7	20.8
	20	60	388.9
	12	98.8	1532
As^{5+}	300	3.7	14.4
	20	60	152.3
	12	98.8	622.7

Adapted with permission from Ref. 8.

as the size of magnetite nanoparticles decreased, the maximum adsorption capacity significantly increased with 12 nm diameter monodisperse Fe_3O_4 nanoparticles having an adsorptive capacity of 1532 $\mu mol\ g^{-1}$ compared to 21 $\mu mol\ g^{-1}$ for 300 nm particles (Table 7.1).[8] A further study by Shen *et al.* found that Fe_3O_4 nanoparticles between 6 and 8 nm were approximately seven times more effective at removing heavy metal ions from water than coarser 18–35-nm sized particles.[9]

However, in terms of practical application in a natural aqueous environment, smaller does not necessarily mean better performance. Larger magnetic fields are required to separate smaller magnetic nanoparticles and also to overcome random Brownian motion, particularly for particles with a diameter less than ~10 nm. Aside from physical limitations, generating sufficiently high magnetic fields and field gradients can be prohibitively costly for an on-site application. Smaller particles also have a greater surface free energy meaning that the energy barrier for deposition on a contacted surface or aggregation into larger particles is lowered. Deposition effectively removes the nanoparticle from the treatment process whilst aggregation greatly reduces the active surface area and consequently adsorptive capacity. Aggregation is also affected by the polydispersity of the nanoparticle size distribution with a greater variance of size leading to enhanced coalescence. Depending on the specific application and the material involved, magnetic nanoparticles with diameters between 10 and 50 nm are typically used in environmental settings which represents an optimum balance between performance, longevity, stability and more practical considerations such as cost and ease of production/implementation.

Overwhelmingly, spherical magnetic nanoparticles are applied to environmental applications in large part due to the ease in which they can be produced in significant quantities. Alternative particle geometries may offer advantages, for example cuboids (Figure 7.3)[10,11] which are attracting

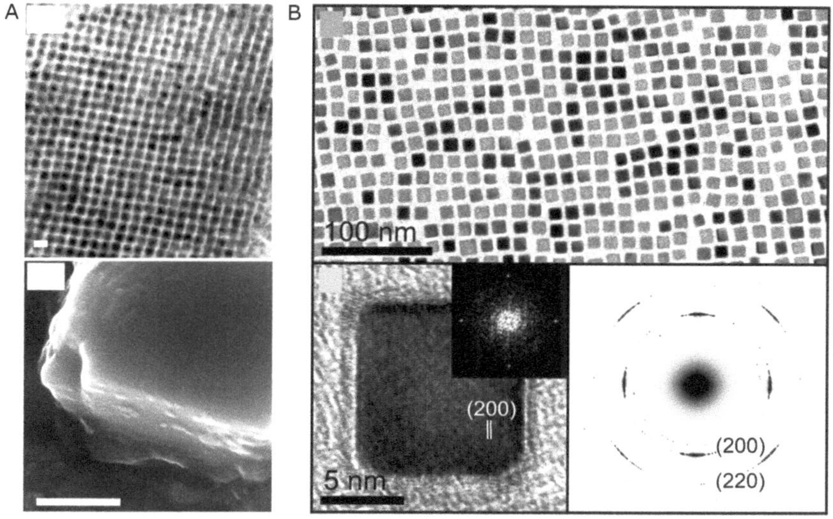

FIGURE 7.3 Synthesis of cuboid magnetic nanoparticles. (A) Transmission (top) and scanning (bottom) electron microscopy images of an assembly of pure iron nanocubes. The bars in the top and bottom figures are 10 and 500 nm, respectively. (B) Transmission electron microscopy image of nickel/nickel oxide core–shell nanocubes and the associated electron diffraction pattern. *(A) adapted with permission from Ref. 10 and (B) reproduced with permission from Ref. 11.*

increased attention in other applications of magnetic nanoparticles such as data storage,[12] catalysis[13] and MRI contrast enhancement.[14] A 25% increase in the surface-to-volume ratio compared to spherical nanoparticles coupled with the ability to select highly reactive facets (typically the {100} orientation for metallic nanoparticles) could lead to significant benefits for environmental applications. Additionally, the possibility of an increased saturation magnetization and a higher density of reactive sites may benefit both sequestration and removal of environmental contaminants. However, it is possible that these potential benefits are mitigated by drawbacks associated with the change in particle shape such as increased aggregation or difficulties in functionalization. This is a fledgling area of nanoparticle research and, to date, there has been no detailed reports in the literature systematically exploring the use of magnetic nanoparticles with a geometry other than spherical in environmental applications.

2.2 Magnetic Properties

Magnetism in a nanoparticle is beneficial to environmental applications as it enables the separation of a contaminant from a solution through the use of an external magnetic field. It many cases, this is a simple and effective approach to the detection and removal of pollutants from contaminated waters, particularly when compared to conventional methods which may involve filtration and centrifugation processes. Most applications involve magnetic nanoparticles that are sufficiently small (<50 nm) that they consist of a single magnetic domain and are superparamagnetic.[15,16] These two finite-size effects lead to higher magnetic moments and hence a stronger interaction with an applied field as the magnetic force experienced by a nanoparticle is proportional to both its magnetic moment and volume. The theoretical critical size below which a spherical nanoparticle is single domain is 15 nm for Fe and hcp Co, 55 nm for Ni and 128 nm for Fe_3O_4,[17] whilst ferrimagnetic or ferromagnetic nanoparticles are generally superparamagnetic for diameters less than 10–20 nm.[16]

Superparamagnetic behaviour occurs above a particular blocking temperature, T_B, at which point the thermal energy is sufficient to overcome the anisotropy energy barrier, that is $k_BT > K_{eff}V$ where K_{eff} is the effective anisotropy constant and V is the particle volume. This results in magnetic nanoparticles acting like giant paramagnets each with a constant magnetic moment that easily aligns to an applied magnetic field. However, once the field is removed, the remanent magnetization of the nanoparticles is zero. This is extremely important in magnetic separation applications as it means that, once the applied magnetic field is effectively switched off, the nanoparticles are redispersed in solution and can be used for further treatment.

Long-range inter-particle magnetic forces can lead to aggregation and, as the blocking temperature is proportional to volume, larger particles are more susceptible to agglomeration than smaller ones. The dominant magnetic interactions contributing to aggregation in a nanoparticulate system arise from anisotropic dipole–dipole forces with direct exchange interactions between particles that are in contact playing a minor role.

An enhanced surface-to-volume ratio is accompanied by a corresponding increase in the number of surface atoms compared to bulk atoms and also in the fraction of the particle volume that is composed of a surface layer. As surface effects become more dominant, the nanoparticle can undergo significant changes to its electronic and magnetic properties, for example a narrowing of the valence band and an associated increase in the density of states at the Fermi level.[18] This can lead to an enhancement of spin and orbital magnetic moments compared to bulk values, as has been observed for 12 nm diameter Fe nanoparticles.[19] As ~23% of the total number of atoms are located at the surface in these particles, the enhancement was attributed to the more dominant role of the orbital moments of the outermost layers.

Coating magnetic nanoparticles can have a large effect on their properties, both individually and as an ensemble. In almost all conditions, pure Fe nanoparticles exposed to atmospheric or aqueous environments will rapidly form a native oxide layer, usually in the inverse-spinel structure of magnetite (Fe_3O_4) and/or maghemite (γ-Fe_2O_3), that is initially several nanometers thick.[20] As well as limiting reactivity, this layer will also reduce the particle magnetization due to the

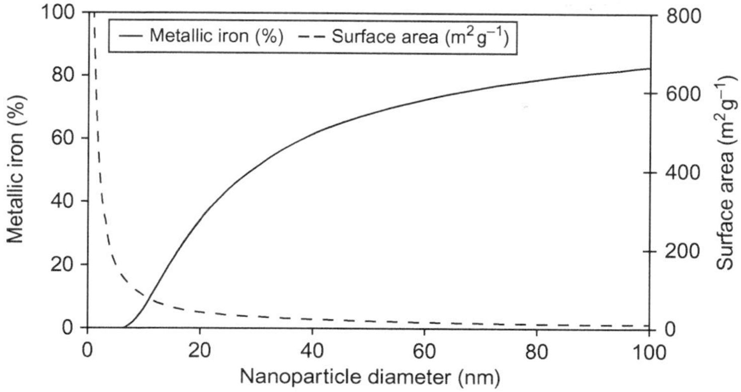

FIGURE 7.4 Plot showing how the surface area and metal iron content of an Fe/Fe oxide nanoparticle changes as a function of particle diameter. A spherical geometry and an oxide layer thickness of 3 nm are assumed and the effects of nanoparticle aggregation are not considered. *Reproduced with permission from Ref. 22.*

lower saturation magnetization, M_s, of Fe oxide (Fe_3O_4: $M_{s,bulk}=92$ emu g^{-1}; 300 K)[14] compared to Fe ($M_{s,\ bulk}=218$ emu g^{-1}; 300 K).[21] The effect is more pronounced for smaller particles as the volumetric fraction of the oxide shell increases (see Figure 7.4).[22] Controlled growth of high-quality Fe/Fe oxide core–shell nanoparticles has been adopted as a way to mitigate the effects of oxidation whilst retaining the properties of the high-moment core.[14] Cheong and coworkers showed that core–shell particles ($M_s=150$ emu g^{-1}) significantly enhanced the contrast in magnetic-resonance imaging when compared to pure Fe oxide nanoparticles ($M_s=40$–70 emu g^{-1}).[14] However, the lack of stability associated with transformation between oxidation states remains an issue for the application of metal/metal oxide nanoparticles to environmental (and biomedical) applications. For example, a recent study of the oxidation of 12 nm cuboid Fe/Fe oxide core–shell nanoparticles revealed a highly crystalline Fe oxide shell but that strain introduced through geometry and confinement effects at the nanoscale led to enhanced ionic transport of Fe^{2+} and O^- and the full oxidation of the nanoparticles within 2 years.[20]

Fully oxidized magnetic nanoparticles, such as those of Fe_3O_4 and γ-Fe_2O_3, are also frequently used in contaminant removal, with the inferior magnetic properties compensated for by greater stability and reusability whilst still retaining versatility of functionalization and ease of production. However, surface effects are still evident as spin canting or a magnetically dead layer at the surface can again reduce the magnetization.[17] Functionalization may also change the magnetic properties at the surface of a ferromagnetic core as the strong bond formed between the binding molecule and the metal modifies the electronic structure. This interaction may also effectively pin the spins of surface atoms so that, over the surface of a spherical particle, their net magnetization is zero.[17]

As well as pre-oxidation to create a protective shell, other approaches to reducing the reactivity of iron nanoparticles include alloying with less-reactive metals or through coating with an inorganic shell, as discussed in the next section.

3 PRODUCTION, COATING AND FUNCTIONALIZATION

The synthesis, coating and functionalization of magnetic nanoparticles are all huge fields in themselves and a detailed discussion of the various methods employed is beyond the scope of this chapter. Here, an overview of the main approaches relevant to environmental applications is given along with representative examples. For more extensive descriptions of the processes employed, the reader is referred to various reviews.[16,17,23–25]

3.1 Production

Chemical synthesis provides inexpensive routes to producing abundant quantities of metallic and metal-oxide magnetic nanoparticles and is the approach generally employed for environmental applications. Chemical pathways to the fabrication of Fe nanoparticles of various geometries are usually based on the thermal decomposition of an iron precursor, usually iron pentacarbonyl ($Fe(CO)_5$) or its derivatives,[10,21] or the reduction of dissolved Fe or Fe oxides.[26] Changing the reaction temperature and time in the thermal decomposition process provides reasonable control over the diameter of the nanoparticles produced which is usually between 2 and 10 nm (Figure 7.5).[21] The thermal decomposition of $Fe[N(SiMe_3)_2]_2$ was shown by Dumestre *et al.* to lead to the formation of cubic nanoparticles (see Figure 7.3A).[10] LaGrow *et al.* described the synthesis of monodisperse Ni nanocubes by varying the surfactant-to-metal precursor ratio to achieve preferential thermodynamic growth of high-surface energy {100} facets (Figure 7.3B).[11] Compared to spherical superparamagnetic nanoparticles, nanocubes displayed a fourfold increase in magnetic saturation. Superparamagnetic nickel nanoparticles were also observed not to magnetically aggregate in solution in the absence of an applied magnetic field.[27] In terms of reductive approaches, ferrous salts are often reduced with borohydride to produce Fe nanoparticles that are highly reactive, but that have a polydisperse particle size distribution.[26] Note that Fe nanoparticles usually form a thin oxide shell immediately after synthesis as can be seen in Figure 7.5.[28] This can also be accompanied by the appearance of voids at the oxide/metal interface which arise due to the Kirkendall effect in which the outwards diffusion of Fe is more rapid than the inwards diffusion of O.[20,29] Hence, especially for smaller particles of Fe and Ni, technically a core–shell structure is usually produced by chemical synthesis.

Fe oxide and other ferrite nanoparticles can also be synthesized using a variety of methods, most often through co-precipitation,[30,31] thermal decomposition[32–34] and microemulsion processes.[24] For example, Zhang *et al.* used

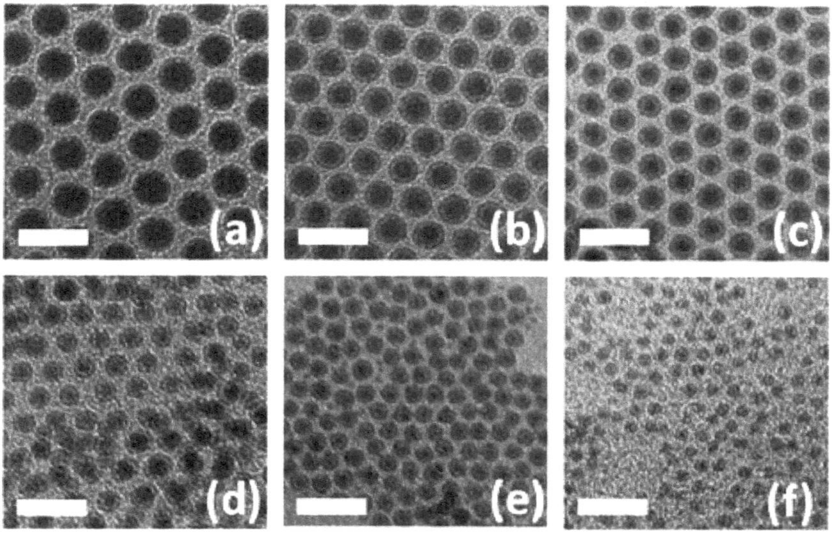

FIGURE 7.5 Transmission electron microscopy images of Fe nanoparticles synthesized through thermal decomposition of a derivative of iron pentacarbonyl, $Fe(CO)_5$, using different reaction temperatures and times. (A) 30 °C for 1 min, (B) 30 °C for 60 min, (C) 30 °C for 180 min, (D) 70 °C for 60 min, (E) 100 °C for 60 min and (F) 130 °C for 60 min. Scale bars: 20 nm. *Reproduced with permission from Ref. 21.*

precursors of $Mn(NO_3)_2$ and $Co(NO_3)_2$ to produce $MnFe_2O_4$ and $CoFe_2O_4$ nanoparticles with high-specific surface areas ($>100\ m^2\ g^{-1}$),[31] whilst Hu and coworkers fabricated Mn, Mg, Zn, Cu, Ni and Co ferrite nanoparticles and systematically investigated their effect on Cr^{6+} removal from wastewater.[30]

3.2 Coating

Encapsulating a magnetic nanoparticle, usually with an inorganic material, provides a method of overcoming some of the issues related to the reactivity and stability of a bare metal or metal oxide particle. Coatings can protect the magnetic core of the resulting core–shell structure, improve dispersion stability in the suspension medium by reducing particle–particle aggregation and enable surface functionalization to target specific metallic or organic substances. Recent studies involving magnetic separation of contaminants, more often than not, use some method of coating. The simplest coatings involve the native oxide layers found on, for example, Fe/Fe oxide core–shell nanoparticles, although, as discussed above these are prone to further reaction and degradation. Instead, magnetic nanoparticles are often coated with other inorganic materials such as silica and carbon. Not only does the coating protect the magnetic core and stabilize the particle, but it can also be used as a base for further functionalization as described below.

Silica (SiO_2) encapsulation is often used as it increases stability in addition to providing control over surface properties. It also allows the dipolar coupling between magnetic nanoparticles to be reduced as this is a function of the distance between particles which can be tuned by changing the thickness of the inert silica shell. Common approaches to producing a silica coating include sol–gel synthesis,[35] surfactant microemulsion[36,37] and the Stöber process.[38–44] The latter method has been used to produce a variety of silica-coated particles including Fe_3O_4@SiO_2 nanoparticles, for example through the hydrolysis and poly-condensation of tetraethoxysilane in ethanol under alkaline conditions,[38] $CoFe_2O_4$@SiO_2 nanoparticles[44] and, by using tetraethyl orthosilicate (TEOS) in water as a silica source, Ni@SiO_2 nanoparticles.[39]

Other inorganic coatings used to encapsulate magnetic cores include alumina[45,46] and carbon[47–49] with Bystrzejewski *et al.* showing that the adsorption capacities of carbon-coated Fe_3O_4 core–shell nanoparticles can be considerably higher than the capacities of activated carbon.[47] Due to their high affinity for bonding with metal ions and bacterial pathogens, organic coatings that activate thiol (—SH), carboxyl (—COOH) and amino (—NH_2) terminal groups are also attractive for contaminant removal.[50]

An alternative approach to producing metal and metal oxide nanoparticles, either coated or uncoated, is by using a gas-aggregation cluster source which enables ultraclean and controlled fabrication of a large variety of nanoparticles under ultrahigh vacuum conditions.[18] This method of growth utilizes magnetron sputtering of a metal target to generate atomic monomers and dimers that effuse into a stream of cool inert gas instigating nucleation of the supersaturated vapour and the formation of metal clusters.[18,28] A recent extension of this method, which could be of interest in terms of environmental applications, enables the production of liquid suspensions of nanoparticles by co-deposition with water vapour into an ice matrix (Figure 7.6). Using this approach, Binns *et al.* produced aqueous suspensions of Fe@Au and Fe@Fe oxide core–shell nanoparticles with diameters in the range 5–15 nm by letting the ice melt after deposition and collecting the resulting hydrosol.[51] To prevent aggregation immediately after melting, a surfactant was added to the water before depositing the ice matrix.

3.3 Functionalization

Functionalization provides an extra level of control over nanoparticle function and performance and is routinely applied to both bare metal/metal oxide nanoparticles and core–shell structures. There is huge variety in the type and properties of available surface functionalization options meaning that magnetic nanoparticles can be tailored for a specific application. Surface modification can also help to improve mobility, further stabilize the nanoparticle suspension against aggregation and degradation, provide increased efficiency in targeted adsorption and provide dual functionality such as chemo- and fluorogenic sensing.

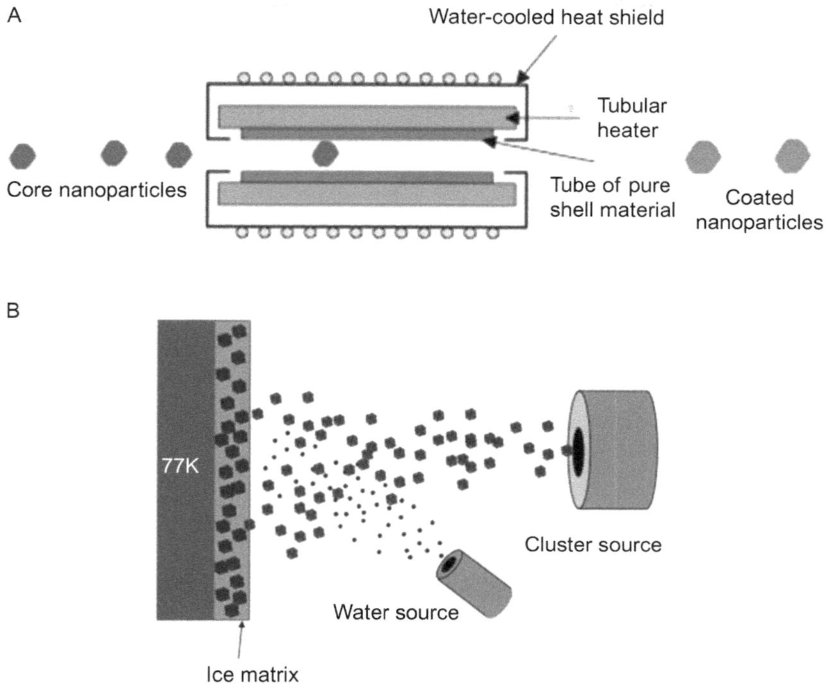

FIGURE 7.6 (A) Schematic showing a method for the ultraclean and controlled production of core–shell nanoparticles using a gas-aggregation cluster source. (B) Co-depositing magnetic nanoparticles with water vapour to form an ice matrix allows hydrosol suspensions to be subsequently collected which could benefit environmental studies. *Reproduced with permission from Ref. 51.*

A good demonstration of how functionalization enables tailoring to a specific contaminant is provided by the work or Warner *et al.* who performed a comprehensive study of the effects of various functionalities on heavy metal sorption (see Figure 7.7).[52] Five different ligands were grafted to Fe_3O_4 nanoparticles stabilized with lauric acid and then their binding affinities for several metals were compared. Thiol-derived functional groups of L-glutathione (GSH), mercaptobutyric acid (MBA), α-thio-ω-(propionic acid) hepta(ethylene glycol) (PEG-SH), *meso*-2,3-dimercaptosuccinic acid (DMSA) in addition to ethylenediamine tetraacetic acid were all chosen due to their known performance in chelating toxic heavy metals. Preference for binding with specific metals showed a large dependence on the type of ligand used.

A large number of synthetic polymers also have the ability to form complexes with metals ions and have therefore been used to coat magnetic nanoparticles and investigate contaminant removal. For example, poly(3,4-ethylenedioxythiophene) (PEDOT) is a polymer that has high conductivity and environmental stability and can be functionalized with thiol groups that

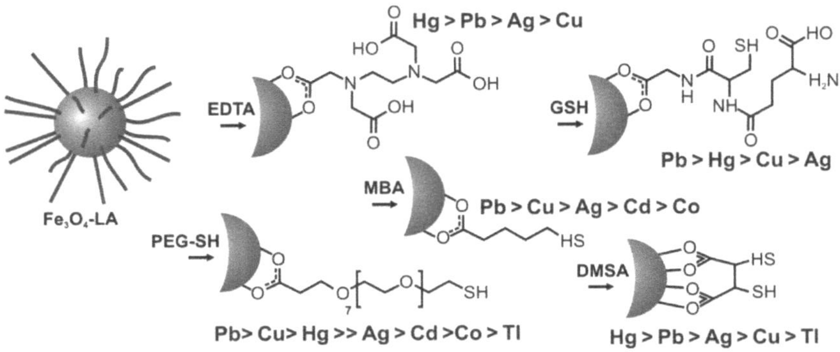

FIGURE 7.7 Functionalization of magnetic nanoparticles. Warner *et al.* grafted five different ligands to the surface of Fe_3O_4 nanoparticles coated with lauric acid (LA) and investigated their performance in removing various metals from aqueous solution. See text for definitions of the abbreviated ligands. *Adapted with permission from Ref. 52.*

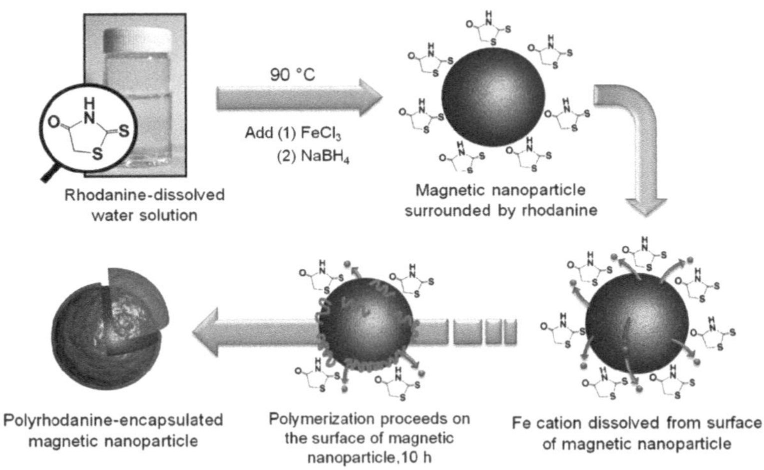

FIGURE 7.8 Schematic showing the synthetic procedure used to produce polyrhodanine-encapsulated γ-Fe_2O_3 nanoparticles used in the removal of metal ions from aqueous solution. *Reproduced with permission from Ref. 54.*

readily conjugate with positively charged metal ions.[53] As an example of polymerization at the surface of magnetic nanoparticles, Figure 7.8 shows the synthetic procedure for the fabrication of polyrhodanine-encapsulated γ-Fe_2O_3 nanoparticles which were shown to be effective at Hg^{2+} removal.[54] When added to an aqueous solution of rhodanine molecules, Fe chloride coordinated with Fe ions so that, after the additional injection of sodium borohydride, magnetic nanoparticles instantly formed with monomers of coordinated rhodanine acting as stabilizers. Upon further reaction, Fe ions redissolved from the surface of the particles leading to the oxidation of the rhodanine monomers and chemical oxidation polymerization. Washing the solution with distilled water allowed the recovery of polyrhodanine-encapsulated γ-Fe_2O_3 nanoparticles.[54]

One of the problems of using bare Fe_3O_4 nanoparticles is their agglomeration in solution. To prevent this, Chen *et al.* exploited the common ion effect by treating Fe_3O_4 nanoparticles with $FeCl_3$ solution which acted as a source of Fe^{3+} ions. These readily adsorbed onto the nanoparticle surface creating a positively charged shell that prevented aggregation.[55] Functionalization was then achieved by using the Fe^{3+} shell to oxidize and polymerize pyrrole monomers to create nanocomposite Fe_3O_4-pyrrole core–shell structures. Other polymers used for coating and functionalization include poly(ethylene glycol), poly(acrylamide), and poly(*N*-isopropylacrylamide),[56] polyacrylic acids,[57] polypyrrole, polystyrene (PS), mercapto polymer,[58] methoxypoly(ethylene glycol)[59] and N-(rhodamine-6G)lactam-ethylenediamine (Rho-en) for its chemosensing properties.[37]

Coating magnetic nanoparticles with naturally occurring nontoxic and biodegradable polymers such as dextran and chitosan in order to stabilize them in solution is also common. Chitosan molecules coordinate reversibly with heavy metal ions and so chitosan functionalization offers an effective means for the magnetic separation of metallic contamination from wastewater.[60,61] Such nanoparticles are efficient in the removal of toxic metals without further functionalization[62] although modification with α-ketoglutaric acid led to an increase in adsorption capacity for Cu^{2+} without significant aggregation.[63] A chitosan coating has also been applied to octadecyl-functionalized Fe_3O_4 nanoparticles to produce compound Fe_3O_4–C_{18}–chitosan structures 20 nm in diameter that were used to remove perfluorocarbons (PFCs) from aqueous solution.[64]

An often used organic coating is DMSA—a highly dispersible sorbent that has a strong affinity for chelation with heavy metals.[50,65–67] DMSA is also an approved United States Food and Drug Administration agent for the treatment of mercury and lead poisoning.[68] Other organic molecules used for functionalization include humic acid,[69] starch,[59] naphthalimide,[70] succinic acid,[50,49] oleic[59,39] and undecanoic acid,[71] sodium dodecyl sulphate,[72] cetyltrimethylammonium bromide,[73] bismuthiol-II[74] and chemisensing molecules such as 4,4-difluoro-4-bora-3a-4a-diaza-s-indacene (BODIPY).[75] Hemimicelles of various compounds including alkyl carboxylates,[76] cetylpyridinium chloride[77] and cetyltrimethylammonium bromides[46,73,77] have also been used to target organic contaminants.

Silica-coated Fe_3O_4 nanoparticles are often coated with chelating agents such as dithiocarbamate (DTC) which has high stability in solution, binds well with heavy metal ions such as Hg^{2+} [41] and has high adsorption capacities (Figure 7.9). Other reported functional compounds grafted to silica-coated Fe_3O_4 include γ-mercaptopropyltrimethoxysilane,[40] triethylenetetramine[78] and amino-activated ligands.[35,79]

In contrast to physically or chemically coating a nanoparticle, another method of protecting its magnetic core and enhancing functionality is to

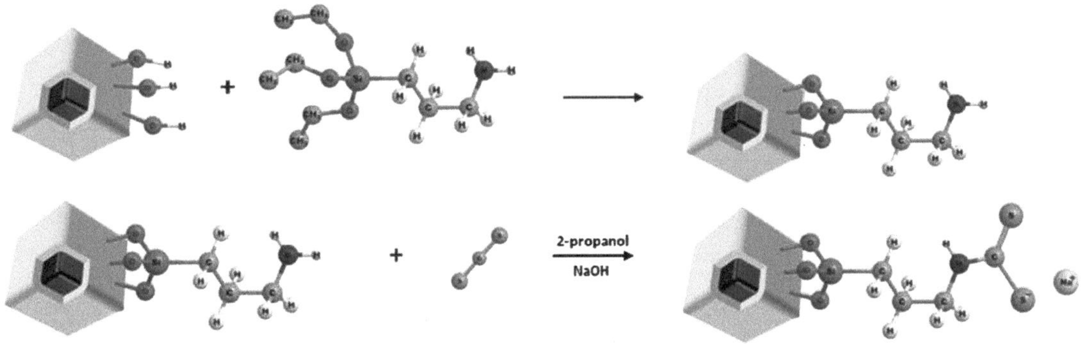

FIGURE 7.9 Silica-coated Fe_3O_4 nanoparticles are often chemically functionalized through the addition of chelating agents as shown here for nanoparticles grafted with dithiocarbamate (DTC) groups which have a high affinity towards heavy metal cations. The silica coating was obtained through alkaline hydrolysis of tetraethyl orthosilicate ($Si(OC_2H_5)_4$, TEOS). *Reproduced with permission from Ref. 41.*

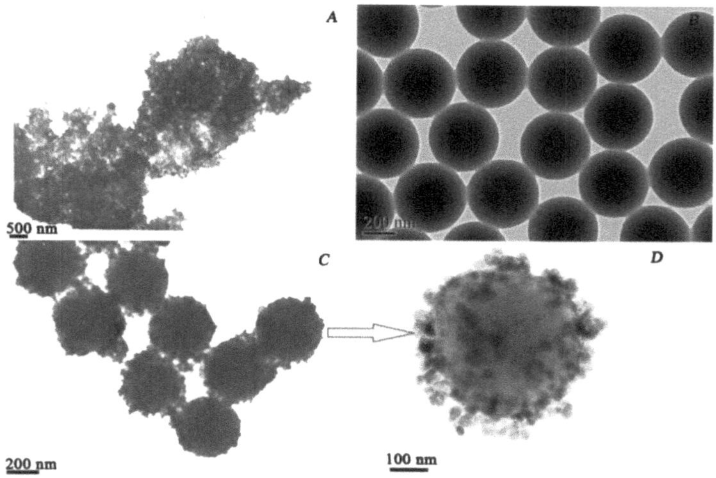

FIGURE 7.10 Magnetically active polymeric particles. To load polystyrene beads with Fe_3O_4 nanoparticles, Jiang *et al.* added drop wise an Fe_3O_4 colloidal solution to a polystyrene emulsion under mechanical stirring. Magnetic separation was used to remove only those particles with a high-density coating of Fe_3O_4 to leave a suspension of polystyrene–Fe_3O_4 particles (Fe_3O_4 21.1 wt%). *Reproduced with permission from Ref. 81.*

embed it in a matrix or supporting network. An example of the effectiveness of this approach is a recent study by Yan *et al.* who found that magnetic composite microspheres consisting of Fe_3O_4 nanoparticles and poly(acrylic acid)-blended chitosan were more effective at removing Cu^{2+} ions from aqueous solutions than microspheres which only contained chitosan-coated Fe_3O_4 nanoparticles.[61] Another example is the preparation of micron-sized magnetically active polymeric particles, in which Fe_3O_4 nanoparticles are dispersed in non-magnetic polymeric beads to provide a polymer-support network capable of removing heavy metals (Figure 7.10).[80,81] Other composite systems demonstrated include $Fe_3O_4@C_{18}$ nanoparticles encaged in barium–alginate,[82] hydrogel networks,[83] Fe_3O_4 nanoparticles supported on graphene sheets,[84] water-soluble multi-walled carbon nanotubes functionalized with Fe nanoparticles[85] and Fe_3O_4 nanoparticles loaded onto multi-walled boron nitride nanotubes.[86]

4 SOIL AND GROUNDWATER REMEDIATION

Remediation involves the removal of contaminants from soil, aquifers and groundwater using various methods in order to reduce toxicity to acceptable levels. Human activities, particularly those related to industry and agriculture, have inevitably led to pollution of soil and groundwater, that is, water present in subsurface porous media such as rock and sand. In the United States alone, billions of gallons of wastewater containing toxic metal ions are annually produced as a result of industrial activity. Although specially designed containers and lagoons are used to store this wastewater, leakages are common leading to contamination of surrounding areas and aquifers. Contamination may be chemical, biological, physical or radioactive in nature and many technologies have been developed to clean up and remediate affected areas.[22] The attractions of nanoscale technology led to the use of nanoporous structures where the vastly increased surface area means their performance is significantly better than conventional sorbent beds. However, the volume flow through such nanomaterials is limited and a dispersible sorbent with an equally large surface area is more suited to contaminant removal. Recently, there has been huge interest in the use of nZVI particles following many years of research into using standard, millimetre-sized zero-valent iron. Since the initial studies on using nZVI particles for treating contaminated aqueous solutions,[26,87] the technology has been applied to remediation studies of numerous pollutants including antibiotics, dyes, chlorinated solvents, pesticides, fertilizers, metalloids such as arsenic, radionuclides and many different metals. A number of review articles exist outlining the mechanisms for the removal of each of these classes of material[22,88–92] and for a comprehensive list of the contaminants treated with nZVI, see, for example, Crane and Scott[22] or O'Carroll *et al.*[91]

The primary attraction of using iron (and other reactive nanoparticles) for remediation arises not from its magnetic properties but from its basic chemistry, specifically its ability to degrade contaminants to less harmful products through reduction. This may be through direct reaction, in which the contaminant is degraded upon interaction with the Fe^0 particle, or through reaction with the corrosion products of Fe^0, specifically Fe^{2+} and hydrogen. These may be produced both with (aerobic) and without (anaerobic) the presence of dissolved oxygen:

$$Fe^0 + 2H_2O \rightarrow Fe^{2+} + H_2 + 2OH^-$$
$$2Fe^0 + O_2 + 2H_2O \rightarrow 2Fe^{2+} + 4OH^-$$

Degradation is particularly effective at reducing organic contaminants such as chlorinated organic solvents (e.g. CCl_4 and C_2CL_4) to harmless substances like CO_2 and H_2O although can be used more generally to remediate a broad range of substances including fertilizers (NO_3^-), pesticides (lindane, DDT) and toxic metals. For direct reaction, after impacting the surface of an Fe^0 particle, the reactant is adsorbed and then chemically degrades before desorbing into the bulk solution. In this way, toxic metal ions and organic compounds can be reduced to more benign species, for example Cr^{6+} to Cr^{3+}[93] and trichloroethene (TCE) to ethane[94]:

$$2Fe^0 + 2H_2CrO_4 + 3H_2O \rightarrow 3(Cr_{0.66}Fe_{0.33})(OH)_3 + FeOOH$$
$$C_2HCl_3 + 4Fe^0 + 5H^+ \rightarrow C_2H_6 + 4Fe^{2+} + 3Cl^-$$

A non-reactive approach to remediation involves sequestration of contaminants through adsorption. In many cases, the reaction of Fe^0 with metallic cations fails to produce insoluble products and so removal through adsorption is usually employed, as discussed in the next section. Here, the focus is on reactive degradation of contaminants in remediation efforts.

The size of nZVI particles used in remediation applications are typically between 10s and 100s of nanometers meaning that they act as environmental colloids. In the form of nanoscale particles, ZVI has a higher specific surface area with more active sites on the nanoparticle surface which leads to higher reactivities. This leads to a greater range of degradation reactions and a faster degradation rate for contaminants whilst also potentially producing less harmful by-products. However, this increased reactivity can lead to rapid deactivation of the nanoparticle and operational inefficiencies, for example in the reaction of nZVI with dissolved oxygen and water to form iron oxides.

The specific mechanism of removal for metals depends on the standard electrode potential, E^0, of the ion relative to that of Fe^0. For metal ions with a more negative E^0, such as for Zn^{2+} ions in which the electrochemical couple is $Fe^0 + Zn^{2+} \rightarrow Zn^0 + Fe^{2+}$, removal by sorption or surface complexation of the ion on Fe^0 occurs. In contrast, for metal ions with a value for E^0 greater than Fe^0, such as Cu^{2+} and Hg^{2+}, chemical reduction results in removal. The corrosive reaction of Fe consumes protons and generates hydroxyl ions leading to an increase in solution pH. Many heavy metals and radionuclides are removed through adsorption rather than being reduced and the long-term ability of Fe^0 to maintain this state to prevent remobilization of the contaminant in to the surrounding environment is one of the major concerns regarding the application of nZVI to remediation. For organic materials, direct degradation through chemical reduction to less harmful products usually takes place. For chlorinated compounds such as TCE and pentachlorophenol (PCP), hydrogen produced during the corrosion process of Fe plays a vital role as it breaks C—Cl bonds releasing Cl as a gas.

Treatment can take place both *in situ*, for example through the injection of nZVI directly into contaminated areas, and *ex situ*, in which the contaminated material is transferred away from the affected area to a treatment plant. The former approach is simpler and therefore more cost-effective; however, the accurate and effective dispersal of the treatment technology has proven difficult to achieve and depends on the nature of the environment involved, that is, whether it primarily consists of soil, sediment or aqueous media. Hence, the injection of low-mobility nZVI may be concentrated on targeted reactive zones, or reactive plumes of more mobile nanoparticles may be generated with the intention that they are transported to the contamination area. Figure 7.11 summarizes various approaches to remediation using nZVI particles.[22]

However, as discussed in more detail in Section 5.2, mobility can be greatly affected by particle–particle and particle–surface interactions which

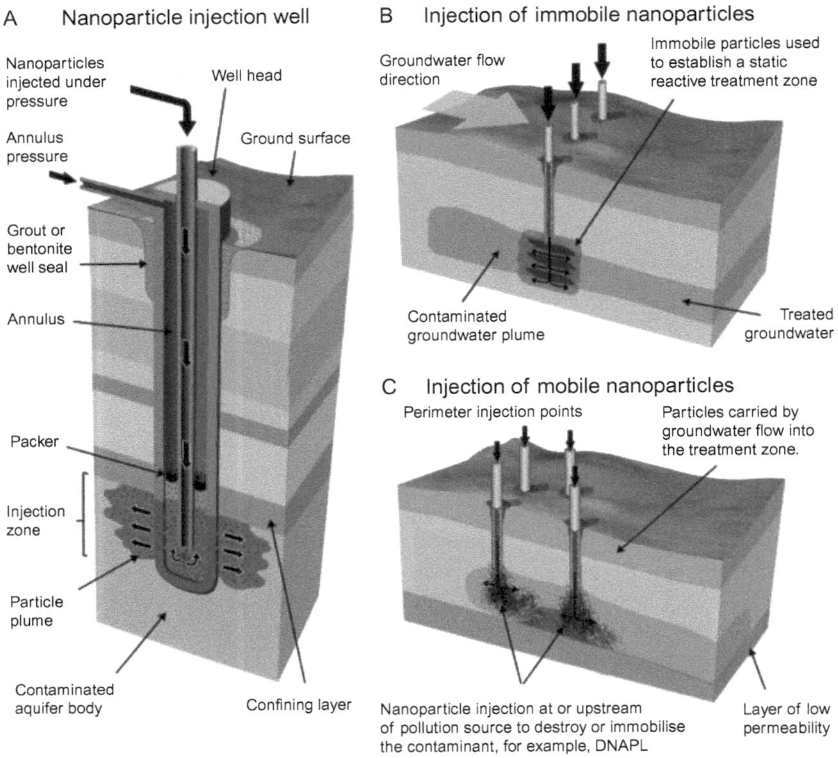

FIGURE 7.11 Comparison of different approaches to remediation using nZVI for groundwater remediation. (A) Schematic of a typical well used to inject nanoparticles into a contaminated aquifer to create reactive treatment zones. Different strategies of injection may be employed depending on whether the contaminant is (B) mobile or (C) immobile as in the case of nonaqueous phase material (DNAPL). *Reprinted with permission from Ref. 22.*

may lead to sedimentation and aggregation. Nanoparticle mobility is a key issue in remediation applications and is determined by the effective sticking coefficient that describes the probability that the nanoparticle will interact with the porous medium and be removed from the fluid, and also by how many collisions the nanoparticle has with that medium. This can be as high as hundreds per metre and so a low-sticking probability is required for high mobility. Brownian motion is the most probable cause for the subsurface collisions, particularly for smaller particles, but gravitational sedimentation and interaction with other substances such as natural organic matter may also play a role. Calculations by Tratnyek and Johnson show that, even for surface-modified iron nanoparticles with low-sticking coefficients (<0.01), mobilities of only several metres can be expected under realistic conditions.[7] Additionally, coating nZVI particles with surfactants, polymers or inorganic materials such as silica and carbon in order to improve stability can effectively buffer contaminant reaction with the nanoparticle surface and reduce performance. If a coating is to be applied, carbon is usually preferred for groundwater applications as it has a high stability over a wide pH range and is biocompatible. Additionally, alloying nZVI with a noble metal such as Pd, Pt or Ag to produce bimetallic particles can increase chemical reactivity[26,91,94] and these were found to be particularly effective in the treatment of chlorinated compounds.[22]

The concept that pure nZVI particles are used in remediation is slightly misleading because, as discussed in Section 2, Fe is so reactive that a native oxide layer several nanometers think will instantaneously form after synthesis. This film is initially porous although reaction further increases the oxide thickness and removes material from the core limiting direct interaction between Fe^0 and the contaminant.[22] Iron oxides/hydroxides themselves act as sorbents and chemical reductants of contaminants thereby contributing to remediation. In fact, pure iron oxide magnetic nanoparticles, specifically those of Fe_3O_4 and γ-Fe_2O_3, have also been investigated for use in remediation applications.[92,95,96]

Approaches to the treatment of soil and groundwater contamination using nZVI particles are sufficiently advanced that the technologies involved have been commercialized in many countries. Fe is abundant, inexpensive and environmentally compatible and nZVI approaches consume less material and energy than conventional ZVI treatment. Colloidal suspension is also relatively easy to deploy via injection at a variety of different locations. This leads to cost-effectiveness,[97] particularly when compared to competing technologies such as pump-and-treat and permeable reactive barriers. However, the effectiveness of nZVI treatment has recently been questioned as simplified laboratory-based studies fail to fully mimic natural environments of contaminants and can lead to an over-estimation of performance, particularly over longer timescales where release of the contaminant may occur, as observed in a study comparing U removal from natural groundwater samples with water only solutions.[98] Additional unknowns in the use of nZVI particles include how surface modification affects the long-term impact on the environment and how rapidly they deactivate and transform through reaction and oxidation. In order to clarify the effectiveness of nZVI in the remediation of contaminated soil and groundwater and its long-term effect on the local environment, Crane and Scott suggested the establishment of a universal testing framework to enable direct comparison of different studies and treatments.[22]

5 MAGNETIC SEPARATION OF ENVIRONMENTAL CONTAMINANTS

Adsorption followed by magnetic separation, or magnetic solid-phase extraction, is an effective approach to the removal of toxic substances such as metal ions and organic and biological materials from wastewater. Techniques based on this principal are generally faster, simpler and more cost-effective than competing technologies such as chemical precipitation, reductive degradation and membrane filtration. The principle of operation is simple: magnetic nanoparticles are used to adsorb a particular contaminant from aqueous effluents before separation from the solution by the application of a magnetic field. A desorption process is then used to strip the contaminant from the nanoparticles for controllable disposal, with the nanoparticles regenerated for further treatment cycles. Secondary contamination issues such as the release of harmful by-products should be avoided. Hence, there is a need for efficient magnetic nanoparticle systems of high capacity that are readily separated from water using practically low field gradients and which are therefore economically viable. They must also operate on a large scale and be recyclable to prevent waste build up and deactivation. The generation of a suitable magnetic field for effective separation at high-throughput volumes is perhaps the biggest complication associated with the technique although the principles involved are well established as magnetic separation is utilized in many other fields such as biochemistry, cell biology, analytical chemistry and mining.[30]

5.1 Separation

Separation of magnetic nanoparticles from a solution is possible due to the force exerted on them by an external magnetic field. This magnetic force, F_m, may generally be expressed as

$$F_m = \mu_0 V_p M_p \nabla H$$

where μ_0 is the permeability constant of vacuum, V_p and M_p are the particle volume and magnetization, respectively and ∇H is the strength of the magnetic field gradient at the location of the particle. M_p is given by $M_p = \chi H$ where χ is the magnetic volume susceptibility, which changes with particle size and shape, and H is the magnetic field strength.

High-gradient magnetic separation (HGMS) is an established concept where a magnetic field is typically applied across an array of magnetically soft metal wires to establish a high-field gradient.[99,100] Magnetic nanoparticles travelling through such a field are attracted to the surfaces of the wires allowing effective collection. However, the operation of such a separator is strongly dependent on the size and magnetic properties of the particles involved and the strength of the magnetic field must be high enough to overcome possible competing colloidal forces acting on a flowing particle suspension arising from fluid drag, gravity, inertia and diffusion.[99]

The gravitational force may be expressed as $F_g = (\rho_p - \rho_g) V_p g$ where ρ_p and ρ_g are the spherical particle and fluid densities, respectively, g is the acceleration due to gravity and V_p is the particle volume. Hydrodynamic fluid drag results in a force according to Stokes' Law: $F_d = 6\pi\eta b(v_f - v_p)$ where η is the dynamic viscosity of the fluid, b is a fluid-dependent constant and v_f and v_p are the fluid and particle velocities, respectively. In many of the laboratory-based *ex situ* demonstrations of removal of a contaminant from its suspension, a handheld magnet is usually strong enough to demonstrate magnetic separation over length scales of a few centimetres. This range may be sufficient for effective separation in, for example, bioseparation, but for industrial-scale *in situ* environmental remediation, a much greater separation distance is usually required necessitating the use of a stronger magnetic field as the dependence of field strength on distance follows an inverse–cube relationship (for a dipole source). Other significant factors acting against magnetic separation include buoyancy and random Brownian motion, both of which are enhanced as the particle becomes smaller. For Fe oxide, the critical diameter of a nanoparticle at which its volume becomes large enough for the magnetic force exerted on it to overcome Brownian motion is ~50 nm, assuming large magnetic field strengths typically used in HGMS are applied.[101]

Yavuz *et al.* first demonstrated magnetic separation with very low fields ($<100\ \mathrm{T\,m^{-1}}$) using monodisperse 12 nm Fe_3O_4 nanoparticles that had previously been applied to the removal of As from water.[101] They demonstrated multiplexed separation of an initially bimodal mixture through the application of different magnetic fields and noted that aggregation, caused by high-field gradients at the surface of the nanoparticles, actually helped the separation process (Figure 7.12).

5.2 Adsorption

Adsorption is the key mechanism governing the effectiveness of removal of a contaminant from an aqueous solution, even more so than the magnetic separation process. An ideal sorbent would have high adsorption capacity with a large surface area, high solution stability and mobility and enable rapid separation in large quantities. Specific adsorbents would also have the ability to target individual metal ions or organic or bacterial substances without deactivation by non-target compounds such as natural organic matter.

Many factors affect the mechanism and efficiency of adsorption including solution pH, initial metal and adsorbent concentrations and contact time. A key feature that characterizes the performance of a nanoparticle in removing a contaminant from solution is the adsorption isotherm which relates the isothermal variation of the concentration of an analyte at the surface of an adsorbent to the concentration of the solution. Two different models are generally accepted when interpreting adsorption isotherms—the linearized Langmuir model and the Freundlich model. The Langmuir isotherm is defined by

$$q_e = \frac{q_m b C_e}{1 + b C_e}$$

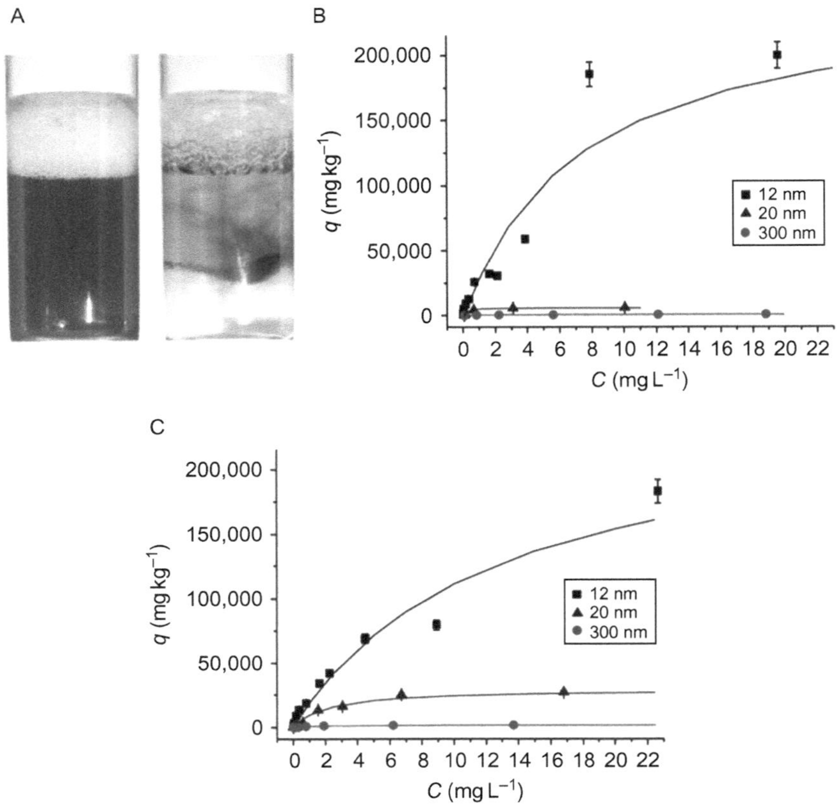

FIGURE 7.12 (A) Magnetic separation of 16-nm water-soluble Fe_3O_4 nanoparticles within several minutes using a low-field gradient of 23.3 T m^{-1}. (B) and (C) Adsorption isotherms comparing the removal of As^{3+} (A) and As^{5+} (B) using Fe_3O_4 nanoparticles of different diameters. Concentrations were measured using inductively coupled plasma-mass spectrometry (ICP-MS). *Adapted with permission from Ref. 101.*

and in linear form by

$$\frac{C_e}{q_e} = \frac{1}{q_m b} + \frac{C_e}{q_m}$$

where q_e (mg g^{-1} or mmol g^{-1}) and C_e (mg L^{-1} or mmol L^{-1}) are the adsorbed and aqueous concentrations at the adsorption equilibrium, respectively, q_m (mg g^{-1} or mmol g^{-1}) is the maximum adsorption capacity and b is the affinity constant that characterizes the bonding energy of adsorption in L mg^{-1} or L mmol^{-1}. This model is built around two assumptions: (i) the adsorbent has adsorption sites which all have the same binding energy and (ii) only one adsorbate can bind to each adsorption site. In contrast, the Freundlich isotherm, given by

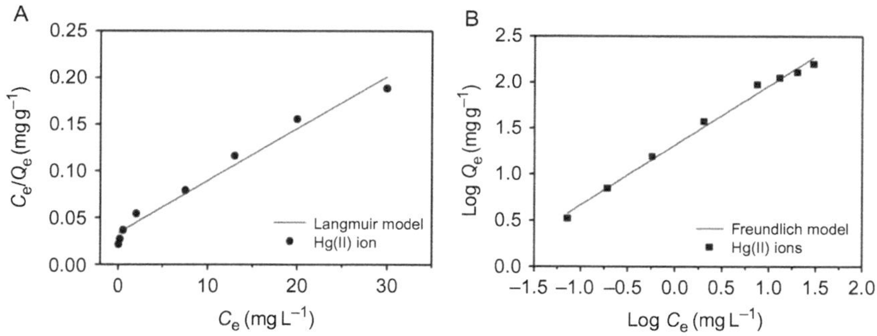

FIGURE 7.13 Isotherms for the adsorption of Hg^{2+} ions onto polyrhodanine-encapsulated γ-Fe_2O_3 nanoparticles fitted with the (A) Langmuir and (B) Freundlich models. The latter provides a better fit due to consideration of the multiple sites available for adsorption at the polymer-coated surface. *Adapted with permission from Ref. 54.*

$$q_e = K_f C_f^{1/n},$$

which can be linearized as

$$\log q_e = \log K_f + \frac{1}{n} C_e$$

where K_f is a constant describing the adsorption capacity of the adsorbent in mg L^{-1} and relates to the adsorption intensity, assumes reversible heterogeneous adsorption takes place with the binding energies of each site not necessarily the same.

In most cases, adsorption isotherm data fit both the Langmuir and Freundlich models reasonably well, although slightly better fitting often arises from the latter, particularly when nanoparticles functionalized with complex molecules are employed. This can be ascribed to the variety of adsorption sites with different binding energies available for these compound nanoparticles compared to simple metal or metal oxide nanoparticles.[54] Figure 7.13 shows typical isotherm data for metal adsorption fitted to both the Langmuir and Freundlich models with the corresponding adsorption parameters shown in Table 7.2.

Analysing adsorption isotherms according to the Langmuir model enables the determination of the maximum adsorption capacity, q_m, which characterizes the quantity of material adsorbed on a nanoparticle for a complete monolayer as a function of its mass. This figure of merit is often used to describe the performance of a nanoparticle in removing a particular contaminant. Another possible way of expressing this is with the adsorption efficiency, that is, the ratio of bound metal ion weight to the weight of the nanoparticle. An example of how q_m can be used to characterize the effect of particle size on removal performance is given by the work of Yean *et al.* who showed that the maximum adsorption capacity of Fe_3O_4 for As^{3+} removal from water dramatically increased from 21 to 1532 μmol g^{-1} as the

TABLE 7.2 Adsorption Parameters Correlating to Figure 7.13 and Linear Regression Coefficient, R^2

Langmuir model				Freundlich model			
q_m ($mg\ g^{-1}$)	b ($L\ mg^{-1}$)	R^2	F-*test*	K_f	n	R^2	F-*test*
179	0.165	0.974	0.98	3.74	1.55	0.992	0.99

Adapted with permission from Ref. 54.

particle size reduced from 300 to 12 nm (Table 7.1).[8] A similar increase in q_m was observed for As^{5+} removal. Adsorption efficiency can be determined by analytical techniques such as inductively coupled plasma-mass spectrometry (ICP-MS) or atomic adsorption spectroscopy. Feng *et al.* also described a rotating disk electrode (RDE) method that could be applied *in situ* without the need to isolate the adsorbent particles from solution.[137] Comparing specific surface areas also helps to understand the performance and behaviour of a nanoparticle in adsorbing a contaminant and to calculate this, Bruauer, Emmett and Teller (BET) analysis is normally used. Analysing the kinetics of the adsorption process is also important in characterizing the performance of a nanoparticle in contaminant removal as it reveals information on the contaminant uptake rate and the adsorption mechanism. A complete investigation of adsorption performance might also involve measuring the pH and temperature dependence, the ability of other contaminants to disrupt adsorption of the target analyte, and the influence of other material present in the solution such as natural organic matter.

5.3 Aggregation

One of the biggest problems hindering the magnetic separation of nanoparticles is aggregation and deposition which may arise due to a number of reasons.[102] An ensemble of nanoparticles with high surface-free-energies can reach a stabilized state through agglomerating into micron-sized aggregates which greatly decreases the total surface area and therefore removal capacity and reactivity. These factors massively increase for smaller particles so that, due to a reduced energy barrier for aggregation, they are more prone to coalesce. Additionally, the magnetic force acting between separated particles is long range compared to, for example, van der Waals and electric double layer interactions, and it may therefore play a dominant role in driving aggregation, particularly for nanoparticles with a high Fe content.

As well as directly reducing the effectiveness of individual nanoparticles, aggregation can greatly reduce mobility and lead to sedimentation and deposition. It has been studied in detail with a number of techniques including dynamic light scattering, small-angle light scattering and fluorescence spectroscopy,[103] whilst deposition onto a surface may be monitored using a packed-bed column or quartz crystal microbalance.[104] Preventing aggregation and producing stable suspensions of magnetic nanoparticles in the aqueous solution to be treated is one of the main motivating factors behind surface coatings as discussed in Section 2. The repulsive steric forces arising from the coating act against the attractive long-range magnetic forces (as well as the attractive van der Waals interaction) between particles thereby stabilizing the colloidal suspension.

Other factors affect nanoparticle transport such as the particle size distribution with polydispersity leading to less stability and consequentially differential sedimentation and aggregation. A higher particle concentration will also lead to increased aggregation due to the increased frequency of particle–particle interactions. Additionally, the physicochemical properties of both of the particles, for example size, composition, surface charge, redox activity, functionalization and also the solution, for example, the presence of natural organic matter, pH, ionic strength and composition, will all influence behaviour. The Derjabuin–Landau–Verwey–Overbeek (DVLO) theory of colloidal stability describes the particle–particle and particle–surface interactions involved in aggregation and deposition that arise from van der Waals and electrical double layer interactions.[105,106] However, non-DVLO forces, aside from the magnetic interaction, also contribute such as steric and electrosteric forces which arise from surface coatings and can lead to repulsion between particles reducing aggregation and deposition. Nanoparticles coated with long-chain organic molecules will be stabilized in solution through steric interactions, whilst electrosteric forces arise when an electrostatic charge is present on these molecules. Petosa *et al.* reviewed the role of physicochemical interactions in the aggregation of nanoparticles in aquatic environments[104] outlining the colloidal forces that determine particle–particle and particle–surface interactions and therefore particle transport.[102]

The particle size and size distribution can also affect the extent and rate of aggregation. Aside from the long-range magnetic force that acts between particles, Brownian diffusion is the dominant process leading to aggregation. An interesting feature of Brownian diffusion is that the collision rate for particles of the same size is independent of that size reflecting the fact that, although larger particles have a lower diffusion constant, they also have a larger interaction cross-section. However, for ultrafine particles with diameters less than approximately 10 nm, higher magnetic fields are required for separation increasing the cost of treatment.[81] Additionally, a polydisperse colloidal suspension tends to aggregate faster than a monodisperse distribution.[104]

5.4 Contaminant Removal

5.4.1 Arsenic

A clear example of where the benefits of magnetic nanoparticles may be utilized to provide new technologies or to enhance existing ones is in the sequestration and removal of As from contaminated water supplies. Arsenic can find its way into groundwater through natural processes such as oxidative weathering of minerals and ores, volcanic ash and geochemical reactions or as a result of uncontrolled industrial processing. This renders arsenic contamination a serious problem affecting millions of people worldwide, but is particularly prevalent on the Indian subcontinent. A wide variety of health problems can result from exposure including cancers, respiratory and cardiovascular diseases, and kidney, liver and skin diseases. Health organizations have responded to this toxic threat by reducing the recommended level in water from 50 to 10 $\mu g\ L^{-1}$ or 10 ppb[107,108] which results in a six in 10,000 lifetime risk of developing skin cancer. This limit was somewhat defined by the limits of testing technology and subsequent data suggest that long periods of As exposure to levels as low as 0.17 $mg\ L^{-1}$ can lead to arsenicosis. This highlights the need for new sensing and removal technologies with extreme sensitivity and effectiveness.

Iron oxides present chemically viable pathways to As removal however, in common with all adsorbents, can lead to sludge and liquid waste formation as a result of inefficient separation from water. Arsenic is commonly found in aqueous environments as either arsenite (As^{3+}) in $H_3AsO_3^0$ or arsenate (As^{5+}) in $H_2AsO_4^{2-}$. Both are toxic although As^{3+} ions are the most damaging to the environment. Fe_3O_4 has a specific affinity for arsenic and, when combined with magnetic separation, therefore shows promise in arsenic sequestration.[8] When loaded onto PS beads, these PS–Fe_3O_4 particles showed enhanced capacity for arsenate adsorption when compared to bulky Fe_3O_4.[81] Effective separation from water was achieved using low magnetic fields of less than 0.035 T. Furthermore, regeneration using NaOH solution allowed the compound beads to be recycled without significant degradation of performance. To provide effective separation and high reactivity on a chemically, thermally and structurally stable support, Chen *et al.* have functionalized boron nitride nanotubes with Fe_3O_4 nanoparticles showing effective As^{5+} removal over a wide range of pH values.[86] The high density of adsorbed Fe_3O_4 also allowed efficient separation with a relatively low magnetic field. For As^{5+} removal, Fe_3O_4 nanoparticles were also modified with cetyltrimethylammonium bromide[109] and by Badruddoza *et al.* who covalently attached phosphonium silane cationic ligands which promoted adsorption of As^{5+} and Cr^{6+} ions through electrostatic attraction and an ion-exchange mechanism.[110] Other Fe oxides such as γ-Fe_2O_3 are effective at removing both As^{3+} and As^{5+} from aqueous media,[111,112] although the reported maximum adsorption capacities are generally not as high as for Fe_3O_4.

Although magnetic separation of As-adsorbed iron oxide nanoparticles is the usual method for treatment, some studies have combined the chemical properties of iron oxide with more traditional separation technologies. For example, Zaspalis *et al.* used ultrafiltrating multilayer ceramic membranes to remove iron oxide nanoparticles exposed to As from wastewater.[113]

The capacities for As adsorption of other ferrite (bimetallic) nanoparticles such as $MnFe_2O_4$ and $CoFe_2O_4$ have been investigated, for example by Zhang *et al.* who measured a significant increase over Fe_3O_4 nanoparticles prepared by the same method (chemical co-precipitation).[31] They measured an adsorption capacity of 94 and 90 mg g^{-1} for $MnFe_2O_4$ and 100 and 74 mg g^{-1} for $CoFe_2O_4$, respectively, for As^{3+} and As^{5+} compared to 50 and 44 mg g^{-1} for Fe_3O_4. Increased surface hydroxyl (M–OH) species were suggested as the cause for this increase in capacity although this was not definitively proven.

5.4.2 Toxic Metals

Over 20 metals are considered toxic with many of these present in sufficient quantities in environmental settings to be considered potentially harmful to human health. Some heavy metals, such as cobalt, copper, manganese and zinc, are vital to the survival of living organisms but only at trace levels and their accumulation in tissue cells are known to be harmful. Even trace levels of other metals such as mercury, chromium and lead can cause significant damage.

Chromium contamination is particularly problematic in areas surrounding electroplating, metal finishing, wood treatment and tannery industries.[114,115] Exposure to species of hexavalent chromium can lead to a wide range of respiratory and skin conditions in humans, whilst it can severely damage local ecosystems.

nZVI nanoparticles have been used to remove Cr^{6+} ions from electroplating wastewater, either supported on bentonite (an absorbent clay),[116] or as pure Fe^0 particles.[117] In the latter case, the superior performance of Fe^0 over Fe_3O_4 nanoparticles in the reductive removal of Cr^{6+} ions from aqueous solution was demonstrated with respective removal efficiencies of 60.4% and 9.13% measured. This was further improved to 99.25% by stabilizing the Fe^0 nanoparticles with starch.

Following initial studies in which nanoscale $MnFe_2O_4$ was shown to have the highest Cr^{6+} adsorption efficiency when compared to other ferrites,[118–120] Hu and coworkers investigated the removal and recovery of Cr^{6+} from synthetic wastewater using 10 nm surface-modified $MnFe_2O_4$ nanoparticles.[121] Various organic and inorganic ligands and different ionic strengths were investigated in a range of pH values with the authors concluding that NH_4^+ ligands enhance the uptake of Cr^{6+} at pH greater than 6.5. This stimulated a systematic study of ferrite nanoparticles prepared by chemical co-precipitation in which it was determined that the Cr^{6+} removal performance followed the order $MnFe_2O_4 > MgFe_2O_4 > ZnFe_2O_4 > CuFe_2O_4 > NiFe_2O_4 > CoFe_2O_4$.[30] As well as having the highest adsorption capacities, $MnFe_2O_4$ nanoparticles were also the most rapid to reach adsorption capacity making it the most promising ferrite adsorbent for the removal of Cr^{6+} (Figure 7.14). It was remarked however that a

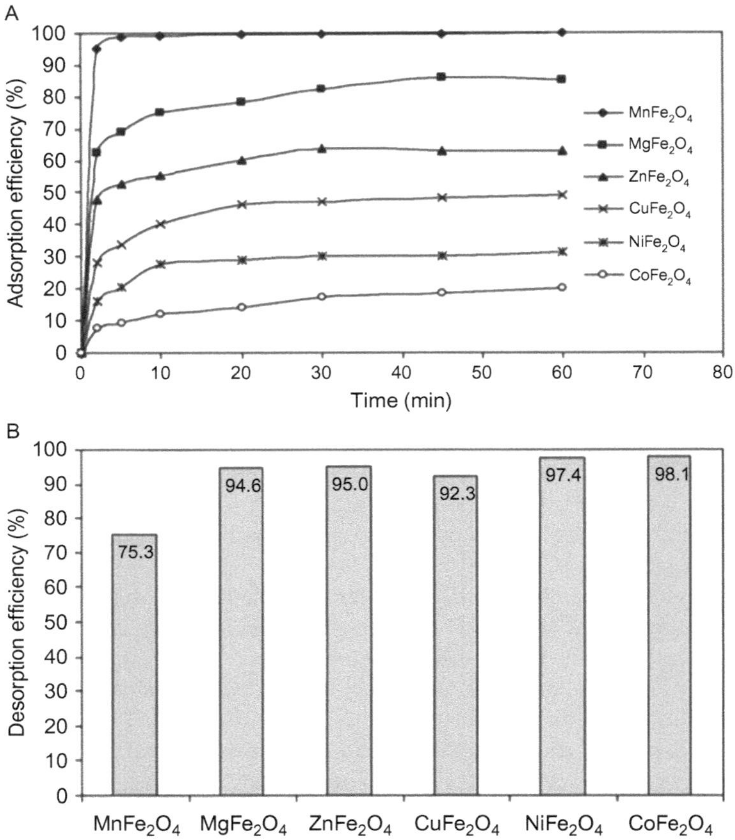

FIGURE 7.14 (A) The effect of contact time on the removal of Cr from synthetic wastewater by various ferrite nanoparticles of ~20 nm diameter. (B) Despite having the highest adsorption capacity and fastest reaction rates, $MnFe_2O_4$ nanoparticles were the least efficient at desorption and therefore recovery. *Adapted with permission from Ref. 30.*

redox reaction between Cr^{6+} and Mn^{2+} led to recovery rates for $MnFe_2O_4$ being lower than for the other ferrites.[30] Surface modification of γ-Fe_2O_3–Fe_3O_4 nanoparticles to enhance the removal of Cr^{6+} from wastewater was performed using a δ-FeOOH coating[122] and water-soluble polyethylenimine (PEI) which acted as a positively charged adsorbent that was effective at low pH.[123] Competition from coexisting ions (K^+, Na^+, Ca^{2+}, Cu^{2+}, Cl^- and NO_3^-) was found to be insignificant. Geng *et al.* used chitosan-stabilized Fe^0 nanoparticles to show that Cr^{6+} ions physically adsorb onto the chitsan-Fe^0 surface before subsequent reduction to Cr^{3+},[124] whilst Shin *et al.* investigated the removal of Cr^{3+} ions using nitrogen-doped magnetic carbon nanoparticles.[125] Wei *et al.* also applied iron-nickel oxide nanoparticles to remove Cr^{6+} from waste nickel liquid.[126]

Mercury and the more toxic organic form, methyl mercury (CH_3Hg^+), are well-known toxic metals. Even at very low concentrations, mercury is very harmful to human health due to its bioaccumulation in the body and toxicity. Hence, if an area is found to be contaminated by mercury, methods are needed to quickly remove it and prevent it from entering the food chain and damaging the local ecosystem. Fe_3O_4 and $Fe_3O_4@SiO_2$ nanoparticles functionalized with 2-mercaptobenzothiazole,[127] naphthalimide,[70] mercapto polymers[58] and thiol[67] have all proved effective in the selective and sensitive adsorption and detection of Hg^{2+}. An interesting recent study by Farrukh *et al.* outlined the use of 'polymer brushes' grown on Fe_3O_4 nanoparticles through a surface-initiated polymerization process to demonstrate the complete removal of Hg^{2+} from water.[128] Dithiocarbamate (DTC) groups on the polymer chains had a high affinity towards Hg^{2+} ions and the higher density of these groups on the polymer brushes, when compared to the monolayer equivalent, led to superior remediation performance (Figure 7.15).

Coated Fe_3O_4 nanoparticles have been applied to the adsorption and magnetic separation of many other toxic metal ions. Silica-coated Fe_3O_4 nanoparticles were functionalized with amino groups which stimulate the complexation of metal ions and were therefore found effective at removing Cu^{2+}, Pb^{2+} and Cd^{2+} ions from aqueous media with effectiveness in that order. Acid treatment enabled the regeneration of the metal-adsorbed nanoparticles and the presence of alkali/earth metal ions such as Na^+, K^+ and Mg^{2+} did not significantly affect the adsorption efficiency. At 25 °C, q_m values of 0.47, 0.37 and 0.20 mmol g^{-1} were measured for Cu^{2+}, Pb^{2+} and Cd^{2+} ions, respectively, all of which increased with temperature indicating an endothermic adsorption mechanism.[79] Liu *et al.* used Fe_3O_4 nanoparticles coated with humic acid to investigate Hg^{2+}, Pb^{2+}, Cd^{2+} and Cu^{2+} removal from natural and

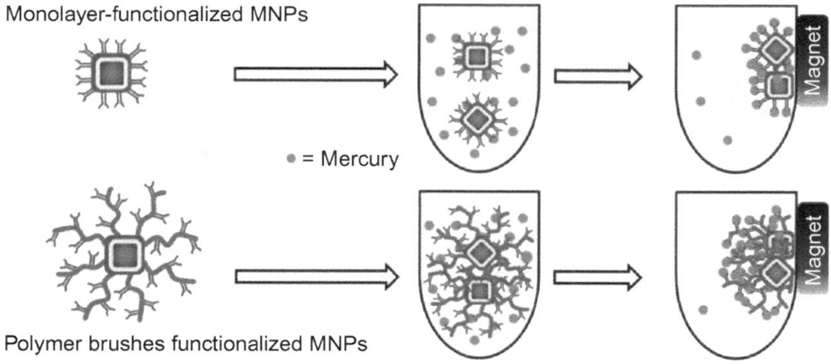

FIGURE 7.15 Fe_3O_4 nanoparticles coated with 'polymer brushes' proved more effective in the removal of Hg^{2+} ions from water than nanoparticles coated with an equivalent monolayer of the same polymer. This was attributed as being due to a higher density of dithiocarbamate groups on the brushes compared to the monolayer. *Reproduced with permission from Ref. 128.*

tap waters.[69] Removal efficiencies of over 95% were measured for these metal ions in less than 15 min with leaching of the heavy metal back into the water sample negligible.

For polymer coatings, Takafuji *et al.* demonstrated that composite poly (1-vinylimidazole)–γ-Fe_2O_3 nanoparticles could selectively bind to divalent metal ions with binding strengths in the order $Cu^{2+} > Ni^{2+} > Co^{2+}$ [56] and Shin *et al.* showed that Fe_3O_4 nanoparticles encapsulated in thiol-containing polymers are efficient at removing Ag^{+}, Hg^{2+} and Pb^{2+}.[53] Liu *et al.* synthesized chitosan-immobilized magnetic nanocomposites from amine-functionalized Fe_3O_4 nanoparticles and demonstrated their effectiveness at removing Pb^{2+}, Cu^{2+} and Cd^{2+} from water. The metal ions could be released from chitosan under weakly acidic conditions in combination with ultrasound exposure.[62]

Carbon-encapsulated magnetic nanoparticles have been demonstrated as effective sorbents for the removal of metal ions such as Pb^{2+},[49] Cr^{6+},[129] Cu^{2+}, Co^{2+} and Cd^{2+}.[47] Wang *et al.* attributed the removal effectiveness of carboxyl-functionalized magnetic carbon nanoparticles for Pb^{2+} from acidic aqueous solutions as being due to a weak electrostatic attraction between the nanoparticles and metal ions together with the porosity of the carbon shell. At a pH of 3, 95% of Pb^{2+} ions were removed from solution at Pb concentrations of 100 mg L^{-1}.[49] A three-step mechanism was proposed as shown in Figure 7.16.

Gupta *et al.* demonstrated an unusual and low-cost method of removing Cd^{2+} ions from simulated electroplating industry wastewater by co-precipitating orange peel powder obtained from agricultural waste with Fe_3O_4 with the resulting compounds showing an 82% removal efficiency.[130] Another interesting study investigated the use of iron sulphide nanoparticles which may be produced by sulphate-reducing bacteria and are efficient adsorbers of a wide range of heavy metals.[131] Adsorption capacities in the range of 100–400 mg g^{-1} were measured for a variety of metals. Iron sulphide may also be strongly magnetic allowing cheap and effective separation.

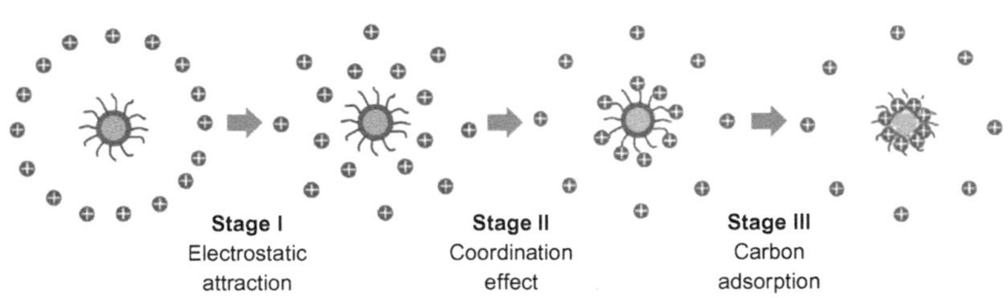

FIGURE 7.16 Removal mechanism of metal ions by carboxyl-functionalized carbon-encapsulated magnetic nanoparticles. Electrostatic attraction causes metal ions to migrate to the nanoparticle where they coordinate with carboxyl groups. Adsorption on the porous carbon shell then occurs. *Adapted with permission from Ref. 49.*

Other magnetic solid-phase extraction studies of the removal of toxic metals from aqueous solution include: Ag^{2+},[53,66] Cd^{2+},[9,40,44,65,66,69,71,74,78,130,132–134] Co^{2+},[56,63] Cu^{2+},[40,43,56,60,69,78,134] Mo^{6+},[135] Ni^{2+},[56] Pb^{2+},[65,67,69,74,134,136] Tl^{2+}[66] and Zn^{2+}.[43,132,134]

5.4.3 Radionuclides

Aside from metal toxicity, radioactive heavy metals such as U, Cs and Sr pose a risk in terms of radiological effects on humans. Uranium is common in both terrestrial and oceanic environments, although concentrated exposure is a serious concern due to use in nuclear energy and its role in enriched form in nuclear weapons. Exposure to uranium, for example, in the form of uranyl metal ions (UO_2^{2+}), can lead to severe health problems. Wang *et al.* showed that Fe_3O_4 nanoparticles functionalized with bisphosphonate are effective at removing U ions from both blood and water with efficiencies of 99% and 69%, respectively.[137] This was the first demonstration of using magnetic separation of functionalized nanoparticles to remove radionuclides from blood.

Crane *et al.* have conducted a series of experiments investigating the use of Fe_3O_4 and ZVI nanoparticles to remediate environmental water contaminated with uranium.[22,98] Oxygen conditions were adjusted to reflect the original environment of the natural water and the chemical evolution of solutions and particulates were tracked over an 84-day reaction period. Their results showed that, within 2 h of reaction, over 98% of uranium was removed from the water by the nZVI nanoparticles to reach a level below that specified by international drinking water regulations. On the other hand, less than 20% of uranium was removed by the magnetite nanoparticles indicating the enhanced reactivity of nanoparticles with a significant Fe^0 core, even if a native surface oxide layer is present. Silica-coated Fe_3O_4 nanoparticles have also been used for the removal of U^{6+} ions from aqueous solution.[138]

5.4.4 Organic Contaminants

Perfluorocarbons (PFCs) are organofluorine compounds that are commonly used in medical and electronic applications and refrigeration but that are also anionic pollutants to environmental waters. Zhang *et al.* investigated the use of chitosan-coated octadecyl-functionalized Fe_3O_4 nanoparticles in the removal of a wide range of PFCs from environmental water samples from various locations determining detection limits between 0.075 and 0.24 ng L^{-1} with recoveries above 56%.[129] The hydrophilic and porous chitosan coating improved dispersibility and allowed the transmission of analyte molecules to the interior of the adsorbent whilst preventing macromolecules of natural organic matter from passing through due to size exclusion and electrostatic repulsion (see Figure 7.17).

Many studies have been conducted on the removal of polycyclic aromatic hydrocarbons from environmental water samples all of which involved fairly advanced functionalization of magnetic nanoparticles. Examples include

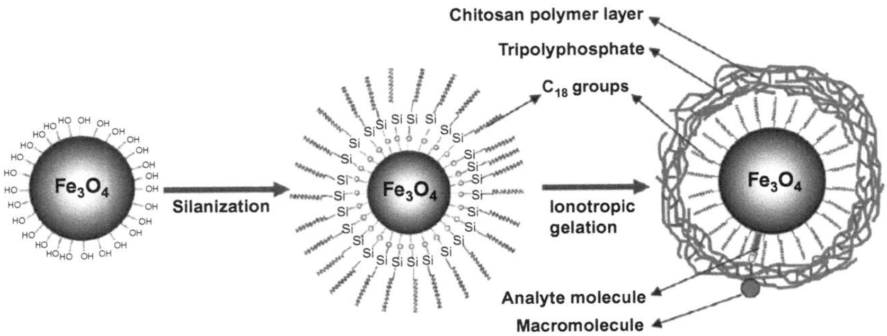

FIGURE 7.17 Synthesis of chitosan-coated octadecyl-functionalized Fe_3O_4 nanoparticles for the removal of perfluorocarbons from environmental water samples. The chitosan coating improves colloidal stability and permits the analyte molecule to pass into the particle interior whilst excluding unwanted natural organic matter. *Adapted with permission from Ref. 64.*

chitosan-coated C_{18}-functionalized Fe_3O_4 nanoparticles for the extraction of phthalate ester compounds,[64] barium–alginate-caged Fe_3O_4@C_{18} nanoparticles for the pre-concentration of polycyclic aromatic hydrocarbons,[82] and hemimicelles of alkyl carboxylates chemisorbed onto magnetic nanoparticles, again for the extraction of polycyclic aromatic hydrocarbons.[76] Fe-nanoparticle-coated multi-walled carbon nanotubes have also been used to remove benzene, toluene, dimethylbenzene and styrene from water.[85]

Zelmanov and Semiat investigated the behaviour of iron-based nanocatalysts for the removal of typical organic matter, for example, ethylene glycol and phenol, from contaminated water.[139] Organic substances were destroyed at room temperature in the presence of hydrogen peroxide in a Fenton-like reaction. Alumina-coated Fe_3O_4 has also been used to remove the organic compounds aminomethylphosphonic acid, glyphosphae[45] and trimethoprim.[46]

Chlorophenols are a class of organic compounds commonly found in pesticides and herbicides. Li *et al.* used mixed hemimicelles composed of Fe_3O_4 nanoparticles coated with cetyltrimethylammonium bromide to remove chlorophenols from environmental water samples.[73] Phenolic compounds were also removed from aqueous solution using Fe_3O_4@SiO_2 nanoparticles.[77,140] A range of sulphonamide compounds have also been removed from water samples using mixed hemimicelles chemisorbed onto Fe_3O_4 nanoparticles.[141]

5.4.5 Bacterial Contaminants

Cyanobacteria present serious risks to the quality of drinking water and can cause disruption to water treatment processes and lead to algal blooms in lakes and ponds. Jiang *et al.* used a composite coagulant prepared using Fe_3O_4 nanoparticles and polyferric chloride to study the removal of cyanobacteria (*Microcystis aeruginosa*) from water.[142] Coagulation is a key process in water treatment and here the authors investigated the coagulation performance of Fe_3O_4 nanoparticle/polyferric chloride composites. The addition of magnetite improved coagulation efficiency leading to higher removal values and a reduced pH dependence. The role of Fe_3O_4 was ascribed to reducing the influence of natural organic matter due to increased settling.

Singh *et al.* explored different surface functionalities of Fe_3O_4 nanoparticles between 6 and 40 nm in diameter in the removal of biological pathogens (*Escherichia coli*) from water. It was found that the capture efficiency of bacteria strongly depended on the concentration of the adsorbents and their inoculation time.[50]

5.4.6 Dyes

Dyes are organic compounds that, as well as often being intrinsically toxic, have a high colour strength which can influence photosynthesis activity in aquatic systems thereby disrupting the microbial activities of organisms found in affected areas. Many dyes used in the textile and paper industries are toxic, such as the carcinogenic Congo red, methylene blue, basic fuchsin and acid black, and find their way into water sources as effluent.

Wang *et al.* investigated the adsorption capability of ferrite nanoparticles for Congo red and found the most efficient performance for $CoFe_2O_4$ which displayed a maximum adsorption capacity of 245 mg g^{-1}.[143] Electrostatic absorption was suggested as the primary removal mechanism. γ-Fe_2O_3 nanoparticles have also been demonstrated to successfully remove Congo red from aqueous samples.[144] For methylene blue removal, $ZnFe_2O_4$ nanoparticles coated with ZnO were shown to be effective, under UV radiation, with higher photodegradation rates than for pure ZnO.[145] Polyacrylic acid-bound iron oxide nanoparticles were also used to remove methylene blue from aqueous solution[57] and a composite consisting of Fe_3O_4 nanoparticles supported on graphene sheets was used for the removal of both methylene blue and Congo red.[84] The use of modified Fe_3O_4 nanoparticles to remove cationic dyes (crystal violet, methylene blue and alkali blue 6B)[146] and basic fuchsin, a magenta dye,[147] has also been reported.

5.4.7 Other Contaminants

Xiong *et al.* coupled magnetic solid-phase extraction using C_{18}-modified Fe_3O_4@SiO_2 nanoparticles with gas-phase chromatography–mass spectrometry to determine the concentration of pesticide residues in environmental water samples.[148] Poisoned pond water was analysed demonstrating the method is rapid, convenient and efficient for the detection of low concentration (0.001–0.008 μg L^{-1}) pesticides in aqueous samples.

Hydrous aluminium oxide embedded with Fe_3O_4 nanoparticles was found to have a specific affinity to fluoride ions and was used to reduce an initial flouride concentration in water 20 mg L^{-1} to 0.3 mg L^{-1} meeting the standards of the World Health Organization for drinking water quality.[149]

6 ENVIRONMENTAL SENSING

In order to be able to acquire measurements of contamination levels on-site, for example at the location of environment incidents such as industrial, chemical and oil spillages, nuclear accidents and natural disasters, there is a drive to develop rapid, accurate and field-deployable sensors. As already outlined in the preceding sections, magnetic nanoparticles are extremely effective at removing a wide range of contaminants either through chemical reaction or through adsorption. Their inherent magnetism then allows the contaminant-coupled nanoparticle to be separated and removed from solution. Much research is now being carried out to add an extra level of functionality that enables contaminant sensing.[150]

Functionalized magnetic nanoparticles can also be used as intrinsic chemosensors of contaminants if chemogenic or fluorogenic molecules are employed. After binding with a target contaminant such as a metal ion, these molecules undergo some physicochemical change that results in a readable signal that can be used for detection. Typically this change will involve light emission or absorption, fluorescence or chemiluminescence.[150] Such sensors have high accuracy and precision, are low-cost and are potentially in-field readable meaning that on-site measurements could easily be taken. Core–shell silica magnetic nanoparticles with diameters in the tens of nanometer range are typically used due to their biocompatibility, stability, single-domain nature and ease of functionalization. Many of these sensors were developed with biomedical applications in mind as the compounds used proved effective in removing metal ions from blood as well as from aqueous solutions. However, there is a clear cross-over with environmental applications.

Yantasee *et al.* used superparamagnetic Fe_3O_4 nanoparticles functionalized with DMSA in an electrochemical sensor capable of detecting river and sea water background levels of <1 ppb of Cd, Pb, Cu and Ag[66]. The sensor consisted of a magnetic or electromagnetic electrode that enabled dispersed metal-bound nanoparticles to be separated from a solution before species identification with a handheld electrochemical detector that allowed voltammetric analysis in an acidic medium (see Figure 7.18). The electromagnetic electrode enabled automation as the coil current could easily be turned on and off with no memory effects whereas the permanent electrode required regeneration after each measurement. The sensor was also effective at detecting the presence of Pb in urine, demonstrating potential for non-invasive biomonitoring of toxic metals and other xenobiotics. For detecting Pb, performance was as good as for inductively coupled plasma-mass spectrometry (ICP-MS) with a greater accuracy and faster analysis time. Other advantages of these sensors stated by the authors include a lack of the need for pH adjustment, high throughput ($\sim$3 min per sample) and economy of use with little waste generation. Table 7.3 shows the binding affinities, K_d, of DMSA–Fe_3O_4 nanoparticles for the adsorption of several metal ions species found

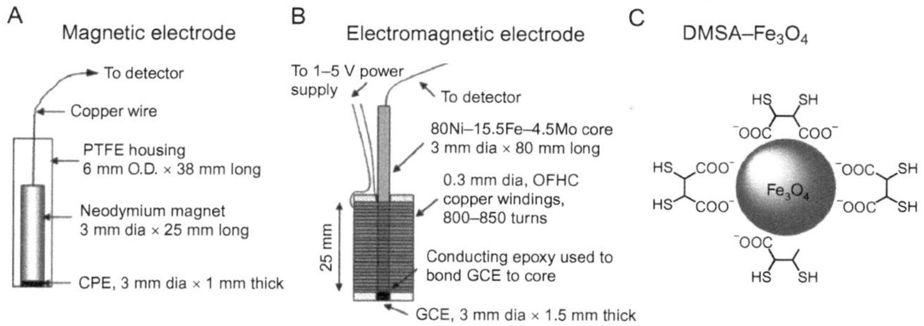

FIGURE 7.18 Electrochemical sensor of contaminants separated from solution using Fe_3O_4 nanoparticles functionalized with dimercaptosuccinic acid (DMSA) (C). Separation was achieved using both (A) a permanent magnet and (B) an electromagnetic electrode. The cylindrical neodymium magnetic electrode produced magnetic field strengths of 1700–180 G at 0–5 mm from the electrode–liquid interface. An advantage of the electromagnetic electrode, consisting of a ferromagnetic round core surrounded by a copper-wire coil, was that it could easily be switched on and off to allow removal of the contaminant and recycling of the nanoparticles. *Reproduced with permission from Ref. 65.*

TABLE 7.3 Binding Affinities of Several Metals Found in River, Ground and Sea Water to DMSA–Fe_3O_4 Nanoparticles

K_d (mL g^{-1})	pH	Pb	Cu	Ag	Hg	Co	Tl	Cd
River water	7.20	280,000	91,000	110,000	390,000	3200	13,000	7400
Ground water	7.77	110,000	63,000	58,000	420,000	6800	11,000	2200
Seawater	7.64	570,000	170,000	2800	16,000	4100	2300	2200

Adapted with permission from Ref. 66.

in river and seawater[66]. DMSA has a high affinity ($K_d > 50{,}000$) for many of the metals present although, for the detection system employed, no Hg signal could be measured due to extremely strong binding to DMSA resulting in Hg ions not being released in the sensor. This work highlights that in sensing contaminants, both the affinity for detection and the contaminant concentration contribute to the signal strength.

In the chromogenic sensing of Pb^{2+} ions, Lee *et al.* used Ni@SiO_2 core–shell nanoparticles functionalized with BODIPY which is a small chemosensing molecule that is strongly UV-adsorbing and which emits fluorescence peaks with high quantum yields.[75] The BODIPY-based receptor that was synthesized, labelled as N-1, was found to undergo UV–vis or fluorescence changes when Pb^{2+} ions were added to the aqueous solution containing the nanoparticles. Usually, the receptor used is nonfluorescent in aqueous solutions but the addition of increasing concentrations of Pb^{2+} resulted in chelation-enhanced fluorescence and an eightfold increase in emission at $\lambda = 510$ nm (see Figure 7.19). The fluorescence change was found to be fully reversible by adding strong base to the acidic solution and could be reproduced over several detection–separation cycles. Interestingly, the addition of other metal ions including Li^{+}, Ca^{2+}, Cu^{2+}, Ag^{2+} and Hg^{2+} did not result in a change and N-1 selectively adsorbed Pb^{2+} when the ions were mixed in the same solution. BODIPY dyes are also insensitive to polarity and pH so that the N-1 compound shows promise for the selective removal of Pb^{2+} ions from aqueous environments. This was demonstrated using ICP-MS measurements which found that 0.1 μM of N-1 can remove 97% of Pb^{2+} at an initial concentration of 15 ppb from water. The same authors also synthesized a receptor that enables dual sensing which provides a system of greater accuracy than single sensors.[151] In this case, BODIPY-functionalized Fe_3O_4@SiO_2 core–shell nanoparticles responded to the addition of metal ions by changing both colour and fluorescence at the same time. Again, the receptor was highly selective in the adsorption of Pb^{2+} in comparison to other metal ions, and was also effective over a large pH range of 3–11. Herrman *et al.* also showed that carbon-coated iron carbide (Fe_3C) core–shell nanoparticles functionalized with a chelating agent of ethylenediaminetetraacetic acid could be used to remove Pb^{2+} from blood and such a receptor could be suitable for environmental applications.[152]

In a dual approach, Tan *et al.* used Fe_3O_4 nanoparticles for the solid-phase extraction of Cu^{2+} from environmental waters before performing a visual test based on functionalized gold nanoparticles that allowed detection of Cu^{2+} ions

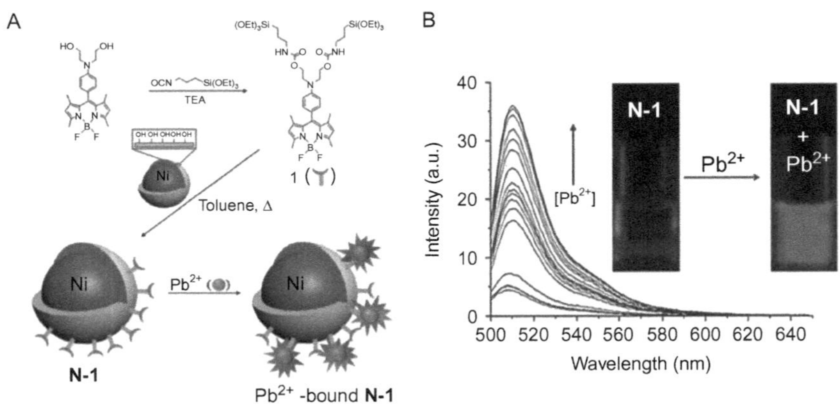

FIGURE 7.19 (A) Schematic diagram showing the synthesis of Ni@SiO_2 nanoparticles functionalized with the fluorogenic sensor BODIPY to produce nanoparticles labelled as N-1. (B) These were found to be highly effective in the selective removal of Pb^{2+} ions from aqueous solution as seen in the fluorescence spectra for increasing Pb^{2+} concentrations, with excitation at 505 nm and emission monitored at 510 nm. *Adapted with permission from Ref. 75.*

at the subparts-per-billion level.[153] Further schemes based on the functionalization of magnetic nanoparticles with receptors enabling chromogenic and/or fluorogenic sensing of toxic metal ions include Ni@SiO_2 nanoparticles functionalized with nitrobenzofurane for Cu^{2+} removal,[154] 3,5-di-*tert*-butyl-2-hydroxybenzaldehyde (DTH)-grafted Fe_3O_4 nanoparticles for the detection of Zn^{2+}[155] and Fe_3O_4 and Fe_3O_4@SiO_2 nanoparticles functionalized with naphthalimide,[70] aminonaphthalimide,[156] porphyrin,[42] thiourea/nitrophenyl,[157] dansyl[158] and Rho-en[37] for the detection of Hg^{2+}. Other functionalizations only demonstrated so far with non-magnetic nanoparticles may also be suitable for use with magnetic nanoparticles, particularly those coated with silica. For example, boehmite (aluminium oxide hydroxide)–silica core–shell nanoparticles with two anthracene–amine systems which proved effective at removing Hg^{2+}.[159]

7 OTHER ENVIRONMENTAL APPLICATIONS

Whilst this chapter has focused on the most widely employed environmental applications of magnetic nanoparticles, that is, groundwater and soil remediation and environmental separation and sensing, many other green technologies exist in which they could have a direct or indirect benefit. For example, iron nanoparticles supported on alumina or carbon nanofibres have recently been demonstrated as effective catalysts for the conversion of synthesis gas, which contains a mixture of H_2 and CO and is used in synthetic natural gas, ammonia and methanol production, to lower olefins (C_2 to C_4).[160] Such a method provides a greener alternative to the usual production process in which crude oil is steam cracked and also addresses supply limitations. Magnetic nanoparticles may also be used as supports for catalysts themselves as in the work of Zhang *et al.* who used a palladium-based catalyst supported on amine-functionalized Fe_3O_4 nanoparticles to achieve fast conversion for various aromatic nitro and unsaturated compounds.[161] Magnetic separation enabled recycling through simple recovery of the catalyst from the reaction mixture.

Takagaki *et al.* demonstrated the use of $CoFe_2O_4$-embedded silica nanoparticles as solid acid catalysts in the hydrolysis of disaccharides and polysaccharides such as cellulose.[162] This is a vital process in the production of the biofuels and bioplastics that are attracting a great deal of attention in terms of biomass utilization. The nanoparticles were found to be highly active catalysts and could be easily recovered from solution using magnetic separation enabling their reuse. Song *et al.* also showed that a photoactive TiO_2 shell surrounding a magnetic SiO_2-coated Fe_3O_4 core could be used as a magnetic photocatalyst, enabling the efficient degradation of methylene blue in aqueous solution with subsequent recovery of the nanoparticle through magnetic separation.[163]

Recently, a number of studies have demonstrated several other novel uses of magnetic nanoparticles for energy applications and environmental clean-up. In a novel study, Wang *et al.* used Fe_3O_4 nanoparticles in the fabrication of polymer solar cells, depositing them as thin films onto a substrate to form a hole extraction layer (Figure 7.20).[164] Aligning the nanoparticles with an external magnetic field led to enhanced electrical conductivity and lower internal resistance, and hence, a corresponding 13% increase in the power conversion efficiency over polymer solar cells incorporating PEDOT:PSS thin films. Fe_3O_4 nanoparticles have also been used in the fabrication of lithium ion batteries which offer higher power densities when compared to conventional batteries. Koo *et al.* investigated the use of hollow Fe_3O_4 nanoparticles as both a cathode and anode material and found that the high concentration of cation vacancies in the Fe_3O_4 could be adopted for reversible Li ion intercalation leading to improved performance with excellent stability.[165]

In terms of water pollution, one of the biggest threats to marine environments is the accidental release of crude oil, as continually demonstrated for many decades. A recent and interesting study demonstrated the first combination of epoxidized natural rubber with magnetite nanoparticles to form a nanocomposite that was used in the absorption of gasoline from an aqueous solution.[166] Natural rubber has poor oil resistance and is commonly used in the form of butyl rubber in the treatment of oil spills. The fabricated magnetically-loaded nanocomposite was demonstrated to have an increased adsorption capacity when compared to butyl rubber with 7 g of oil removed for every gram of adsorbent. As well as improving porosity, the epoxidized rubber also allowed a high density of magnetite nanoparticles to be incorporated enabling magnetic separation and recovery.

In household and commercial water systems, magnetic nanoparticles have also been shown to be effective in characterizing the corrosion products typically formed in municipal water pipes made of iron and steel.[167] Senftle *et al.*

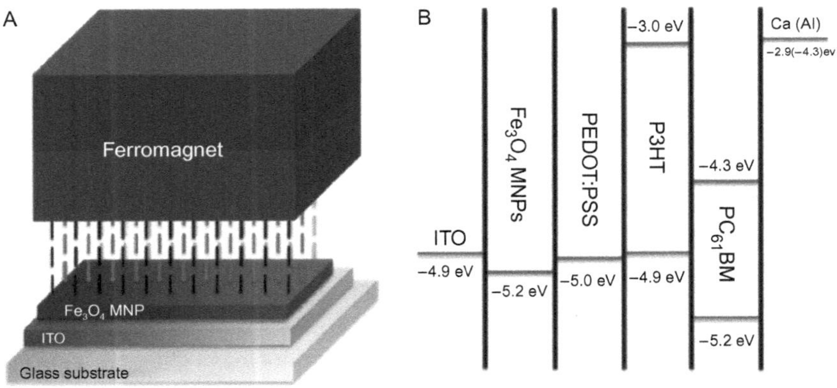

FIGURE 7.20 Use of magnetic nanoparticles to produce a hole extraction layer in the fabrication of polymer solar cells. (A) An external magnetic field was used to align the Fe_3O_4 nanoparticles in the hole extraction layer, changing the film morphology and leading to enhanced efficiency and stability. (B) Shows the energy-level alignment in the fabricated polymer solar cell. *Reproduced with permission from Ref. 164.*

found that superparamagnetic particles <10 nm in diameter exist along with multidomain ferromagnetic particles in samples extracted from flowing and stagnant tap water.[167] However, both types of particle were trapped in household filters.

8 SUMMARY AND FUTURE PERSPECTIVES

It is clear that the inherent properties of magnetic nanoparticles can provide enormous benefits to many technologies which are focused on improving the quality of the environment in which we live. As reflected in this chapter, the majority of research to date incorporating magnetic nanoparticles for environmental applications has been in the treatment of water whether in the remediation of soil and groundwater or through the magnetic separation and/or sensing of contaminants present in various aqueous systems. Both the use of nZVI for contaminant degradation and functionalized magnetic nanoparticles for removal through adsorption and separation are now well-established fields in which the basic principles of operation have been established and demonstrated many times for various systems.[168] What is now needed is a greater focus on translating the success of many laboratory-based studies to field applications in which conditions can vary greatly. In this regard, the recent development of many magnetic nanoparticle systems functionalized with chemosensing molecules shows great promise in providing rapid and accurate onsite readings of contamination levels. However, various issues, such as the true long-term effectiveness of nZVI in remediating contaminated areas, or how practical and costly the use of magnetic fields are for the large-scale separation of contaminant-adsorbed nanoparticles, still require attention. Factors such as aggregation, rapid on-stream deactivation and contaminant rerelease could all severely check the many advantages offered by magnetic nanoparticles and need further understanding.

As a subclass of engineered nanomaterials, magnetic nanoparticles have been subject to intense discussion in terms of biological and ecological toxicity.[169–178] According to the Project on Emerging Nanotechnologies,[179] more than 1600 nanotechnology-based consumer products are currently on the market, a number that has approximately doubled in the last 5 years.[176] The *source-pathway-receptor-impact* paradigm has arisen to describe the lifecycle of an engineered nanomaterial in the ecosystem.[169] The source of magnetic nanoparticle introduction may be direct, for example through controlled injection at a contamination site or accidental release, or indirect as occurs at sewage and waste treatment plants and landfill sites.[173,175] The pathway in the case of magnetic nanoparticles is predominantly subsurface transport and groundwater flow.[170] Murphy has also pointed out that with the increasing interest and use of nanoparticles across a wide array of applications, 'green' synthesis techniques that reduce and recycle waste with room temperature processing are required for sustainability.[174]

Other, less mature fields that could provide environmental benefits are beginning to explore the use of magnetic nanoparticles, for example energy and catalytic technologies. After some very recent and exciting studies, this can be expected to multiply. Treatment of less common environmental and geopolitical incidents such as oceanic oil spills or chemical weapon contamination may be further demonstrated. A noteworthy use of magnetic nanoparticles is in the removal of radionuclides such as uranium from soil and water systems and following the recent Fukushima Daiichi nuclear disaster, it is timely to wonder if a similar approach could in some way help in the environmental clean-up which is expected to take decades to complete. Coupled with the many applications already demonstrated and outlined in this chapter, these exciting prospects will ensure that the use of magnetic nanoparticles to improve the quality of our environment will continue to attract great interest.

REFERENCES

1. Bakir F, Damluji SF, Amin-Zaki L, Murtadha M, Khalidi A, Al-Rawi NY, et al. Methylmercury poisoning in Iraq. *Science* 1973;**181**:230–41.
2. Hassan A, Velasquez E, Belmar R, Coye M, Drucker E, Landrigan PJ, et al. Mercury poisoning in Nicaragua: a case study of the export of environmental and occupational health hazards by a multinational corporation. *Int J Health Serv* 1981;**11**:221–6.
3. Eto K. Pathology of Minamata disease. *Toxicol Pathol* 1997;**25**:614–23.
4. Johnson LL, Lomax DP, Myers MS, Olson OP, Sol SY, O'Neill SM, et al. Xenoestrogen exposure and effects in English sole (Parophyrys vetulus) from Puget Sounds, WA. *Aquat Toxicol* 2008;**88**:29–38.
5. Peck KA, Lomax DP, Olson OP, Sol SY, Swanson P, Johnson LL. Development of an enzyme-linked immunosorbent assay for quantifying vitellogenin in Pacific salmon and assessment of field exposure to environmental estrogens. *Environ Toxicol Chem* 2011;**30**:477–86.
6. Lawrence Berkeley National Laboratory, Earth Sciences Division. Environmental Remediation Subsurface Science. Retrieved October 30th 2013 from: http://esd.lbl.gov/research/programs/erwr/research_areas/erss.html.
7. Tratnyek PG, Johnson RL. Nanotechnologies for environmental cleanup. *Nano Today* 2006;**1**:44–8.
8. Yean S, Cong L, Yavuz CT, Mayo JT, Yu WW, Kan AT, et al. Effect of magnetite particle size on adsorption and desorption of arsenite and arsenate. *J Mater Res* 2005;**20**:3255–64.
9. Shen YF, Tang J, Nie ZH, Wang YD, Ren Y, Zuo L. Tailoring size and structural distortion of Fe_3O_4 nanoparticles for the purification of contaminated water. *Bioresour Technol* 2009;**100**:4139–46.
10. Dumestre F, Chaudret B, Amiens C, Renaud P, Fejes P. Superlattices of iron nanocubes synthesized from $Fe[N(SiMe_2)_2]_2$. *Science* 2004;**303**:821–3.
11. LaGrow AP, Ingham B, Cheong S, Williams GVM, Dotzler C, Toney MF, et al. Synthesis, alignment, and magnetic properties of monodisperse nickel nanocubes. *J Am Chem Soc* 2012;**134**:855–8.
12. Terris BD, Thomson T. Nanofabricated and self-assembled magnetic structures as data storage media. *J Phys D Appl Phys* 2005;**38**:R199–222.
13. Jia C-J, Schüth F. Colloidal metal nanoparticles as a component of designed catalyst. *Phys Chem Chem Phys* 2011;**13**:2457–87.

14. Cheong S, Ferguson P, Feindel KW, Hermans IF, Callaghan PT, Meyer C, et al. Simple synthesis and functionalization of iron nanoparticles for magnetic resonance imaging. *Angew Chem Int Ed* 2011;**50**:4206–9.
15. Batlle X, Labarta A. Finite-size effects in fine particles: magnetic and transport properties. *J Phys D Appl Phys* 2002;**35**:R15–42.
16. Huber DL. Synthesis, properties, and applications of iron nanoparticles. *Small* 2005;**1**:482–501.
17. Lu A-H, Salabas EL, Schuth F. Magnetic nanoparticles: synthesis, protection, functionalization, and application. *Angew Chem Int Ed* 2007;**46**:1222–4.
18. Binns C. Magnetism in nanoclusters and cluster-assembled thin films. *J Nanosci Nanotechnol* 2001;**1**:243–62.
19. Bansmann J, Baker SH, Binns C, Blackman JA, Bucher J-P, Dorantes-Dávila J, et al. Magnetic and structural properties of isolated and assembled clusters. *Surf Sci Rep* 2005;**56**:189–275.
20. Pratt A, Lari L, Hovorka O, Shah A, Woffinden C, Tear SP, et al. Enhanced oxidation of nanoparticles through strain-mediated ionic transport. *Nat Mater* 2013;**13**:26–30.
21. Kura H, Takahashi M, Ogawa T. Synthesis of monodisperse iron nanoparticles with a high saturation magnetization using an $Fe(CO)_x$ – oleylamine reacted precursor. *J Phys Chem C* 2010;**114**:5835–8.
22. Crane RA, Scott TB. Nanoscale zero-valent iron: future prospects for an emerging water treatment technology. *J Hazard Technol* 2012;**211–212**:112–25.
23. Li L, Fan M, Brown RC, Van Leeuwen JH, Wang J, Wang W, et al. Synthesis, properties, and environmental applications of nanoscale iron-based materials: a review. *Crit Rev Environ Sci Technol* 2006;**36**:405–31.
24. Wu W, He Q, Jiang C. Magnetic iron oxide nanoparticles: synthesis and surface functionalization strategies. *Nanoscale Res Lett* 2008;**3**:397–415.
25. Rao CNR, Matte HSSR, Voggu R, Govindaraj A. Recent progress in the synthesis of inorganic nanoparticles. *Dalton Trans* 2012;**41**:5089–120.
26. Wang CB, Zhang WX. Synthesizing nanoscale iron particles for rapid and complete dechlorination of TCE and PCBs. *Environ Sci Technol* 1997;**31**:2154–6.
27. Lee IS, Lee N, Park J, Kim BH, Yi Y-W, Kim T, et al. Ni/NiO core/shell nanoparticles for selective binding and magnetic separation of histidine-tagged proteins. *J Am Chem Soc* 2006;**128**:10658–9.
28. Pratt A, Woffinden C, Kröger R, Tear S, Binns C. Surface spin polarization of Fe nanoclusters. *IEEE Trans Magn* 2010;**46**:1660.
29. Yin Y, Rious RM, Erdonmez CK, Hughes S, Somojai GA, Alivisatos AP. Formation of hollow nanocrystals through the nanoscale Kirkendall effect. *Science* 2004;**304**:711–4.
30. Hu J, Lo IMC, Chen GH. Comparative study of various magnetic nanoparticles for Cr(VI) removal. *Sep Purif Technol* 2007;**56**:249–56.
31. Zhang S, Niu H, Cai Y, Zhao X, Shi Y. Arsenite and arsenate adsorption on coprecipitated bimetal oxide magnetic nanomaterials: $MnFe_2O_4$ and $CoFe_2O_4$. *Chem Eng J* 2010;**158**:599–607.
32. Sun S, Zeng H. Size-controlled synthesis of magnetite nanoparticles. *J Am Chem Soc* 2002;**124**:8204–5.
33. Shavel A, Rodríguez-González B, Spasova M, Farle M, Liz-Marzán LM. Synthesis and characterization of iron/iron oxide core/shell nanocubes. *Adv Funct Mater* 2007;**17**:3870–6.
34. Abdulwahab K, Malik MA, O'Brien P, Govender K, Muryn CA, Timco GA, et al. Synthesis of monodispersed magnetite nanoparticles from iron pivalate clusters. *Dalton Trans* 2013;**42**:196–206.

35. Zhang J, Zhai S, Li S, Xiao Z, Song Y, An Q, et al. Pb(II) removal of Fe_3O_4@SiO_2–NH_2 core–shell nanomaterials prepared via a controllable sol–gel process. *Chem Eng J* 2013;**215–216**:461–71.
36. Vestal CR, Zhang ZJ. Synthesis and magnetic characterization of Mn and Co spinel ferrite-silica nanoparticles with tunable magnetic core. *Nano Lett* 2003;**3**:1739–43.
37. Peng X, Wang Y, Tang X, Liu W. Functionalized magnetic core–shell Fe_3O_4@SiO_2 nanoparticles as selectivity-enhanced chemosensor for Hg(II). *Dyes Pigments* 2011;**91**:26–32.
38. Sun Y, Duan L, Guo Z, DuanMu Y, Ma M, Xu L, et al. An improved way to prepare superparamagnetic magnetite-silica core-shell nanoparticles for possible biological application. *J Magn Magn Mater* 2005;**285**:65–70.
39. Lee S-Y, Harris MT. Surface modification of magnetic nanoparticles capped by oleic acids: characterization and colloidal stability in polar solvents. *J Colloid Interface Sci* 2006;**293**:401–8.
40. Huang C, Hu B. Silica-coated magnetic nanoparticles modified with γ-mercaptopropyltrimethoxysilane for fast and selective solid phase extraction of trace amounts of Cd, Cu, Hg, and Pb in environmental and biological samples prior to their determination by inductively coupled plasma mass spectrometry. *Spectrochim Acta B* 2008;**63**:437–44.
41. Girginova PI, Daniel-da-Silva AL, Lopes CB, Figueira P, Otero M, Amaral VS, et al. Silica coated magnetite particles for magnetic removal of Hg^{2+} from water. *J Colloid Interface Sci* 2010;**345**:234–40.
42. Sun L, Li Y, Sun M, Wang H, Xu S, Zhang C, et al. Porphyrin-functionalized Fe_3O_4@SiO_2 core/shell magnetic colorimetric material for detection, adsorption and removal of Hg^{2+} in aqueous solution. *New J Chem* 2011;**35**:2697–704.
43. Zhai Y, He Q, Han Q, Duan S. Solid-phase extraction of trace metal ions with magnetic nanoparticles modified with 2,6-diaminopyridine. *Microchim Acta* 2012;**178**:405–12.
44. Wang Y, Tian T, Wang L, Hu X. Solid-phase preconcentration of cadmium(II) using amino-functionalized magnetic-core silica-shell nanoparticles, and its determination by hydride generation atomic fluorescence spectrometry. *Microchim Acta* 2013;**180**:235–42.
45. Hsu CC, Whang CW. Microscale solid phase extraction of glyphosate and aminomethylphosphonic acid in water and guava fruit extract using alumina-coated iron oxide nanoparticles followed by capillary electrophoresis and electrochemiluminescence detection. *J Chromatogr A* 2009;**1216**:8575–80.
46. Sun L, Zhang CZ, Chen LG, Liu J, Jin HY, Xu HY, et al. Preparation of alumina-coated magnetite nanoparticle for extraction of trimethoprim from environmental water samples based on mixed hemimicelles solid-phase extraction. *Anal Chim Acta* 2009;**638**:162–8.
47. Bystrzejewski M, Pyrzyńska K, Huczko A, Lange H. Carbon-encapsulated magnetic nanoparticles as separable and mobile sorbents of heavy metal ions from aqueous solutions. *Carbon* 2009;**47**:1201–4.
48. Zhang D, Wei S, Kaila C, Su X, Wu J, Karki AB, et al. Carbon-stabilized iron nanoparticles for environmental remediation. *Nanoscale* 2010;**2**:917–9.
49. Wang H, Yu Y-F, Chen Q-W, Cheng K. Carboxyl-functionalized nanoparticles with magnetic core and mesopore carbon shell as adsorbents for the removal of heavy metal ions from aqueous solution. *Dalton Trans* 2011;**40**:559–63.
50. Singh S, Barick KC, Bahadur D. Surface engineered magnetic nanoparticles for removal of toxic metal ions and bacterial pathogens. *J Hazard Mater* 2011;**192**:1539–47.
51. Binns C, Prieto P, Baker S, Howes P, Dondi R, Burley G, et al. Preparation of hydrosol suspensions of elemental and core–shell nanoparticles by co-deposition with water vapour from the gas-phase in ultra-high vacuum conditions. *J Nanopart Res* 2012;**14**:1136.

52. Warner CL, Addleman RS, Cinson AD, Droubay TC, Engelhard MH, Nash MA, et al. High-performance, superparamagnetic, nanoparticle-based heavy metal sorbents for removal of contaminants from natural waters. *ChemSusChem* 2010;**3**:749–57.
53. Shin S, Jang J. Thiol containing polymer encapsulated magnetic nanoparticles as reusable and efficiently separable adsorbent for heavy metal ions. *Chem Commun* 2007;4230–2.
54. Song J, Kong H, Jang J. Adsorption of heavy metal ions from aqueous solution by polyrhodanine-encapsulated magnetic nanoparticles. *J Colloid Interface Sci* 2011;**359**:505–11.
55. Chen A, Wang H, Zhao B, Li X. The preparation of polypyrrole–Fe_3O_4 nanocomposites by the use of common ion effect. *Synth Met* 2003;**139**:411–5.
56. Takafuji M, Ide S, Ihara H, Xu Z. Preparation of poly(1-vinylimidazole)-grafted magnetic nanoparticles and their application for removal of metal ions. *Chem Mater* 2004;**16**:1977–83.
57. Mak S-Y, Chen D-H. Fast adsorption of methylene blue on polyacrylic acid-bound iron oxide magnetic nanoparticles. *Dyes Pigments* 2004;**61**:93–8.
58. Pan S, Zhang Y, Shen H, Hu M. An intensive study on the magnetic effect of mercapto-functionalized nano-magnetic Fe_3O_4 polymers and their adsorption mechanism for the removal of Hg(II) from aqueous solution. *Chem Eng J* 2012;**210**:564–74.
59. Kim DK, Mikhaylova M, Zhang Y, Muhammed M. Protective coating of superparamagnetic iron oxide nanoparticles. *Chem Mater* 2003;**15**:1617–27.
60. Chang YC, Chen DH. Preparation and adsorption properties of monodisperse chitosan-bound Fe_3O_4 magnetic nanoparticles for removal of Cu(II) ions. *J Colloid Interface Sci* 2005;**283**:446–51.
61. Yan H, Yang L, Yang Z, Yang H, Li A, Cheng R. Preparation of chitosan/poly(acrylic acid) magnetic composite microspheres and applications in the removal of copper(II) ions from aqueous solutions. *J Hazard Mater* 2012;**229–230**:371–80.
62. Liu X, Hu Q, Fang Z, Zhang X, Zhang B. Magnetic chitosan nanocomposites: a useful recyclable tool for heavy metal ion removal. *Langmuir* 2009;**25**:3–8.
63. Zhou Y-T, Nie H-L, Branford-White C, He Z-Y, Zhu L-M. Removal of Cu^{2+} from aqueous solution by chitosan-coated magnetic nanoparticles modified with α-ketoglutaric acid. *J Colloid Interface Sci* 2009;**330**:29–37.
64. Zhang X, Niu H, Zhang S, Cai Y. Preparation of a chitosan-coated C_{18}-functionalized magnetite nanoparticle sorbent for extraction of phthalate ester compounds from environmental water samples. *Anal Bioanal Chem* 2010;**397**:791–8.
65. Yantasee W, Hongsirikarn K, Warner CL, Choi D, Sangvanich T, Toloczko MB, et al. Direct detection of Pb in urine and Cd, Pb, Cu, and Ag in natural waters using electrochemical sensors immobilized with DMSA functionalized magnetic nanoparticles. *Analyst* 2007;**133**:348–55.
66. Yantasee W, Warner CL, Sangvanich T, Addleman RS, Carter TG, Wiacek RJ, et al. Removal of heavy metals from aqueous systems with thiol functionalized superparamagnetic nanoparticles. *Environ Sci Technol* 2007;**41**:5114–9.
67. Zhang S, Zhang Y, Liu J, Xu Q, Xiao H, Wang X, et al. Thiol modified $Fe_3O_4@SiO_2$ as a robust, high effective, and recycling magnetic sorbent for mercury removal. *Chem Eng J* 2013;**226**:30–8.
68. Miller AL. Dimercaptosuccinic acid (DMSA), a non-toxic, water-soluble treatment for heavy metal toxicity. *Altern Med Rev* 1998;**3**:199–207.
69. Liu J-F, Zhao Z-S, Jiang G-B. Coating Fe_3O_4 magnetic nanoparticles with humic acid for high efficient removal of heavy metals in water. *Environ Sci Technol* 2008;**42**:6949–54.
70. Zhu B, Zhao J, Yu H, Liangguo Y, Ain W, Bin D. Naphthalimide-functionalized $Fe_3O_4@SiO_2$ core/shell nanoparticles for selective and sensitive adsorption and detection of Hg^{2+}. *Chem Eng J* 2013;**219**:411–8.

71. Hai B, Wu J, Chen X, Protasiewicz JD, Scherson DA. Metal-ion adsorption on carboxyl-bearing self-assembled monolayers covalently bound to magnetic nanoparticles. *Langmuir* 2005;**21**:3104–5.
72. Wang Y, Luo X, Tang J, Hu X, Xu Q, Yang C. Extraction and preconcentration of trace levels of cobalt using functionalized magnetic nanoparticles in a sequential injection lab-on-valve system with detection by electrothermal atomic absorption spectrometry. *Anal Chim Acta* 2012;**713**:92–6.
73. Li J, Zhao X, Shi Y, Cai Y, Mou S, Jiang G. Mixed hemimicelles solid-phase extraction based on cetyltrimethylammonium bromide-coated nano-magnets Fe_3O_4 for the determination of chlorophenols in environmental water samples coupled with liquid chromatography/spectrophotometry detection. *J Chromatogr A* 2008;**1180**:24–31.
74. Suleiman JS, Hu B, Peng H, Huang C. Separation/preconcentration of trace amounts of Cr, Cu and Pb in environmental samples by magnetic solid-phase extraction with Bismuthiol-II-immobilized magnetic nanoparticles and their determination by ICP-OES. *Talanta* 2009;**77**:1579–83.
75. Lee HY, Bae DR, Park JC, Song H, Han WS, Jung JH. A selective fluoroionophore based on BODIPY-functionalized magnetic silica nanoparticles: removal of Pb^{2+} from human blood. *Angew Chem Int Ed* 2009;**48**:1239–43.
76. Ballesteros-Gomez A, Rubio S. Hemimicelles of alkyl carboxylates chemisorbed onto magnetic nanoparticles: study and application to the extraction of carcinogenic polycyclic aromatic hydrocarbons in environmental water samples. *Anal Chem* 2009;**81**:9012–20.
77. Zhao X, Shi Y, Wang T, Cai Y, Jiang G. Preparation of silica-magnetite nanoparticle mixed hemimicelle sorbents for extraction of several typical phenolic compounds from environmental water samples. *J Chromatogr A* 2008;**1188**:140–7.
78. Mahmoud ME, Abdelwahab MS, Fathallah EM. Design of novel nano-sorbents based on nano-magnetic iron oxide-bound-nano-silicon oxide-immobilized-triethylenetetramine for implementation in water treatment of heavy metals. *Chem Eng J* 2013;**223**:318–27.
79. Wang J, Zheng S, Shao Y, Liu J, Xu Z, Zhu D. Amino-functionalized Fe_3O_4@SiO_2 core–shell magnetic nanomaterial as a novel adsorbent for aqueous heavy metals removal. *J Colloid Interface Sci* 2010;**349**:293–9.
80. Cumbal L, Greenleaf J, Leun D, SenGupta AK. Polymer supported inorganic nanoparticles: characterization and environmental applications. *React Funct Polym* 2003;**54**:167–80.
81. Jiang W, Chen X, Niu Y, Pan B. Spherical polystyrene-supported nano-Fe_3O_4 of high capacity and low-field separation for arsenate removal from water. *J Hazard Mater* 2012;**243**:319–25.
82. Zhang SX, Niu HY, Cai YQ, Shi YL. Barium alginate caged Fe_3O_4@C_{18} magnetic nanoparticles for the pre-concentration of polycyclic aromatic hydrocarbons and phthalate esters from environmental water samples. *Anal Chim Acta* 2010;**665**:167–75.
83. Ozay O, Ekici S, Baran Y, Aktas N, Sahiner N. Removal of toxic metal ions with magnetic hydrogels. *Water Res* 2009;**43**:4403–11.
84. Yao Y, Miao S, Liu S, Ma P, Sun H, Wang S. Synthesis, characterization, and adsorption properties of magnetic Fe_3O_4@graphene nanocomposite. *Chem Eng J* 2012;**184**:326–32.
85. Jin J, Li R, Wang H, Chen H, Liang K, Ma J. Magnetic Fe nanoparticle functionalized water-soluble multi-walled carbon nanotubules towards the preparation of sorbent for aromatic compounds removal. *Chem Commun* 2007;386–8.
86. Chen R, Zhi C, Yang H, Bando Y, Zhang Z, Sugiur N, et al. Arsenic (V) adsorption on Fe_3O_4 nanoparticle-coated boron nitride nanotubes. *J Colloid Interface Sci* 2011;**359**:261–8.

87. Glavee GN, Klabunde KJ, Sorenson CM, Hadjipanayis GC. Chemistry of borohydride reduction of iron(II) and iron(III) ions in aqueous and nonaqueous media. Formation of nanoscale Fe, FeB, and Fe_2B powders. *Inorg Chem* 1995;**34**:28–32.
88. Zhang W-X. Nanoscale iron particles for environmental remediation: a review. *J Nanopart Res* 2003;**5**:323–32.
89. Li XQ, Elliot DW, Zhang WX. Zero-valent iron nanoparticles for abatement of environmental pollutants: materials and engineering aspects. *Crit Rev Solid State Mater Sci* 2006;**31**:111–22.
90. Cundy AB, Hopkinson L, Whitby RLD. Use of iron-based technologies in contaminated land and groundwater remediation: a review. *Sci Total Environ* 2008;**400**:42–51.
91. O'Carroll D, Sleep B, Krol M, Boparai H, Kocur C. Nanoscale zero valent iron and bimetallic particles for contaminated site remediation. *Adv Water Res* 2013;**51**:104–22.
92. Tang SCN, Lo IMC. Magnetic nanoparticles: essential factors for sustainable environmental applications. *Water Res* 2013;**47**:2613–32.
93. Ponder SM, Darab JG, Mallouk TE. Remediation of Cr(VI) and Pb(II) aqueous solutions using supported nanoscale zero-valent iron. *Environ Sci Technol* 2000;**34**:2564–9.
94. Elliott DW, Zhang W-X. Field assessment of nanoscale bimetallic particles for groundwater treatment. *Environ Sci Technol* 2001;**35**:4922–6.
95. Yang GCC, Wu M-Y. Injection of nanoscale Fe_3O_4 slurry coupled with the electrokinetic process for remediation of NO_3^- in saturated soil: remediation performance and reaction behavior. *Sep Purif Technol* 2011;**79**:272–7.
96. Xu P, Zeng GM, Huang DL, Feng CL, Hu S, Zhao MH, et al. Use of iron oxide nanomaterials in wastewater treatment: a review. *Sci Total Environ* 2012;**424**:1–10.
97. Masciangioli T, Zhang WX. Environmental technologies at the nanoscale. *Environ Sci Technol* 2003;**37**:102A–8A.
98. Crane RA, Dickinson M, Popescu IC, Scott TB. Magnetite and zero-valent iron nanoparticles for the remediation of uranium contaminated environmental water. *Water Res* 2011;**45**:2931–42.
99. Ebner AD, Ritter JA, Ploehn HJ. Feasibility and limitations of nanolevel high gradient magnetic separation. *Sep Purif Technol* 1997;**11**:199–210.
100. Wang Q, Guan Y, Ren X, Yang M, Liu X. Application of magnetic extractant for the removal of hexavalent chromium from aqueous solution in high gradient magnetic separator. *Chem Eng J* 2012;**183**:339–48.
101. Yavuz CT, Mayo JT, Yu WW, Prakash A, Falkner JC, Yean S, et al. Low-field magnetic separation of monodisperse Fe_3O_4 nanocrystals. *Science* 2006;**304**:964–7.
102. Elimelech M, Gregory J, Jia X, Williams RA. *Particle deposition and aggregation: measurement, modeling, and simulation*. Oxford: Butterworth-Heinemann; 1995.
103. Tiede K, Boxall ABA, Tear SP, Lewis J, David H, Hassellov M. Detection and characterization of engineered nanoparticles in food and the environment. *Food Addit Contam Part A Chem Anal Control Expo Risk Assess* 2008;**25**:795–821.
104. Petosa AR, Jaisi DP, Quevedo IR, Elimelech M, Tufenkji N. Aggregation and deposition of engineered nanomaterials in aquatic environments: role of physicochemical interactions. *Environ Sci Technol* 2010;**44**:6532–49.
105. Derjaguin BV, Landau LD. Theory of stability of strongly charged lyophobic sols and the adhesion of strongly charged particles in solutions of electrolytes. *Acta Physicochim URSS* 1941;**14**:633–62.
106. Verwey EJW, Overbeek JTG. *Theory of the stability of lyophobic colloids*. Amsterdam: Elsevier; 1948.

107. United States Environmental Protection Agency. List of drinking water contaminants for various substances in the US. Retrieved [October 28th 2013], from http://water.epa.gov/drink/contaminants/index.cfm#List.
108. World Health Organization. Guidelines for drinking-water quality, 4th edition. Retrieved [October 28th 2013], from http://www.who.int/water_sanitation_health/publications/2011/dwq_guidelines/en/index.
109. Jin Y, Liu F, Tong M, Hou Y. Removal of arsenate by cetyltrimethylammonium bromide modified magnetic nanoparticles. *J Hazard Mater* 2012;**227–228**:461–8.
110. Badruddoza AZM, Shawon ZBZ, Rahman MT, Hao KW, Hidajat K, Uddin MS. Ionically modified magnetic nanomaterials for arsenic and chromium removal from water. *Chem Eng J* 2013;**225**:607–15.
111. Tuutijärvi T, Lu J, Sillanpää M, Chen G. As(V) adsorption on maghemite nanoparticles. *J Hazard Mater* 2009;**166**:1415–20.
112. Chowdhury SR, Yanful EK. Arsenic and chromium removal by mixed magnetite–maghemite nanoparticles and the effect of phosphate on removal. *J Environ Manage* 2010;**91**:2238–47.
113. Zaspalis V, Pagana A, Sklari S. Arsenic removal from contaminated water by iron oxide sorbents and porous ceramic membranes. *Desalination* 2007;**217**:167–80.
114. Brown PA, Gill SA, Allen SJ. Metal removal from wastewater using peat. *Water Res* 2002;**34**:3907–16.
115. Demir A, Arisoy M. Biological and chemical removal of Cr(VI) from waste water: cost and benefit analysis. *J Hazard Mater* 2007;**147**:275–80.
116. Shi L-N, Zhang X, Chen Z-L. Removal of chromium (VI) from wastewater using bentonite-supported nanoscale zero-valent iron. *Water Res* 2011;**45**:886–92.
117. Alidokht L, Khataee AR, Reyhanitabar A, Oustan S. Reductive removal of Cr(VI) by starch-stabilized Fe^0 nanoparticles in aqueous solution. *Desalination* 2011;**270**:105–10.
118. Hu J, Lo IMC, Chen GH. Removal of Cr(VI) by magnetite nanoparticles. *Water Sci Technol* 2004;**50**:139–46.
119. Hu J, Lo IMC, Chen G. Fast removal and recovery of Cr(VI) using surface-modified Jacobsite ($MnFe_2O_4$) nanoparticles. *Langmuir* 2005;**21**:11173–9.
120. Hu J, Lo IMC, Chen GH. *International symposium on nanotechnology in environmental protection and pollution*. Bangkok: Asia Pacific Nanotechnology Forum (APNF); 2005.
121. Hu J, Chen GH, Lo IMC. Removal and recovery of Cr(VI) from wastewater by maghemite nanoparticles. *Water Res* 2005;**39**:4528–36.
122. Hu J, Lo IMC, Chen GH. Performance and mechanism of chromate (VI) adsorption by δ-FeOOH-coated maghemite (γ-Fe_2O_3) nanoparticles. *Sep Purif Technol* 2007;**58**:76–82.
123. Pang Y, Zeng G, Tang L, Zhang Y, Liu Y, Lei X, et al. Preparation and application of stability enhanced magnetic nanoparticles for rapid removal of Cr(VI). *Chem Eng J* 2011;**175**:222–7.
124. Geng B, Jin Z, Li T, Qi X. Kinetics of hexavalent chromium removal from water by chitosan-Fe^0 nanoparticles. *Chemosphere* 2009;**75**:825–30.
125. Shin KY, Hong JY, Jang J. Heavy metal ion adsorption behavior in nitrogen-doped magnetic carbon nanoparticles: isotherms and kinetic study. *J Hazard Mater* 2011;**190**:36–44.
126. Wei L, Yang G, Wang R, Ma W. Selective adsorption and separation of chromium (VI) on the magnetic iron–nickel oxide from waste nickel liquid. *J Hazard Mater* 2009;**164**:1159–63.
127. Parham H, Zargar B, Shiralipour R. Fast and efficient removal of mercury from water samples using magnetic iron oxide nanoparticles modified with 2-mercaptobenzothiazole. *J Hazard Mater* 2012;**205–206**:94–100.

128. Farrukh A, Akram A, Ghaffar A, Hanif S, Hamid A, Duran H, et al. Design of polymer-brush-grafted magnetic nanoparticles for highly efficient water remediation. *Appl Mater Interfaces* 2013;**5**:3784–93.
129. Zhang X, Nui H, Pan Y, Shi Y, Cai Y. Chitosan-coated octadecyl-functionalized magnetite nanoparticles: preparation and application in extraction of trace pollutants from environmental water samples. *Anal Chem* 2010;**82**:2363–71.
130. Gupta VK, Nayak A. Cadmium removal and recovery from aqueous solutions by novel adsorbents prepared from orange peel and Fe_2O_3 nanoparticles. *Chem Eng J* 2012;**180**:81–90.
131. Watson JHP, Cressey BA, Roberts AP, Ellwood DC, Charnock JM, Soper AK. Structural and magnetic studies on heavy-metal-adsorbing iron sulphide nanoparticles produced by sulphate-reducing bacteria. *J Magn Magn Mater* 2000;**214**:13–30.
132. Feng Y, Gong J-L, Zeng G-M, Niu Q-Y, Zhang H-Y, Niu C-G, et al. Adsorption of Cd (II) and Zn (II) from aqueous solutions using magnetic hydroxyapatite nanoparticles as adsorbents. *Chem Eng J* 2010;**162**:487–94.
133. Feng Z, Zhu S, De Godoi DRM, Samia ACS, Scherson D. Adsorption of Cd^{2+} on carboxyl-terminated superparamagnetic iron oxide nanoparticles. *Anal Chem* 2012;**84**:3764–70.
134. Ge F, Li M-M, Ye H, Zhao B-X. Effective removal of heavy metal ions Cd^{2+}, Zn^{2+}, Pb^{2+}, Cu^{2+} from aqueous solution by polymer-modified magnetic nanoparticles. *J Hazard Mater* 2012;**211–212**:366–72.
135. Afkhami A, Norooz-Asl R. Removal, preconcentration and determination of Mo(VI) from water and wastewater samples using maghemite nanoparticles. *Colloid Surf A* 2009;**346**:52–7.
136. Kong J, Coolahan K, Mugweru A. Manganese based magnetic nanoparticles for heavy metal detection and environmental remediation. *Anal Methods* 2013;**5**:5128–33.
137. Wang L, Yang Z, Gao J, Xu K, Gu H, Zhang B, et al. A biocompatible method of decorporation: biphosphonate-modified magnetite nanoparticles to remove uranyl ions from blood. *J Am Chem Soc* 2006;**128**:13358–9.
138. Fan FL, Qin Z, Bai J, Rong WD, Fan FY, Tian W, et al. Rapid removal of uranium from aqueous solutions using magnetic $Fe_3O_4@SiO_2$ composite particles. *J Environ Radioact* 2012;**106**:40–6.
139. Zelmanov G, Semiat R. Iron(3) oxide-based nanoparticles as catalysts in advanced organic aqueous oxidation. *Water Res* 2008;**42**:492–8.
140. Ji YS, Lui XY, Guan M, Zha CD, Huang HY, Zhang HX, et al. Preparation of functionalized magnetic nanoparticulate sorbents for rapid extraction of biphenolic pollutants from environmental samples. *J Sep Sci* 2009;**32**:2139–45.
141. Sun L, Chen LG, Sun X, Du XB, Yue YS, He DQ, et al. Analysis of sulfonamides in environmental water samples based on magnetic mixed hemimicelles solid-phase extraction coupled with HPLC-UV detection. *Chemosphere* 2009;**77**:1306–12.
142. Jiang C, Wang R, Ma W. The effect of magnetic nanoparticles on Microcytis aeruginosa removal by a composite coagulant. *Colloid Surf A* 2010;**369**:260–7.
143. Wang L, Li J, Wang Y, Zhao L, Jiang Q. Adsorption capability for Congo red on nanocrystalline MFe_2O_4 (M = Mn, Fe, Co, Ni) spinel ferrites. *Chem Eng J* 2012;**181–182**: 72–9.
144. Afkhami A, Moosavi R. Adsorptive removal of Congo red, a carcinogenic textile dye, from aqueous solutions by maghemite nanoparticles. *J Hazard Mater* 2010;**174**:398–403.
145. Shao R, Sun L, Tang L, Chen Z. Preparation and characterization of magnetic core–shell $ZnFe_2O_4@ZnO$ nanoparticles and their application for the photodegradation of methylene blue. *Chem Eng J* 2013;**217**:185–91.

146. Ge F, Ye H, Li M-M, Zhao B-X. Efficient removal of cationic dyes from aqueous solution by polymer-modified magnetic nanoparticles. *Chem Eng J* 2012;**198–199**:11–7.
147. Zargar B, Parham H, Hatamie A. Modified iron oxide nanoparticles as solid phase extractor for spectrophotometric determination and separation of basic fuchsin. *Talanta* 2009;**77**:1328–31.
148. Xiong Z, Zhang L, Zhang R, Zhang Y, Chen J, Zhang W. Solid-phase extraction based on magnetic core–shell silica nanoparticles coupled with gas chromatography–mass spectrometry for the determination of low concentration pesticides in aqueous samples. *J Sep Sci* 2012;**35**:2430–7.
149. Zhao X, Wang J, Wu F, Wang T, Cai Y, Shi Y, et al. Removal of fluoride from aqueous media by Fe_3O_4@$Al(OH)_3$ magnetic nanoparticles. *J Hazard Mater* 2010;**173**:102–9.
150. Jung JH, Lee JH, Shinkai S. Functionalized magnetic nanoparticles as chemosensors and adsorbents for toxic metal ions in environmental and biological fields. *Chem Soc Rev* 2011;**40**:4464–74.
151. Son H, Lee HY, Lim JM, Kang D, Han WS, Lee SS, et al. A highly sensitive and selective turn-on fluorogenic and chromogenic sensor based on BODIPY-functionalized magnetic nanoparticles for detecting lead in living cells. *Chem Eur J* 2010;**16**:11549–53.
152. Herrman IK, Urner M, Koehler FM, Hasler M, Roth-Z'Graggen B, Grass RN, et al. Blood purification using functionalized core/shell nanomagnets. *Small* 2010;**6**:1388–92.
153. Tan Z, Liu J, Jiang G. Visual test of subparts per billion-level copper(II) by Fe_3O_4 magnetic nanoparticle-based solid phase extraction coupled with a functionalized gold nanoparticle probe. *Nanoscale* 2012;**4**:6735–8.
154. Park M, Seo S, Lee SJ, Jung JH. Functionalized Ni@SiO_2 core/shell magnetic nanoparticles as a chemosensor and adsorbent for Cu^{2+} ion in drinking water and human blood. *Analyst* 2010;**135**:2802–5.
155. Wang Y, Peng X, Shi J, Tang X, Jiang J, Liu W. Highly selective fluorescent chemosensor for Zn^{2+} derived from inorganic–organic hybrid magnetic core/shell Fe_3O_4@SiO_2 nanoparticles. *Nanoscale Res Lett* 2012;**7**:86.
156. Park M, Seo S, Lee IS, Jung JH. Ultraefficient separation and sensing of mercury and methylmercury ions in drinking water by using aminonaphthalimide-functionalized Fe_3O_4@SiO_2 core/shell magnetic nanoparticles. *Chem Commun* 2010;**46**:4478–80.
157. Lin Q, Fu Y-P, Chen P, Wei T-B, Zhang Y-M. Colorimetric chemosensors designed to provide high sensitivity for Hg^{2+} in aqueous solutions. *Dyes Pigments* 2013;**96**:1–6.
158. Dong Z, Tian X, Chen Y, Guo Y, Ma J. Dansyl derivative functionalized Fe_3O_4@SiO_2 fluorescent probe for detection and removal of Hg^{2+} in aqueous solution. *RSC Adv* 2013;**3**:1082–8.
159. Delgado-Pinar E, Montoya N, Galiana M, Albelda MT, Frísa JC, Jiménez HR, et al. Preparation of Hg^{2+} selective fluorescent chemosensors based on surface modified core-shell aluminosilicate nanoparticles. *New J Chem* 2010;**34**:567–70.
160. Torres Galvis HM, Bitter JH, Khare CB, Ruitenbeek M, Dugulan AI, de Jong KP. Supported iron nanoparticles as catalysts for sustainable production of lower olefins. *Science* 2012;**335**:835–8.
161. Zhang F, Jin J, Zhong X, Shuwen L, Niu J, Li R, et al. Pd immobilized on amine-functionalized magnetite nanoparticles: a novel and highly active catalyst for hydrogenation and Heck reactions. *Green Chem* 2011;**13**:1238–43.
162. Takagaki A, Nishimura M, Nishimura S, Ebitani K. Hydrolysis of sugars using magnetic silica nanoparticles with sulfonic acid groups. *Chem Lett* 2011;**40**:1195–7.
163. Song X, Gao L. Fabrication of bifunctional titania/silica-coated magnetic spheres and their photocatalytic activities. *J Am Ceram Soc* 2007;**90**:4015–9.

164. Wang K, Ren H, Yi C, Liu C, Wang H, Huang L, et al. Solution-processed Fe_3O_4 magnetic nanoparticle thin film aligned by an external magnetostatic field as a hole extraction layer for polymer solar cells. *Appl Mater Interfaces* 2013;**5**:10325–30.
165. Koo B, Xiong H, Slater MD, Parkapenka VB, Balasubramanian M, Podsiadlo P, et al. Hollow iron oxide nanoparticles for application in lithium ion batteries. *Nano Lett* 2012;**12**:2429–35.
166. Venkatanarasimhan S, Raghavachari D. Epoxidized natural rubber-magnetite nanocomposites for oil spill recovery. *J Mater Chem A* 2013;**1**:868–76.
167. Senftle FE, Thorpe AN, Grant JR, Barkatt A. Superparamagnetic nanoparticles in tap water. *Water Res* 2007;**41**:3005–11.
168. Stephens JR, Beveridge JS, Williams ME. Analytical methods for separating and isolating magnetic nanoparticles. *Phys Chem Chem Phys* 2012;**14**:3280–9.
169. Nowack B. Occurrence, behavior and effects of nanoparticles in the environment. *Environ Pollut* 2007;**150**:5–22.
170. Wiesner MR, Bottero J-Y. *Environmental nanotechnology*. New York: The McGraw-Hill Companies; 2007.
171. Owen R, Handy R. Formulating the problems for the environmental risk assessment of nanomaterials. *Environ Sci Technol* 2007;**41**:5582–8.
172. Klaine SJ, Alvarez PJJ, Batley GE, Fernandes TF, Handy RD, Lyon DY, et al. Nanomaterials in the environment: behavior, fate, bioavailability, and effects. *Environ Toxicol Chem* 2008;**27**:1825–51.
173. Mueller NC, Nowack B. Exposure modeling of engineered nanoparticles in the environment. *Environ Sci Technol* 2008;**42**:4447–53.
174. Murphy CJ. Sustainability as an emerging design criterion in nanoparticle synthesis and applications. *J Mater Chem* 2008;**18**:2173–6.
175. Farré M, Gajda-Schrantz K, Kantiani L, Barceló D. Ecotoxicity and analysis of nanomaterials in the aquatic environment. *Anal Bioanal Chem* 2009;**393**:81–95.
176. Posner JD. Engineered nanomaterials: where they go, nobody knows. *Nano Today* 2009;**4**:114–5.
177. Wiesner MR, Lowry GV, Jones KL, Hochella Jr MF, DiGiulio RT, Casman E, et al. Decreasing uncertainties in assessing environmental exposure, risk, and ecological implications of nanomaterials. *Environ Sci Technol* 2009;**43**:6458–62.
178. Brar SK, Verma M, Tyagi RD, Surampalli RY. Engineered nanoparticles in wastewater and wastewater sludge—evidence and impacts. *Waste Manage* 2010;**30**:504–20.
179. Project on Emerging Nanotechnologies (2013). Consumer Products Inventory. Retrieved [October 28th 2013], from http://www.nanotechproject.org/cpi.

Index

Note: Page numbers followed by "*f*" indicate figures, and "*t*" indicate tables.

T

U

V

W

X

FSC
www.fsc.org
MIX
Papier | Fördert
gute Waldnutzung
FSC® C083411

Kolibri 360 GmbH
Ferdinand-Jühlke-Straße 7
99095 Erfurt, Deutschland
produktsicherheit@kolibri360.de